GENERALIZED FUNCTIONS AND PARTIAL
DIFFERENTIAL EQUATIONS

Mathematics and Its Applications

*A Series of Monographs and Texts Edited by
Jacob T. Schwartz, Courant Institute of Mathematical Sciences,
New York University*

Additional volumes in preparation

Generalized Functions and Partial Differential Equations

GEORGI E. SHILOV
Professor of Mathematics,
Moscow State University

Authorized English Edition
Revised by the Author

Translated and Edited by
Bernard D. Seckler

G|B

GORDON AND BREACH

NEW YORK · LONDON · PARIS   SCIENCE PUBLISHERS INC.

COPYRIGHT © 1968 BY GORDON AND BREACH, *Science Publishers, Inc.*
150 Fifth Avenue, New York, New York 10011

Library of Congress Catalog Card Number 67–28235

Editorial Office for Great Britain:
Gordon and Breach Science Publishers Ltd., 8 Bloomsbury Way, London WC1, England

Editorial Office for France:
7–9 rue Emile Dubois, Paris 14ᵉ

Distributed in France by:
Dunod Editeur, 92 rue Bonaparte, Paris 6ᵉ

Distributed in Canada by:
The Ryerson Press, 299 Queen Street West, Toronto 2B, Ontario

Printed in Germany at VEB Leipziger Druckhaus

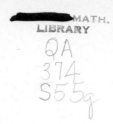
Contents

PART 2

Problems in the General Theory of Partial Differential Equations

Preface

This special course in mathematical analysis presents the fundamentals of the theory of generalized functions with applications to partial differential equations. The author has given this course a number of times in the Mathematics-Mechanics Department of Moscow University under the title *Analysis 4*.

The first half of the book is devoted to the elements of generalized function theory. The Sobolev-Schwartz definition is taken as a basis (a generalized function is a continuous linear functional on the space of infinitely differentiable functions with compact support). The material selected for inclusion from generalized functions has been dictated mainly by the needs of the second half. The general theory of partial differential equations to which the second half is devoted already embodies a great number of important results. We have chosen two topics for consideration in the book: the theory of fundamental functions (and Hörmander's theory of hypoellipticity which is related to it) and questions concerning well-posed problems for a half-space. One of the important reasons for choosing precisely these topics was the possibility of using comparatively simple analytical tools.

A second important reason was that these topics are not elaborated upon in the familiar series of books *Generalized Functions* (I. M. Gelfand *et al.*). But above all, of course, was the fact that these two topics reflect the ideas of the general theory of partial differential equations very graphically. Important and deep results are established without having to make any special demands as to type and order.

As in the first course, the development is accompanied by exercises. They deal with interesting questions that do not lie on the direct path of the main theory (in particular everything concerning S', the space containing functions of power growth and their derivatives).

The reader will find the material readily accessible if he has had a general course in mathematical analysis encompassing such topics as metric and Hilbert spaces, theory of the integral, and Fourier transforms. Some of these are covered, for example, in the author's book, *Mathematical Analysis:* 2nd edition, Fizmatgiz, Moscow, 1961

G. E. SHILOV

Part 1

GENERALIZED FUNCTIONS

Chapter 1

Elementary Theory of Generalized Functions

1.1. Problem of Extending the Collection of Ordinary Functions

The problems of mathematical analysis are couched mostly in classical function-theoretic language formulated at the beginning of the nineteenth century. Depending on the nature of the problem, one considers functions that are analytic, differentiable, continuous, (Lebesgue) integrable, and so on. Despite the vast arsenal of tools created in the profound development of the various parts of function theory, it has become clear nowadays that the classical apparatus is often inadequate and requires a certain extension.

In order to clarify this thought, let us agree to label a real function $y = f(x)$ *ordinary* if it is defined (almost everywhere) for $-\infty < x < \infty$ and (Lebesgue) integrable in each finite interval $a \leq x \leq b$. Thus, for example, every continuous function and every bounded (measurable) function are among the ordinary functions. As customary, we shall consider $f_1(x)$ and $f_2(x)$ to be the same ordinary function if $f_1(x) = f_2(x)$ almost everywhere.

The set of all ordinary functions will be denoted by E. Ordinary functions may be added together and multiplied by real numbers, and so they form a real linear space.

Various modes of convergence can also be defined within the space E. For example, if $f_1, f_2, \ldots, f_\nu, \ldots$ is a sequence of ordinary functions converging to $f(x)$ almost everywhere such that

$$|f_\nu(x)| \leq f_0(x)$$

for all ν, where $f_0(x)$ is a fixed ordinary function, then by the familiar Lebesgue theorem, the limit function $f(x)$ is locally integrable. In other words, $f(x)$ is also an ordinary function.

1*

3

On the other hand, an operation of such importance to analysis as differentiation is far from being applicable to every ordinary function. Certain ordinary functions and even ones that are continuous have no derivatives at all as is exemplified by the familiar Weierstrass function.† Certain other ordinary functions have a derivative but it is not an ordinary function (for instance, $y = 1/\sqrt{|x|}$). Sometimes the derivative of an ordinary function exists (almost everywhere) and is an ordinary function, but it is such that when integrated the original function is not recovered (for example, the Cantor function $y = C(x)$). Thus, its derivative is of no use. It is known‡ that an absolutely continuous function (and only such a function) posseses an ordinary derivative $f'(x)$ such that

$$f(x) = f(a) + \int_a^x f'(\xi)\, d\xi.$$

(Cantor's function $C(x)$ is continuous but not absolutely continous.) Finally, the fact that a sequence $f_\nu(x)$ converges to $f(x)$ is a long way off from always implying the convergence of the derivatives $f_\nu'(x)$ to $f'(x)$, even when they exist.

There is a very familiar class of functions in which none of the undesirable things mentioned above occur, namely, the class of analytic functions. However, it is too narrow for applications. It would be desirable to retain the use of at least all ordinary functions. Strange as it may seem at first glance, it turns out to be possible to expand rather than to contract the class of ordinary functions as a means of eliminating the difficulties with derivatives.

In this chapter, we shall extend E to a new collection in which it will be possible to give a reasonable definition of differentiation and moreover so that it is continuous under convergence.

To begin with, we observe the following. Suppose $\varphi(x)$ is a bounded (measurable) function having compact support. The latter means that $\varphi(x)$ vanishes outside some finite interval $a \leqq x \leqq b$. With each ordinary function $f(x)$, we associate the number

$$(f, \varphi) = \int_{-\infty}^{\infty} f(x)\, \varphi(x)\, dx.$$

---

† Or the similar function of van der Waerden. See E. C. Titchmarsh, *Theory of Functions*, Oxford University Press, 2nd ed. 1935, Chapt. XI, Sec. 11.23.

‡ See E. C. Titchmarsh, *ibid.*, Chapt. XI, Sec. 11.8.

The integral in this actually extends only over the interval $[a, b]$. If $f(x)$ is absolutely continuous and has an ordinary derivative $f'(x)$, we can similarly form the expression

$$(f', \varphi) = \int_{-\infty}^{\infty} f'(x) \, \varphi(x) \, dx.$$

Suppose now that $\varphi(x)$ is also absolutely continuous and has a bounded derivative $\varphi'(x)$. Then integration by parts can be applied to the last integral. This leads to the relation

$$(f', \varphi) = f(x) \, \varphi(x) \Big|_{-\infty}^{\infty} - \int_{-\infty}^{\infty} f(x) \, \varphi'(x) \, dx = - \int_{-\infty}^{\infty} f(x) \, \varphi'(x) \, dx.$$

The integrated term vanishes since $\varphi(x)$ has compact support. Therefore, for the case in question, we have

$$(f', \varphi) = - \int_{-\infty}^{\infty} f(x) \, \varphi'(x) \, dx = -(f, \varphi').$$

But even when $f'(x)$ does not exist, the expression

$$\int_{-\infty}^{\infty} f(x) \, \varphi'(x) \, dx \tag{1}$$

is nevertheless meaningful for each bounded function $\varphi(x)$ of compact support with a bounded derivative $\varphi'(x)$. Thus, although there may exist no function $f'(x)$ as such, if we needed the value of the integral of $f'(x)$ multiplied by a bounded function of compact support having a bounded derivative, we could always obtain one as if $f'(x)$ existed by taking (1) with the opposite sign.

It is this observation that permits us to arrive at a suitable extension of the concept of function. Before, we required a function to have a definite value at every point (or almost every point). What interests us now are merely the values of the integrals of a given function multiplied by certain "test" functions. If the *only* thing known about a function are such integrals, then we say that it is a *generalized function*. The foregoing discussion shows that if we choose our fund of test functions properly, then every generalized function will have a derivative, which will likewise be a generalized function.

But how can the integral of the product of a generalized function and a test function be defined without having the values of the function at individual points to work with? The answer is simple. *In this instance, the integral should be defined axiomatically rather than constructively.* This will be described in a subsequent section.

Test functions may be selected in a variety of ways and different classes of generalized functions thereby result. The above discussion indicates that the existence of the derivative of a generalized function can be assured if the (ordinary) derivative of each test function is also a test function. Hence, it follows that each test function must be taken to be infinitely differentiable. On the other hand, if we want an ordinary function with an arbitrary rate of growth at $\pm\infty$ to be among the generalized functions, i.e., if we want the integral

$$\int_{-\infty}^{\infty} f(x)\, \varphi(x)\, dx$$

to exist for *any ordinary function*, then we must assume that a test function $\varphi(x)$ has *compact support*.

With these heuristic considerations, we can begin a systematic presentation of the subject.

1.2. Test Functions of One Variable

We say that a real function $\varphi(x)$ defined over $-\infty < x < \infty$ is a *test function* if it is continuous, infinitely differentiable (in the usual sense) and has compact support, or in other words, vanishes outside a finite interval.

In particular, if a test function $\varphi(x)$ vanishes outside the interval $[a, b]$, we shall say that *it is carried in the interval* $[a, b]$ and we shall call $[a, b]$ the *support* of $\varphi(x)$ (notation: $[a, b] = \operatorname{supp} \varphi$).

Test functions may be added and multiplied by real numbers, and so they form a linear space. The space of all test functions will be denoted by K.

Multiplication of a test function $\varphi(x)$ by an *arbitrary* infinitely differentiable function $g(x)$ (with any kind of behavior at infinity) can also be defined in K. The operation is distributive over addition.

In many questions, it is necessary to consider complex-valued test functions of the form $\varphi_1(x) + i\varphi_2(x)$, where $\varphi_1(x)$ and $\varphi_2(x)$ are real test functions. Besides, the indicated operations, the space of complex-valued test functions (denoted as before by K) admits multiplication by a complex number and multiplication by an infinitely differentiable complex-valued function.

Of basic importance is the notion of convergence in K. We shall say that a sequence of test functions $\varphi_1(x), ..., \varphi_\nu(x), ...$ *converges to zero in K* if the functions all vanish outside the same bounded interval $[a, b]$ and converge uniformly to zero together with their derivatives of any order.

EXAMPLE 1°. The function

$$\varphi(x; a) = \begin{cases} e^{-\frac{a^2}{a^2-x^2}} & \text{for} \quad |x| < a, \\ 0 & \text{for} \quad |x| \geqq a \end{cases}$$

is a test function (Fig. 1). The sequence $\varphi_\nu(x) = \dfrac{1}{\nu} \varphi(x; a)$

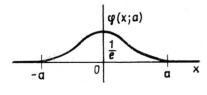

FIG. 1

converges to zero in K. The sequence $\varphi_\nu(x) = \dfrac{1}{\nu} \varphi\left(\dfrac{x}{\nu}; a\right)$ converges uniformly to zero over the whole line $-\infty < x < \infty$ together with all its derivatives. But it does *not* converge to zero in K since there is no one interval outside of which all $\varphi_\nu(x)$ vanish.

A sequence of test functions $\varphi_1(x), ..., \varphi_\nu(x), ...$ is said to *converge in K to* $\varphi(x)$ if $\varphi(x)$ belongs to K and the difference $\varphi(x) - \varphi_\nu(x)$ converges to zero in K.

Addition and multiplication by a number or an infinitely differentiable function are continuous operations under this convergence. In other words, *if* $\varphi_\nu \to \varphi$ *in K and* $\psi_\nu \to \psi$ *in K, then* $\alpha\varphi_\nu + \beta\psi_\nu \to \alpha\varphi + \beta\psi$ *in K for any real* α *and* β, *and* $g(x)\varphi_\nu(x) \to g(x)\varphi(x)$ *in K for any infinitely differentiable function* $g(x)$.

REMARK. The space K is *non-metrizable*. Stated otherwise, no distance function $\varrho(\varphi, \psi)$ with the standard properties† can be introduced in K so that the convergence of a sequence φ_ν to φ is equivalent to the fulfillment of the relation $\varrho(\varphi, \varphi_\nu) \to 0$.

---

† See J. L. Kelley, *General Topology*, D. van Nostrand, Princeton, 1955, Chapt. 4.

Indeed, if a metric space contains a system of sequences

$$\varphi_1^{(1)}, ..., \varphi_\nu^{(1)}, ... \to \varphi^{(1)}$$

$$\cdot \quad \cdot \quad \cdot \quad \cdot \quad \cdot \quad \cdot \quad \cdot \quad \cdot$$

$$\varphi_1^{(m)}, ..., \varphi_\nu^{(m)}, ... \to \varphi^{(m)}$$

$$\cdot \quad \cdot \quad \cdot \quad \cdot \quad \cdot \quad \cdot \quad \cdot \quad \cdot \quad \cdot$$

and the limits $\varphi^{(1)}, ..., \varphi^{(m)}, ...$ themselves form a convergent sequence with limit φ, then by choosing one element from each row in the proper way, we can always construct a sequence

$$\varphi_{\nu_1}^{(1)}, ..., \varphi_{\nu_m}^{(m)}, ...$$

also converging to φ.

However, this is not the sort of picture in K. For example, let

$$\varphi_\nu^{(m)} = \frac{1}{\nu}\varphi(x; m) \quad (\nu = 1, 2, ...; \; m = 1, 2, ...),$$

where $\varphi(x; m)$ is the function of Example 1°. For each fixed m, the sequence $\varphi_\nu^{(m)}$, converges to zero in K as in Example 1°. But no sequence of the form $\varphi_{\nu_m}^{(m)}$ converges to zero in K since there is no common closed interval outside of which all $\varphi_{\nu_m}^{(m)}$ vanish.

Thus, the convergence in K cannot be specified by a metric. However, it can be specified by a certain topology. See Prob. 7 of this section.

Problems

1. Show that K is *complete* under the convergence defined in the space. Completeness here means the following: If $\varphi_1, ..., \varphi_\nu, ...$ is a sequence of test functions carried in the same interval $[a, b]$ and the sequence of derivatives $\varphi_\nu^{(q)}(x)$ converges uniformly to the limit $\psi_q(x)$ for each $q = 0, 1, 2, ...$, then $\psi_0(x) = \lim_{\nu \to \infty} \varphi_\nu(x)$ is a test function, $\psi_q(x) = \psi_0^{(q)}(x)$, and the sequence $\varphi_\nu(x)$ converges to $\psi_0(x)$ in K.

2. Let $\varphi_\nu(x)$, $\nu = 1, 2, ...$ be a given sequence of test functions which vanish outside a fixed interval $[a, b]$. Show that if the constants b_ν tend to 0 sufficiently fast, then the sequence $b_\nu \varphi_\nu(x)$ converges to zero in K and the series $\sum_{\nu=1}^{\infty} b_\nu \varphi_\nu(x)$ converges in K.

Hint. The sequence b_ν^{-1} has to approach infinity faster than any sequence $m_{\nu p} = \max |\varphi_\nu^{(p)}(x)|$ (for each fixed $p = 0, 1, 2, ...$).

3. Let $\{\varphi(x; t)\}$ be a family of test functions carried in the same interval $[a, b]$ and depending continuously on a parameter t varying over the interval $[c, d]$. Continuous dependence on t here means the following: If $t_\nu \to t$, then

$$\varphi(x; t_\nu) \to \varphi(x; t) \quad \text{in} \quad K.$$

Show that the family $\{\varphi(x; t)\}$ can be integrated with respect to t and that the integral is again a test function carried in $[a, b]$. The integral is to be understood to be the limit (in K) of the customary type sums,

$$\sum_{k=0}^{m} \varphi(x; t_k)\, \Delta t_k.$$

Hint. Use Prob. 1. The function $\varphi(x; t)$ is uniformly continuous in x and t, as are all its derivatives with respect to x.

4. Consider the linear space C_0 of all continuous functions on the real line having compact support. Introduce a topology in C_0 by defining a neighborhood of zero to be the collection of all $\varphi \in C_0$ such that

$$|\varphi(x)| \leq \gamma(x),$$

with $\gamma(x)$ any everywhere positive continuous function.

Show that the convergence corresponding to the topology is as follows: a sequence $\varphi_\nu(x) \in C_0$ converges to zero in C_0 if all $\varphi_\nu(x)$ vanish outside the same closed interval and tend uniformly to zero as $\nu \to \infty$.

Hint. Two facts have to be proved: a) the elements φ_ν lie in a neighborhood of zero from some point onward, and b) if the functions $\varphi_\nu(x)$ lie in a neighborhood of zero from some point onward, they tend to zero in the indicated sense (prove by contradiction).

5. Show that the topological space of Prob. 4 is non-metrizable.
Hint. See the remark on p.7.

6. Consider a countable set A of elements of C_0 of the form

$$\frac{\varphi_0(x)}{m} + \frac{\varphi_m(x)}{k},$$

where $k = 1, 2, \ldots$, $m = 1, 2, \ldots$, $\varphi_0(0) \neq 0$, and $\varphi_m(x) = \varphi_0\left(\dfrac{x}{m}\right)$.

Show that $\varphi(x) \equiv 0$ belongs to the closure of the set A and yet is not the limit of any sequence of elements of A converging to 0 (O. G. Smolyanov).

7. Assign a topology to K in the following fashion. Take $m + 1$ positive continuous functions $\gamma_0(x), \ldots, \gamma_m(x)$, for any given m. A neighborhood of zero consists of all test functions $\varphi(x)$ for which

$$|\varphi(x)| \leq \gamma_0(x), \ldots, |\varphi^{(m)}(x)| \leq \gamma_m(x).$$

Show that convergence in K corresponds to this topology.
Hint. Cf. Prob. 4.

1.3. Generalized Functions of One Variable

1.3.1. With each ordinary function $f(x)$ (Sec. 1.1), there is associated a functional on the space K through

$$(f, \varphi) = \int_{-\infty}^{\infty} f(x)\, \varphi(x)\, dx. \tag{1}$$

The functional is linear. That is to say, for any φ_1 and φ_2 in K and any two real numbers α_1 and α_2,

$$(f, a_1\varphi_1 + \alpha_2\varphi_2) = \alpha_1(f, \varphi_1) + \alpha_2(f, \varphi_2).$$

Moreover, it is continuous in the following sense. If $\varphi_1, \ldots, \varphi_\nu, \ldots$ is a sequence of test functions converging to 0 in K, then

$$\lim_{\nu \to \infty} (f, \varphi_\nu) = 0.$$

This is a consequence of the customary properties of the integral.

Thus, the expression (1) is a continuous linear functional on K.

There exist other continuous linear functionals on K that do not reduce to the form (1).

Consider the example of the functional δ which assigns to each test function $\varphi(x)$ the number $\varphi(0)$. Such a functional is clearly linear and continuous. On the other hand, we assert that no functional of the form (1) can assign the number $\varphi(0)$ to each $\varphi(x)$. In fact, suppose that

$$\int_{-\infty}^{\infty} f(x)\, \varphi(x)\, dx = \varphi(0)$$

for some ordinary function $f(x)$ and every $\varphi(x)$ in K. In particular, let $\varphi(x) = \varphi(x; a)$ (Sec. 1.2, 1°). We must then have

$$\int_{-\infty}^{\infty} f(x)\, \varphi(x; a)\, dx = \varphi(0; a) = \frac{1}{e}. \tag{2}$$

But when $a \to 0$, the integral on the left-hand side approaches 0, which contradicts (2).

We define a *generalized function* to be a continuous linear functional on K, i.e., any functional f satisfying the conditions

(a) $(f, \alpha_1\varphi_1 + \alpha_2\varphi_2) = \alpha_1(f, \varphi_1) + \alpha_2(f, \varphi_2)$ for all φ_1 and φ_2 in K and any real numbers α_1 and α_2;

(b) if $\varphi_\nu \to 0$ in K, then $(f, \varphi_\nu) \to 0$.

If we can represent the functional in the form (1), we call it *regular* (or a *regular function*). If we cannot represent it in the form (1), we call it a *singular* functional (or *singular function*).

The functional

$$(\delta, \varphi) = \varphi(0)$$

just considered is singular.

Each ordinary function has a generalized function associated with it through formula (1). For example, corresponding to the ordinary function $f(x) \equiv C$ is the generalized function given by

$$(f, \varphi) \equiv (C, \varphi) \equiv C \int_{-\infty}^{\infty} \varphi(x)\, dx.$$

This generalized function is naturally called the *constant C*. In particular, the generalized function *unity* acts according to the formula

$$(1, \varphi) = \int_{-\infty}^{\infty} \varphi(x)\, dx.$$

1.3.2. Two generalized functions f_1 and f_2 are considered to be *equal* if the values of their corresponding functionals coincide on each test function:

$$(f_1, \varphi) \equiv (f_2, \varphi),$$

and to be *distinct* if their values are different on at least one φ_0 in K:

$$(f_1, \varphi_0) \neq (f_2, \varphi_0).$$

Let us show that *the generalized functions corresponding to two unequal ordinary functions $f_1(x)$ and $f_2(x)$ are distinct*. Suppose $f_1(x)$ and $f_2(x)$ determine the same functional through formula (1), so that for all $\varphi(x) \in K$,

$$\int_{-\infty}^{\infty} f_1(x)\, \varphi(x)\, dx = \int_{-\infty}^{\infty} f_2(x)\, \varphi(x)\, dx.$$

Set $f(x) = f_1(x) - f_2(x)$; $f(x)$ is also ordinary and we have

$$\int_{-\infty}^{\infty} f(x)\,\varphi(x)\,dx = 0$$

for every test function $\varphi(x) \in K$. Consider at first just those test functions $\varphi(x)$ with support in a fixed interval $[a, b]$. For any such function, we likewise have

$$\int_a^b f(x)\,\varphi(x)\,dx = 0. \tag{3}$$

The function $F(x) = \int_a^x f(\xi)\,d\xi$ is continuous. Integrating (3) by parts, we find

$$\int_a^b F(x)\,\varphi'(x)\,dx = 0. \tag{4}$$

By the fundamental lemma of the calculus of variations,† this implies that $F(x)$ is a constant. Since $F(a) = 0$, $F(x)$ vanishes identically. Hence, $f(x)$ which is equal almost everywhere to the derivative of its indefinite integral is equal to zero almost everywhere on $[a, b]$.

Since the interval $[a, b]$ is arbitrary, $f(x)$ is equal to zero almost everywhere on the whole line $-\infty < x < \infty$. Thus, $f_1(x)$ and $f_2(x)$ are equal almost everywhere, as asserted.

For completeness, we shall present a proof of the fundamental lemma as it applies to our case. Some of the considerations involved in it will be used later on.

Assume that $F(x)$ is not constant. Then there are points, say x_1 and x_2, where $F(x_1) < F(x_2)$. We shall show that there exists a test function $\varphi(x)$ for which the equation (4) does not hold. Choose an arbitrary number A between $F(x_1)$ and $F(x_2)$. Since $F(x)$ is a continous function, one can find disjoint intervals Δ_1 and Δ_2 containing x_1 and x_2, respectively, such that for any $x' \in \Delta_1$ and $x'' \in \Delta_2$,

$$F(x') < A < F(x'').$$

As $\varphi'(x)$, we choose any infinitely differentiable function which is positive on Δ_1, negative on Δ_2, vanishes outside Δ_1 and Δ_2, and is such that

$$\int_{-\infty}^{\infty} \varphi'(x)\,dx = \int_{\Delta_1} \varphi'(x)\,dx + \int_{\Delta_2} \varphi'(x)\,dx = 0.$$

† See G. Bliss, *Lectures on the Calculus of Variations*, Chicago University Press, 1946, Sec. 5.

Of course, we define $\varphi(x)$ itself by

$$\varphi(x) = \int_{-\infty}^{x} \varphi'(\xi)\, d\xi;$$

$\varphi(x)$ clearly belongs to K. We then have

$$\int_{-\infty}^{\infty} [F(x) - A]\, \varphi'(x)\, dx = \int_{\Delta_1} + \int_{\Delta_2} < 0,$$

since both terms are negative. But this implies that

$$\int_{-\infty}^{\infty} F(x)\, \varphi'(x)\, dx = \int_{-\infty}^{\infty} [F(x) - A]\, \varphi'(x)\, dx + A \int_{-\infty}^{\infty} \varphi'(x)\, dx < 0.$$

Thus, for our $\varphi'(x)$, the equation (4) is not satisfied. For later application, we mention that it is enough to require that (3) hold for just a countable set of test functions rather than for all of them. Such a set would be, for example, countably many $\varphi(x)$ each constructed according to the above rule for each pair of disjoint intervals Δ_1 and Δ_2 with rational endpoints.

1.3.3. Thus, the collection of all ordinary functions may be considered to be a subset of the collection of generalized functions. The collection of all generalized functions will be denoted by K'.

Let us discuss a few further examples of generalized functions which do not lead to ordinary functions.

1°. *The shifted delta-function* $\delta(x - a)$. This functional assigns to each test function $\varphi(x)$ the value $\varphi(a)$.

2°. *The functional* $1/x$. The function $1/x$ is not ordinary (it is not integrable in any neighborhood of the origin). Nevertheless, for any test function, one can compose the expression

$$\left(\frac{1}{x}, \varphi\right) = \int_{-\infty}^{\infty} \frac{\varphi(x)}{x}\, dx,$$

wherein the integral is understood to be the Cauchy principal value, i.e.,

$$\lim_{\varepsilon \to 0} \left\{ \int_{-\infty}^{-\varepsilon} \frac{\varphi(x)}{x}\, dx + \int_{\varepsilon}^{\infty} \frac{\varphi(x)}{x}\, dx \right\}.$$

By changing the variable in the first integral from x to $-x$, we obtain

$$\left(\frac{1}{x}, \varphi\right) = \lim_{\varepsilon \to 0} \left\{ -\int_{\varepsilon}^{\infty} \frac{\varphi(-x)\,dx}{x} + \int_{\varepsilon}^{\infty} \frac{\varphi(x)}{x}\,dx \right\}$$

$$= \lim_{\varepsilon \to 0} \int_{\varepsilon}^{\infty} \frac{\varphi(x) - \varphi(-x)}{x}\,dx = \int_{0}^{\infty} \frac{\varphi(x) - \varphi(-x)}{x}\,dx.$$

The integral is now absolutely convergent. The resulting functional can easily be seen to be linear and continuous.

3°. *A functional with order of singularity $\leq p$.* Such a functional has the form

$$(f, \varphi) = \sum_{k=0}^{p} \int_{-\infty}^{\infty} f_k(x)\,\varphi^{(k)}(x)\,dx = \sum_{k=0}^{p} (f_k(x), \varphi^{(k)}(x)), \tag{5}$$

where $f_0(x), \ldots, f_p(x)$ are ordinary functions.

The functions $f_k(x)$ in the representation (5) are no longer uniquely determined in contrast to (1) (see Prob. 5 of Sec. 1.4). A functional f is said to have order of singularity exactly p if it can be expressed as (5) for this p but cannot be so expressed for a smaller p. Ordinary functions have order of singularity 0.

The functional δ has order of singularity not exceeding 1 since it is possible to write

$$(\delta, \varphi) = \varphi(0) = -\int_{0}^{\infty} \varphi'(x)\,dx = \int_{-\infty}^{\infty} [-\theta(x)]\,\varphi'(x)\,dx,$$

where $\theta(x)$ is the ordinary function equal to 1 for $x > 0$ and to 0 for $x < 0$. Since it has been shown that $\delta(x)$ is not an ordinary function, its order of singularity is exactly equal to 1.

The functional f which assigns the number $\varphi(0) + \varphi'(1) + \cdots + \varphi^{(k)}(k) + \cdots$ to a test function $\varphi(x)$ does not have a finite order of singularity. In such instances, we shall say that the order of singularity of the functional f is $+\infty$

1.3.4. Let us pause to consider the question of functionals in the space of complex-valued test functions. The definition of a generalized function as a continuous linear functional remains the same as before. The only change is the possibility of the functional taking on complex values.

A functional may then be of the *first kind*, i.e., it satisfies the condition

$$(f, \alpha\varphi) = \alpha(f, \varphi),$$

for every complex number α, or of the *second kind*:

$$(f, \alpha\varphi) = \bar{\alpha}(f, \varphi),$$

where $\bar{\alpha}$ is the complex conjugate of α.

It is possible to associate four linear functionals with each ordinary complex function $f(x)$, namely,

$$(f, \varphi)_1 = \int_{-\infty}^{\infty} f(x)\,\varphi(x)\,dx; \qquad (6)$$

$$(f, \varphi)_2 = \int_{-\infty}^{\infty} \overline{f(x)}\,\varphi(x)\,dx; \qquad (7)$$

$$(f, \varphi)_3 = \int_{-\infty}^{\infty} f(x)\,\overline{\varphi(x)}\,dx; \qquad (8)$$

$$(f, \varphi)_4 = \int_{-\infty}^{\infty} \overline{f(x)}\,\overline{\varphi(x)}\,dx. \qquad (9)$$

The first two are functionals of the first kind and the last two are functionals of the second kind.

The multiplication of a functional f by a number α is *always* defined by the formula

$$(\alpha f, \varphi) = \alpha(f, \varphi).$$

Therefore, for the first and third functionals, the multiplication of $f(x)$ by α goes over into multiplying the corresponding functional by the same number. In the second and fourth cases, the multiplication of the function by α corresponds to multiplying the functional by the complex conjugate of α. Thus, one's choice of any of the functionals (6)–(9) is connected with a different way of imbedding the ordinary functions in the space of generalized functions.

If $f(x)$ is an ordinary function and $\bar{f}(x)$ is its complex conjugate, then it is easy to verify for $j = 1, 2, 3,$ and 4 that $(\bar{f}, \varphi) = \overline{(f, \bar{\varphi})}$. This motivates us to define the complex conjugate $\bar{f}$ of a given functional f by the formula

$$(\bar{f}, \varphi) = \overline{(f, \bar{\varphi})}. \qquad (10)$$

The functional $\bar{f}$ *defined* by (10) is obviously linear and continuous. Moreover, the transition from f to $\bar{f}$ for an ordinary function $f(x)$ *always* corresponds to taking the complex conjugate of $f(x)$ regardless of which of the four ways the set of ordinary functions is imbedded in the space of generalized functions.

The theorems that are proved for one of the types (6)–(9) are easily converted into theorems for functionals of any other type by the insertion and removal of complex conjugate signs. To simplify matters, we shall (until Sec. 2.7) always have the functional (6) in mind when considering the complex case.

Problems

1. Let $\varphi(x; t)$ be a test function depending on a parameter t, $c \leq t \leq d$, which is continuous with respect to t in K (Sec. 1.2, Prob. 3). Then according to Prob. 3 of Sec. 1.2.

$$\varphi(x) = \int_c^d \varphi(x; t)\, dt \tag{1}$$

is a test function. Show that for any $f \in K'$,

$$(f, \varphi) = \left(f, \int_c^d \varphi(x; t)\, dt \right) = \int_c^d (f, \varphi(x; t))\, dt.$$

Hint. The integral (1) is the limit of its Riemann sums in the sense of convergence in K.

2. Show that $\dfrac{1}{x}$ (Sec. 1.3.3, 2°) is a singular functional.

Hint. Assume the representation

$$\left(\frac{1}{x}, \varphi \right) = \int_{-\infty}^{\infty} f(x)\, \varphi(x)\, dx$$

to be possible, where $f(x)$ is an ordinary function. Then using test functions φ with support outside of the origin, show that $f(x)$ has to coincide with $\dfrac{1}{x}$ for $x \neq 0$.

3. Show that the functional $\dfrac{1}{x}$ has order of singularity 1.
Hint.

$$\left(\frac{1}{x}, \varphi \right) = -\int_{-\infty}^{\infty} \log |x|\, \varphi'(x)\, dx.$$

4. *The regularization problem.* Let $f(x)$ be locally integrable everywhere except in a neighborhood of $x = 0$ where it has at most a power singularity. Thus, for some m and sufficiently small $|x|$,

$$|f(x)| \leqq \frac{C}{|x|^m}.$$

Show that there exists a functional $f \in K'$ which operates on all test functions equal to zero in a neighborhood of $x = 0$ according to the formula

$$(f, \varphi) = \int_{-\infty}^{\infty} f(x)\,\varphi(x)\,dx.$$

Hint. For example,

$$(f, \varphi) = \int_{-\infty}^{\infty} f(x) \left\{ \varphi(x) - \left[\varphi(0) + x\varphi'(0) + \cdots \right. \right.$$

$$\left. \left. \cdots + \frac{x^{m-1}}{(m-1)!}\varphi^{(m-1)}(0) \right] h(x) \right\} dx,$$

where $h(x)$ is equal to 1 for $|x| \leqq 1$ and to 0 for $|x| > 1$.

5. Show that Prob. 4 is unsolvable for a function $f(x)$ which satisfies the estimate

$$|f(x)| > \frac{A_m}{|x|^m} \qquad (0 < x < x_0), \tag{1}$$

for any $m = 0, 1, 2, \ldots$ ("no regularization exists for the function in K'".)
Hint. Let $\varphi(x)$ be the function $\varphi(x; a)$ of Sec. 1.2. Then for sufficiently small ε_ν, the functions

$$\varphi_\nu(x) = \varepsilon_\nu \varphi\left(\nu\left(x - \frac{2}{\nu} \right) \right)$$

tend to 0 in K. Making use of condition (1), one can choose the numbers ε_ν so that

$$(f, \varphi_\nu) = \int_{-\infty}^{\infty} f(x)\,\varphi_\nu(x)\,dx \to \infty.$$

1.4. Operations on Generalized Functions of One Variable

1.4.1. Various operations are defined in the class of ordinary functions (though sometimes partially, as in the case of differentiation). Our problem now is to carry them over to the entire class of generalized functions.

2 Shilov

Addition and multiplication by a real number are defined by the formula

$$(\alpha_1 f_1 + \alpha_2 f_2, \varphi) = \alpha_1(f_1, \varphi) + \alpha_2(f_2, \varphi).$$

It is easy to verify that the functional $\alpha_1 f_1 + \alpha_2 f_2$ which it determines is again linear and continuous. Moreover, if f_1 and f_2 are regular functionals corresponding to functions $f_1(x)$ and $f_2(x)$, then $\alpha_1 f_1 + \alpha_2 f_2$ is also a regular functional corresponding to $\alpha_1 f_1(x) + \alpha_2 f_2(x)$.

Of course, the product of a generalized function and any infinitely differentiable function $\alpha(x)$ may also be defined. For if $f(x)$ is an ordinary function, then so is $\alpha(x) f(x)$, and thus

$$(\alpha(x) f(x), \varphi) = \int_{-\infty}^{\infty} \alpha(x) f(x) \varphi(x)\, dx = \int_{-\infty}^{\infty} f(x) \alpha(x) \varphi(x)\, dx = (f, \alpha\varphi),$$

where $\alpha(x) \varphi(x)$ is again a test function. If f is now an arbitrary generalized function, the product $\alpha(x) f$ may be *defined* by the formula

$$(\alpha(x)f, \varphi) = (f, \alpha(x) \varphi).$$

The functional $\alpha(x) f$ is clearly linear and continuous together with the functional f.

If $\alpha(x)$ is not infinitely differentiable, there is no reasonable way of defining in K' the product of $\alpha(x)$ with a generalized function (see Prob. 3 of Sec. 2.2).

1.4.2. We now proceed to define differentiation. In accordance with our discussion of Sec. 1.1, we set

$$(f', \varphi) = (f, -\varphi')$$

for any generalized function f. Since $\varphi'(x)$ is a test function together with $\varphi(x)$, the expression on the right-hand side is well-defined. The resulting functional is obviously linear and continuous as well: if $\varphi_\nu \to 0$ in K, then $\varphi_\nu' \to 0$ in K also, and therefore

$$(f', \varphi_\nu) = (f, -\varphi_\nu') \to 0.$$

Thus, *every generalized function has a derivative* (which likewise is a generalized function). Since the process of differentiation may be applied further, *every generalized function has derivatives of all orders.*

In this connection,

$$(f'', \varphi) = (f', -\varphi') = (f, \varphi''),$$

and similarly for any q,

$$(f^{(q)}, \varphi) = (f, (-1)^q \varphi^{(q)}) = (-1)^q (f, \varphi^{(q)}). \qquad (1)$$

It is easy to see that the derivative possesses the usual linearity property. If f_1 and f_2 are two generalized functions and α_1 and α_2 are constants, then

$$\begin{aligned}((\alpha_1 f_1 + \alpha_2 f_2)', \varphi) &= (\alpha_1 f_1 + \alpha_1 f_2, -\varphi') \\ &= \alpha_1(f_1, -\varphi') + \alpha_2(f_2, -\varphi') \\ &= \alpha_1(f_1', \varphi) + \alpha_2(f_2', \varphi) = (\alpha_1 f_1' + \alpha_2 f_2', \varphi).\end{aligned}$$

Hence,

$$(\alpha_1 f_1 + \alpha_2 f_2)' = \alpha_1 f_1' + \alpha_2 f_2'.$$

Furthermore, if $\alpha(x)$ is an infinitely differentiable function, then

$$\begin{aligned}((\alpha(x)f)', \varphi) &= (\alpha f, -\varphi') = (f, -\alpha\varphi') = (f, -(\alpha\varphi)') + (f, \alpha'\varphi) \\ &= (f', \alpha\varphi) + (\alpha'f, \varphi) = (\alpha f' + \alpha'f, \varphi)\end{aligned}$$

and therefore

$$(\alpha f)' = \alpha f' + \alpha' f.$$

If an ordinary function $f(x)$ has an ordinary derivative and it can be recovered from it by integration (in other words, $f(x)$ is absolutely continuous on each finite interval), then as we saw in Sec. 1.1, the derivative of $f(x)$ as a generalized function coincides with $f'(x)$. Conversely, if $f(x)$ is an ordinary function and its derivative f' in K' is also an ordinary function, i.e., $f' = f'(x)$, then $f'(x)$ is the derivative of $f(x)$ in the usual sense (almost everywhere and indeed everywhere if $f'(x)$ is continuous; see Sec. 1.5.1). But in the general case, the derivative of an ordinary function $f(x)$ is no longer an ordinary function.

1.4.3. Let us consider a few examples involving the determination of the derivative.

1°. We first find the derivative of the ordinary function

$$\theta(x) = \begin{cases} 1 & \text{for } x > 0, \\ 0 & \text{for } x < 0. \end{cases}$$

By definition,

$$(\theta', \varphi) = (\theta, -\varphi') = -\int_0^\infty \varphi'(x)\, dx = \varphi(0) = (\delta, \varphi),$$

whence,

$$\theta' = \delta.$$

2*

In similar fashion, we have

$$\theta'(x - a) = \delta(x - a)$$

for any a. (Observe that the derivative of $\theta(x)$ in the usual sense is equal to 0 for $x \neq 0$ and does not exist for $x = 0$.)

2°. $\theta(x)$ is the simplest discontinuous function. Let us look at the derivative of a piecewise absolutely continuous function $f(x)$ with piecewise

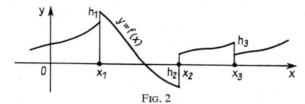

FIG. 2

continuous derivative $f'(x)$. The discontinuities are at $x_1, x_2, \ldots$, and the corresponding jumps are $h_1, h_2, \ldots$ (Fig. 2). Introduce the function

$$f_1(x) = f(x) - \sum_k h_k \theta(x - x_k).$$

The function $f_1(x)$ is absolutely continuous and can be recovered from its derivative $f_1'(x)$, which coincides with $f'(x)$ everywhere except at the discontinuities of $f(x)$ where $f'(x)$ does not exist. Therefore, $f_1'(x)$ is the derivative of the generalized function f_1 in K'. On the other hand, by the result of Example 1°,

$$f_1'(x) = f' - \sum_k h_k \delta(x - x_k), \tag{2}$$

wherein f' is the derivative of the generalized function $f(x)$. From (2), we obtain

$$f' = f_1'(x) + \sum_k h_k \delta(x - x_k). \tag{3}$$

Thus, the derivative of the generalized function $f(x)$ is composed of its ordinary derivative and a sum of delta-functions at its discontinuities with the corresponding jumps as coefficients.

In particular, if $f(x)$ vanishes everywhere outside the interval $[a, b]$ and

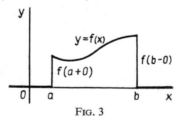

FIG. 3

is absolutely continuous inside it (Fig. 3), then

$$f' = f_1'(x) + f(a + 0)\,\delta(x - a) - f(b - 0)\,\delta(x - b). \qquad (4)$$

3°. Let us find the derivative of the ordinary function $y = \log|x|$. Its usual derivative $y' = 1/x$ is not an ordinary function. Applying the general rule, we obtain

$$((\log|x|)', \varphi) = (\log|x|, -\varphi') = -\int_{-\infty}^{\infty} \log|x|\,\varphi'(x)\,dx$$

$$= -\lim_{\varepsilon \to 0}\left\{\int_{-\infty}^{-\varepsilon} \log|x|\,\varphi'(x)\,dx + \int_{\varepsilon}^{\infty} \log|x|\,\varphi'(x)\,dx\right\}$$

$$= -\lim_{\varepsilon \to 0}\left\{\log\varepsilon\varphi(-\varepsilon) - \int_{-\infty}^{-\varepsilon}\frac{\varphi(x)}{x}\,dx - \log\varepsilon\varphi(\varepsilon) - \int_{\varepsilon}^{\infty}\frac{\varphi(x)}{x}\,dx\right\}$$

$$= -\lim_{\varepsilon \to 0}\left\{\log\varepsilon[\varphi(-\varepsilon) - \varphi(\varepsilon)] - \int_{|x|\geq\varepsilon}\frac{\varphi(x)}{x}\,dx\right\}$$

$$= \lim_{\varepsilon \to 0}\int_{|x|\geq\varepsilon}\frac{\varphi(x)}{x}\,dx = \left(\frac{1}{x}, \varphi\right),$$

where $1/x$ is the functional defined in Sec. 1.3.3, 2°.

4°. The derivative of the generalized function $\delta(x)$ is the functional operating through the formula

$$(\delta', \varphi) = (\delta, -\varphi') = -\varphi'(0).$$

Similarly, for any q,

$$(\delta^{(q)}, \varphi) = (\delta, (-1)^q\,\varphi^{(q)}(x)) = (-1)^q\,\varphi^{(q)}(0).$$

5°. A functional of order of singularity not exceeding p (Sec. 1.3.3, 3°) can be expressed by means of (1) in the form

$$(f, \varphi) = \sum_{k=0}^{p}(-1)^k(f_k^{(k)}, \varphi) = \left(\sum_{k=0}^{p}(-1)^k f_k^{(k)}, \varphi\right).$$

In other words, it is a sum of derivatives of ordinary functions of at most p-th order. Conversely, every such sum is clearly a functional of order of singularity not exceeding p.

Problems

1. Find the derivatives of the functions

a) $y = x_+^\lambda = \begin{cases} x^\lambda & \text{for } x > 0, \\ 0 & \text{for } x < 0, \quad -1 < \lambda < 0; \end{cases}$

b) $y = \log x_+ = \begin{cases} \log x & \text{for } x > 0, \\ 0 & \text{for } x < 0. \end{cases}$

Ans.

$$\text{a) } (y', \varphi) = \int_0^\infty \lambda x^{\lambda-1} \, [\varphi(x) - \varphi(0)] \, dx;$$

$$\text{b) } (y', \varphi) = \int_0^\infty \frac{1}{x} \, [\varphi(x) - \varphi(0) \, \theta(1 - x)] \, dx.$$

2. Derive the formula

$$\alpha(x) \, \delta^{(k)}(x) = \sum_{j=0}^{k} (-1)^{j+k} \binom{k}{j} \alpha^{(k-j)}(0) \, \delta^{(j)}(x)$$

for any infinitely differentiable function $\alpha(x)$.

3. Find the q-th derivative of the function x_+^k ($k = 0, 1, 2, ...$).
Ans.

$$(x_+^k)^{(q)} = \begin{cases} k(k - 1) \cdots (k - q + 1) x_+^{k-q} & q \leq k, \\ k! \, \delta^{(q-k-1)}(x) & q > k. \end{cases}$$

4. What is the derivative of the Cantor function $C(x)$? ($C(x)$ is a continuous monotone function on $[0, 1]$. Its value increases from 0 up to 1 while remaining constant on each interval contiguous to Cantor's set.)†
Ans.

$$(C'(x), \varphi) = \sum_j C_j[\delta(x - \alpha_j) - \delta(x - \beta_j)],$$

where C_j is the value of $C(x)$ in the contiguous interval (α_j, β_j).

5. If

$$\int_{-\infty}^\infty f(x) \, \varphi(x) \, dx = 0$$

for every test function $\varphi(x)$, then $f(x) \equiv 0$ (Sec. 1.3.2). What conclusions

† See E. C. Titchmarsh, *ibid.*, Chapt. XI, Sec. 11.72.

may be drawn about the ordinary functions $f_0(x)$, $f_1(x)$, ..., $f_m(x)$ if it is known that

$$\int_{-\infty}^{\infty} [f_0(x)\, \varphi(x) + f_1(x)\, \varphi'(x) + \cdots + f_m(x)\, \varphi^{(m)}(x)]\, dx = 0$$

for any test function $\varphi(x)$?

Ans. There exist ordinary functions $g_0(x)$, ..., $g_m(x)$ such that

$$f_0 = Ig_0 + g_1, f_1 = Ig_1 + g_2, ..., f_{m-1} = Ig_{m-1} + g_m, f_m = Ig_m,$$

where I denotes indefinite integral.

1.5. Ordinary Differential Equations

The operations which have been defined for generalized functions, namely, differentiation, multiplication by a function, and addition, permit us to construct differential expressions of the form

$$\alpha_0(x)\, y^{(m)} + a_1(x)\, y^{(m-1)} + \cdots + a_m(x)\, y - b(x),$$

where $a_0(x)$, $a_1(x)$, ..., $a_m(x)$ are prescribed infinitely differentiable functions and y and $b(x)$ are generalized functions. Equating such an expression to zero, we obtain an ordinary m-th order linear differential equation for the generalized function y. This raises the question of describing the set of all solutions of such an equation.

1.5.1. Consider first the simplest equation

$$y' = 0. \tag{1}$$

We shall show that *the general solution of this equation in the class of generalized functions is $y = C$* (a constant).

Equation (1) is equivalent to the equation

$$(y', \varphi) = (y, -\varphi') = 0 \tag{2}$$

for each test function φ. The functional y is thereby already defined on the collection K_0 of test functions that can be represented as derivatives of other test functions. We must clarify in what way the functional y can be extended from K_0 to the whole space K.

It is easy to verify that a test function $\varphi_0(x)$ can be represented as the derivative of another test function if and only if

$$\int_{-\infty}^{\infty} \varphi_0(x)\, dx = 0. \tag{3}$$

Indeed, if $\varphi_0(x) = \varphi_1'(x)$, then

$$\int_{-\infty}^{\infty} \varphi_0(x)\, dx = \varphi_1(x)\; \Big|_{-\infty}^{\infty} = 0.$$

On the other hand, if condition (3) holds, we set

$$\varphi_1(x) = \int_{-\infty}^{x} \varphi_0(\xi)\, d\xi.$$

It merely remains to be shown that $\varphi_1(x)$ is a test function. But this is obvious since $\varphi_1(x)$ is infinitely differentiable together with $\varphi_0(x)$ and it has compact support on account of condition (3).

Now let $\varphi_1(x)$ be a fixed test function with the property

$$\int_{-\infty}^{\infty} \varphi_1(x)\, dx = 1.$$

For any test function $\varphi(x)$, we can write down the relation

$$\varphi(x) - \varphi_1(x) \int_{-\infty}^{\infty} \varphi(x)\, dx = \varphi_0(x),$$

where $\varphi_0(x)$ obviously satisfies the condition (3). Hence, it is apparent that if we prescribe the value of the required functional y on the test function $\varphi_1(x)$, then its value on any φ will be uniquely determined by

$$(y, \varphi) = (y, \varphi_1) \int_{-\infty}^{\infty} \varphi(x)\, dx. \tag{4}$$

Let the value of (y, φ_1) be an arbitrary fixed number, for instance, C_1. Then (4) yields

$$(y, \varphi) = C_1 \int_{-\infty}^{\infty} \varphi(x)\, dx = \int_{-\infty}^{\infty} C_1 \varphi(x)\, dx,$$

or in other words, the generalized function y is a constant C_1, as asserted.

As a corollary, we obtain the following:

If $f' = g'$ for two generalized functions f and g, then f and g differ by a constant.

As an example, let f be a functional having a derivative which is an ordinary function $f'(x)$. Then f is an ordinary function, moreover absolutely continuous, and its derivative (in the usual sense) coincides with $f'(x)$. To show this, consider the indefinite integral

$$g(x) = \int_{0}^{x} f'(\xi)\, d\xi.$$

The function $g(x)$ is defined everywhere and is absolutely continuous. Its usual derivative coincides with $f'(x)$ almost everywhere and its derivative in K' is likewise $f'(x)$. Hence, $f = g(x) + C$, as asserted.

1.5.2. Consider now an arbitrary homogeneous linear system of p equations in p unknown functions,†

$$\left.\begin{array}{l} y_0' = a_{00}(x)\, y_0 + \cdots + a_{0,\,p-1}(x)\, y_{p-1}, \\ \cdots\cdots\cdots\cdots\cdots\cdots\cdots\cdots\cdots\cdots\cdots\cdots \\ y_{p-1}' = a_{p-1,\,0}(x)\, y_0 + \cdots + a_{p-1,\,p-1}(x)\, y_{p-1}, \end{array}\right\} \tag{5}$$

where $a_{00}(x), \ldots, a_{p-1,p-1}(x)$ are infinitely differentiable functions. The system has a certain number of ordinary solutions, which are also infinitely differentiable. Let us show that it has no other solutions in the space K' besides the ordinary ones. As we know, the classical solutions form a p-dimensional linear space for which any set of p linearly independent solutions is a basis. If we write down the components of the separate solutions in columns, we obtain the fundamental matrix corresponding to such a basis,

$$U = \left\|\begin{array}{ccc} u_{00}(x) & \cdots & u_{0,\,p-1}(x) \\ \cdot\cdot\cdot & \cdot\cdot\cdot & \cdot\cdot\cdot \\ u_{p-1,\,0}(x) & \cdots & u_{p-1,\,p-1}(x) \end{array}\right\|.$$

The determinant of the matrix (the Wronskian) does not vanish. If the system (5) is written as

$$y' = Ay, \tag{6}$$

where y is an unknown p-dimensional vector in K' and $A = \|a_{jk}(x)\|$ is the coefficient matrix, then each of the columns of the matrix U considered as a vector will be a solution of (6). This fact can be expressed by the matrix equation

$$U' = AU. \tag{7}$$

Let $y = Uz$, where z is a new unknown vector (in K'). Substituting in (6) and making use of (7), we obtain

$$U'z + Uz' = AUz = U'z,$$

or

$$Uz' = 0.$$

---

† The advantage in numbering from 0 to $p-1$ (rather than from 1 to p) will become clear as we go along.

Multiplication by U^{-1} then yields the uncoupled system

$$z' = 0.$$

By our earlier result, z is a constant. Hence, the vector $y = Uz$ is a linear combination of the vectors in the fundamental matrix and is thereby a classical solution.

1.5.3. Consider the simplest non-homogeneous equation

$$g' = f. \tag{8}$$

Here, f is a prescribed generalized function and g is an unknown one.

THEOREM. *The equation (8) with any right-hand side f has a solution in the class of generalized functions.*

The solution is naturally called the *primitive* or *indefinite integral* of the generalized function f.

Proof. Equation (8) is equivalent to the equation

$$(g, -\varphi') = (g', \varphi) = (f, \varphi) = \left(f, \int_{-\infty}^{x} \varphi'(\xi)\, d\xi \right)$$

for any test function φ. As a consequence, *the functional g is already defined on any test function φ_0 which is the derivative of some other test function φ*, i. e., it is defined on the manifold K_0 considered in Sec. 1.5.1. We must extend the functional g to the whole space K. This may be accomplished as follows. As in Sec. 1.5.1. we introduce a test function φ_1 for which

$$\int_{-\infty}^{\infty} \varphi_1(x)\, dx = 1.$$

By means of φ_1, we can represent any test function φ in the form

$$\varphi(x) = \varphi_1(x) \int_{-\infty}^{\infty} \varphi(x)\, dx + \varphi_0(x),$$

where φ_0 belongs to K_0. We have thereby associated with each test function φ its "projection" on the subspace K_0.

If a sequence of functions $\varphi_1, \varphi_2, \dots$ converges to 0 in K, then the sequence of its projections $\varphi_{10}, \varphi_{20}, \dots$ also converges to 0 in K since

$$\int_{-\infty}^{\infty} \varphi_\nu(x)\, dx = (1, \varphi_\nu) \to 0.$$

We now set

$$(g_0, \varphi) = (g, \varphi_0) = -\left(f, \int_{-\infty}^{x} \varphi_0(\xi)\, d\xi \right) \tag{9}$$

for any $\varphi(x) \in K$. It is easy to verify that the functional g_0 so defined is linear and continuous. The general solution of (8) is obtained by adding to the particular solution just found the general solution of the homogeneous equation, which by Sec. 1.5.1 is an arbitrary constant.

Thus, all the solutions of equation (8) are described by the formula

$$g = g_0 + C,$$

where g_0 is the functional given by (9).

If $f = f(x)$ is an ordinary function, its ordinary indefinite integral $F(x)$ is already a solution of (8) and the general solution of the non-homogeneous equation (8) is the sum of the particular solution $F(x)$ and the general solution of the homogeneous equation $g_0' = 0$. The latter is once more the constant function and so the general solution of (8) is $F(x) + C$. We see that when f is an ordinary function, all of the solutions of (8) in K' are classical solutions.

1.5.4. The determination of the general solution of a non-homogeneous system

$$
\left.
\begin{aligned}
y_0' - a_{00}(x)\, y_0 - \cdots - a_{0,\,p-1}(x)\, y_{p-1} &= f_0, \\
\cdots \quad \cdots \quad \cdots \quad \cdots \quad \cdots \quad \cdots \quad \cdots \quad & \\
y_{p-1}' - a_{p-1,\,0}(x)\, y_0 - \cdots - a_{p-1,\,p-1}(x)\, y_{p-1} &= f_{p-1},
\end{aligned}
\right\} \quad (10)
$$

where the f_j are generalized functions and the $a_{jk}(x)$ are ordinary infinitely differentiable functions, can be reduced to solving equations of the form (8).

In fact, if we make the substitution $y = Uz$ applied before, where U is the fundamental matrix of solutions of the corresponding homogeneous system ($f_j = 0$), then we obtain $Uz' = f$ or $z' = U^{-1}f$. The unknowns in this system have been "separated" and each equation in it is of the form (8).

Finally, a higher order non-homogeneous differential equation

$$y^{(p)} + a_{p-1}(x)\, y^{(p-1)} + \cdots + a_0(x)\, y = f, \qquad (11)$$

where the $a_j(x)$ are infinitely differentiable functions and f is any generalized function, can be transformed into a system of the form (10) by means of the substitution $y_0 = y,\ y_1 = y',\ \ldots,\ y_{p-1} = y^{(p-1)}$. Hence, the determination of the general solution of an equation such as (11) can also be reduced to solving equations of the form (8).

In every one of these instances, if the right-hand sides are ordinary functions, all of the solutions in K' also turn out to be ordinary functions, namely, the classical solutions.

REMARK. Equations with singular coefficients generally behave in a more complicated fashion. They may have more linearly independent solutions than the order of the equation (see Prob. 1) or a fewer number than the order (see Prob. 2, where a first order equation has no non-trivial solutions at all in K').

1.5.5. We shall say that a generalized function f is *equal to zero for* $x < 0$, if $(f, \varphi) = 0$ for each test function φ which vanishes for $x \geq 0$ (that is to say, which is different from zero for $x < 0$ at the most). Every ordinary function equal to zero for $x < 0$ (in the usual sense) has this property. Then, $\delta(x)$, for example, has this property as do all of its derivatives.

LEMMA. *If $f \in K'$ and is equal to zero for $x < 0$, then it has a primitive $g \in K'$ also equal to zero for $x < 0$ which is unique.*

Proof. The uniqueness of a primitive equal to zero for $x < 0$ follows from the fact that two primitives differ by a constant. Since the constant equals zero for $x < 0$, it is identically equal to zero. As our required primitive g_0, we can take the one determined by formula (9) wherein

$$\varphi_0(x) = \varphi(x) - \varphi_1(x) \int_{-\infty}^{\infty} \varphi(x)\, dx, \quad \int_{-\infty}^{\infty} \varphi_1(x)\, dx = 1$$

and $\varphi_1(x)$ vanishes for $x > 0$. If in addition $\varphi(x) = 0$ for $x > 0$, then $\varphi_0(x)$ also vanishes for $x > 0$ and hence so does its primitive. Therefore,

$$(g_0, \varphi) = (g, \varphi_0) = -\left(f, \int_{-\infty}^{x} \varphi_0(\xi)\, d\xi \right) = 0,$$

which proves the lemma.

Consider a system of differential equations given in matrix form by

$$y' - Ay = f, \tag{12}$$

where the generalized vector function f is equal to zero for $x < 0$.

Let us show that the system has a unique solution $y = y(x)$ which is equal to zero for $x < 0$.

In fact, as before, the substitution $y = Uz$ transforms the system (12) into the form

$$z' = U^{-1}f, \tag{13}$$

where $U^{-1}f$ is also evidently equal to zero for $x < 0$. Applying the lemma, we find that the system (13) has a unique solution z whose components are all equal to zero for $x < 0$. Hence, $y = Uz$ is equal to zero for $x < 0$, as asserted.

1.5.6. When $f_0(x), ..., f_{p-1}(x)$ are ordinary functions, the solution $y(x)$ $= \{y_0(x), ..., y_{p-1}(x)\}$ to the system of first order linear differential equations (10) is uniquely determined by the assignment of an initial vector $y(0) = \{y_0(0), ..., y_{p-1}(0)\}$. This raises the question of how to pose the corresponding problem in the class of generalized vector functions considering that it makes no sense to talk about the values of a generalized function at isolated points.

If $f = \{f_0(x), ..., f_{p-1}(x)\}$ is an ordinary vector function, then as proved earlier every solution $y(x)$ is also an ordinary vector function. Let $Y(x)$ denote a generalized vector function equal to zero for $x < 0$ and to the required (ordinary) solution $y(x)$ for $x > 0$. Further, let $Y_1(x)$ denote a generalized vector function equal to zero for $x < 0$ and to $y'(x)$ for $x > 0$. The system (10) may now be expressed as

$$Y_1 - AY = F, \tag{14}$$

where $A = \|a_{jk}\|$ is the coefficient matrix of (10) and F is a generalized function equal to zero for $x < 0$ and to $f(x)$ for $x > 0$.

The function Y_1 is generally speaking not the derivative of Y in K'. What holds is the relation

$$Y' = Y_1 + y(0) \delta(x), \tag{15}$$

which follows from formula (3) of Sec. 1.4.

Combining (14) and (15), we end up with a system of equations

$$Y' - AY = F + y(0) \delta(x) \tag{16}$$

for Y in K'.

The general solution of (16) is the sum of any available particular solution Y and the general solution of the homogeneous system $Y' - AY = 0$, shown before to be the classical solution. A unique solution Y is determined from the general solution by imposing the additional condition that

$$Y = 0 \quad \text{for} \quad x < 0. \tag{17}$$

As we have already said, (17) has the following meaning for a generalized function Y: For each test function $\varphi(x)$ which is different from zero for at most $x < 0$, the relation $(Y, \varphi) = 0$ holds.

Thus, we have arrived at the following result:

Let there be given a system of equations

$$Y' - AY = F(x) + y(0)\,\delta(x),$$

where F is a generalized vector function equal to zero for $x < 0$ and $y(0)$ is a prescribed vector.

Then the system has a solution $Y \in K'$, which is unique and vanishes for $x < 0$.

If $F(x)$ is an ordinary vector function, then the corresponding solution is the ordinary solution of the system

$$y'(x) - Ay(x) = F(x) \quad (x \geqq 0)$$

equaling $y(0)$ at $x = 0$.

We next formulate the appropriate result for a p-th order differential equation

$$a_p y^{(p)} + a_{p-1} y^{(p-1)} + \cdots + a_0 y = f(x). \tag{18}$$

Introducing the new unknown functions

$$y_0 = y, \; y_1 = y', \ldots, y_{p-1} = y^{(p-1)},$$

we write (18) as a system

$$\hat{y}' - A\hat{y} = \hat{f}, \tag{19}$$

with a coefficient matrix A of the special form

$$A = \begin{Vmatrix} 0 & 1 & 0 & \cdots & 0 \\ 0 & 0 & 1 & \cdots & 0 \\ & & \cdots & & \\ 0 & 0 & 0 & \cdots & 1 \\ -\dfrac{a_0}{a_p} & -\dfrac{a_1}{a_p} & -\dfrac{a_2}{a_p} & \cdots & -\dfrac{a_{p-1}}{a_p} \end{Vmatrix}$$

and a right-hand side $\hat{f} = \left(0, 0, \ldots, \dfrac{1}{a_p} f(x)\right)$.

To (19), there corresponds the following system in generalized functions;

$$
\begin{array}{llll}
Y_0' & - Y_1 & = y_0(0)\,\delta(x), \\
Y_1' & - Y_2 & = y_1(0)\,\delta(x), \\
\end{array}
$$

$$\cdot\ \cdot\ \cdot\ \cdot\ \cdot\ \cdot\ \cdot\ \cdot\ \cdot\ \cdot\ \cdot\ \cdot\ \cdot\ \cdot\ \cdot\ \cdot$$

$$
Y_{p-2}' \qquad\qquad\qquad - Y_{p-1} = y_{p-2}(0)\,\delta(x),
$$

$$
Y_{p-1}' + \frac{a_0}{a_p}\,Y_0 + \frac{a_1}{a_p}\,Y_1 + \cdots + \frac{a_{p-1}}{a_p}\,Y_{p-1} = \frac{F(x)}{a_p} + y_{p-1}(0)\,\delta(x).
$$

This system is in turn equivalent to a p-th order equation which results when the unknowns $Y_1, \ldots, Y_{p-1}$ are eliminated. Namely, we have

$$
Y_1 = Y_0' - y_0(0)\,\delta(x),
$$

$$
Y_2 = Y_1' - y_1(0)\,\delta(x) = Y_0'' - y_0(0)\,\delta'(x) - y_1(0)\,\delta(x),
$$

$$\cdot\ \cdot\ \cdot\ \cdot\ \cdot\ \cdot\ \cdot\ \cdot\ \cdot\ \cdot\ \cdot\ \cdot\ \cdot\ \cdot\ \cdot\ \cdot$$

$$
Y_{p-1} = Y_0^{(p-1)} - y_0(0)\,\delta^{(p-2)}(x) - y_1(0)\,\delta^{(p-3)}(x) - \cdots - y_{p-2}(0)\,\delta(x)
$$

Hence,

$$
Y_{p-1}' + \sum_{j=0}^{p-1} \frac{a_j}{a_p}\,Y_j = Y_0^{(p)} - y_0(0)\,\delta^{(p-1)}(x) - y_1(0)\,\delta^{(p-2)}(x)
$$

$$
\cdots - y_{p-2}(0)\,\delta'(x)
$$

$$
+ \frac{a_0}{a_p}\,Y_0 + \frac{a_1}{a_p}\,(Y_0' - y_0(0)\,\delta(x)) + \cdots
$$

$$
\cdots + \frac{a_{p-1}}{a_p}\,(Y_0^{(p-1)} - y_0(0)\,\delta^{(p-2)}(x) -
$$

$$
\cdots - y_{p-2}(0)\,\delta(x)) = y_{p-1}(0)\,\delta(x) + \frac{1}{a_p}\,F(x),
$$

or equivalently,

$$
\sum_{k=0}^{p} a_k \left[Y^{(k)} - \sum_{j=0}^{k-1} y_j(0)\,\delta^{(k-1-j)}(x) \right] = F(x). \tag{20}
$$

By virtue of the foregoing discussion, we have obtained the following result.

THEOREM. *The equation* (20) *has a solution* $Y \in K'$, *which is unique and vanishes for* $x < 0$. *If* $f(x)$ *is an ordinary function, the indicated solution* *is the classical solution of equation (18) for $x \geq 0$ which at $x = 0$ satisfies the conditions $y(0) = y_0(0), \ldots, y^{(p-1)}(0) = y_{p-1}(0)$.*

1.5.7. Let

$$P[y] \equiv \sum_{k=0}^{p} a_k y^{(k)} \tag{21}$$

be a linear differential operator. In certain questions, an important role is played by the fundamental functions for the operator $P[y]$. By definition, a function $\mathscr{E}(x)$ is a *fundamental function* for $P[y]$ if it satisfies the equation

$$P[\mathscr{E}] \equiv \sum_{k=0}^{p} a_k \mathscr{E}^{(k)} = \delta(x). \tag{22}$$

Of course, a fundamental function is uniquely determined by (22) only to within an additive term. Any solution of the homogeneous equation

$$\sum_{k=0}^{p} a_k y^{(k)}(x) = 0 \tag{23}$$

may be added to it.

A fundamental function can be uniquely determined by requiring in addition that it vanish for $x < 0$. As is apparent from formula (20), the fundamental function $\mathscr{E}(x)$ is then an ordinary function equal to zero for $x < 0$ and to the solution of (23) for $x > 0$ satisfying the initial data

$$y_0(0) = \cdots = y_{p-2}(0) = 0, \quad y_{p-1}(0) = \frac{1}{a_p}.$$

If we sum this ordinary function with any solution of (23) for all x, we obtain a description of all fundamental functions for the operator (21):

Every fundamental function $\mathscr{E}(x)$ is an ordinary solution of (23) both for $x < 0$ and for $x > 0$ such that

$$y_0(0+) = y_0(0-), \ldots, y_{p-2}(0+) = y_{p-2}(0-),$$

$$y_{p+1}(0+) = y_{p-1}(0-) + \frac{1}{a_p}.$$

Problems

1. Construct $k + q$ linearly independent solutions of the q-th order equation

$$x^k y^{(q)}(x) = 0$$

in K'.

Ans.
For $q \geqq k$,

$$y_1 = x^{q-1}, \ y_2 = x^{q-2}, \ ..., \ y_q = 1, \ y_{q+1} = x_+^{q-1}, \ ..., \ y_{q+k} = x_+^{q-k};$$

$$\text{for} \quad q \leqq k, \ y_1 = x^{q-1}, \ ..., \ y_q = 1, \ y_{q+1} = x_+^{q-1}, \ ..., \ y_{2q} =$$

$\theta(x), \ y_{2q+1} = \delta(x), \ ..., \ y_{q+k} = \delta^{(k+q-1)}(x)$.

2. Show that the equation

$$-x^3 y' = 2y \tag{1}$$

has no non-trivial solutions in K'.

Hint. The classical solution of equation (1) is Ce^{1/x^2}. Show that a solution in K' would have to operate on a $\varphi(x) \in K$ vanishing in a neighborhood of $x = 0$ according to the formula

$$(y, \varphi) = \int_{-\infty}^{\infty} Ce^{\frac{1}{x^2}} \varphi(x) \, dx.$$

Use Prob. 5 of Sec. 1.3.

3. Prove that the equation

$$xy' = f$$

has a solution for any generalized function $f \in K'$.

Hint. Use the method of Sec. 1.5.3. A test function $\psi(x)$ can be represented in the form $(x\varphi)'$ with $\varphi \in K$ if and only if

$$\int_{-\infty}^{\infty} \psi(x) \, dx = 0, \quad \int_{-\infty}^{0} \psi(x) \, dx = 0.$$

4. Formulate in terms of generalized functions the analogue of the classical boundary value problem,

$$y'' + p(x) \, y' + q(x) \, y = f(x),$$

$$y(a) = y_a, \ y(b) = y_b.$$

Ans.

$$Y'' + y_a \delta(x - a) - y_b \delta'(x - b) + y_a' \delta(x - a) - y_b' \delta(x - b)$$

$$+ p(x) \left[Y' + y_a \delta(x - a) - y_b \delta(x - b) \right] + q(x) \, Y = F(x).$$

Here, $F(x)$ is equal to $f(x)$ for $x \in [a, b]$ and to zero for $x \notin [a, b]$. The constants y_a' and y_b' are determined by the condition: $Y(x) = 0$ exterior to $[a, b]$.

3 Shilov

1.6. Test Functions and Generalized Functions of Several Variables

1.6.1. The definitions of the preceding sections carry over to the case of several variables without any especial difficulties.

Let R_n be real n-dimensional space and let $x = (x_1, ..., x_n)$. By definition, a function is *ordinary* if it is defined almost everywhere in R_n and is (Lebesgue) integrable in each bounded block $B = \{x: a_1 \leqq x_1 \leqq b_1, ..., a_n \leqq x_n \leqq b_n\}$.

Every *test* function is infinitely differentiable and vanishes outside some sufficiently large block.

A block or generally a set outside of which a test function $\varphi(x)$ and all its derivatives vanish is the *support* of $\varphi(x)$ and is denoted by supp φ.

The linear space of *all* test functions in R_n is denoted by K (or $K(R_n)$ or K_n when the dimension n needs to be indicated). A sequence $\varphi_\nu(x) \in K$, $\nu = 1, 2, ...$, *converges to* 0 *in* K, if the functions $\varphi_\nu(x)$ all vanish outside the same block and tend uniformly to 0 together with their derivatives of any order.

Any continuous linear functional on K is a *generalized function*. The collection of all generalized functions is denoted by K' (or $K'(R_n)$ or K'_n). Each ordinary function $f(x)$ determines a functional on K by means of the formula

$$(f(x), \varphi(x)) = \int_{R_n} f(x)\, \varphi(x)\, dx. \tag{1}$$

Such functionals are called *regular*.

Just as in the case $n = 1$, *an ordinary function $f(x)$ is uniquely determined by the quantities (f, φ).*

It suffices to show that if $(f, \varphi) = 0$ for every $\varphi \in K$, then $f(x)$ is equal to zero almost everywhere. We proved this in Sec. 1.3 for $n = 1$, and we proceed in the general case by induction. As one test function $\varphi(x)$, we take the product of $\varphi_1(x_1)$ and $\varphi_{n-1}(x_2, ..., x_n)$, where φ_1 and φ_{n-1} are test functions of one and $n - 1$ independent variables, respectively. By Fubini's theorem,

$$0 = \int_{R_1} \int_{R_{n-1}} f(x)\, \varphi(x)\, dx = \int_{R_{n-1}} \left\{ \int_{-\infty}^{\infty} f(x_1, ..., x_n)\, \varphi_1(x_1)\, dx_1 \right\}$$
$$\times \varphi_{n-1}(x_2, ..., x_n)\, dx_2 \,...\, dx_n,$$

the inner integral being an integrable function of $x_2, ..., x_n$ over the $(n - 1)$-dimensional block in which φ_{n-1} is carried. Since φ_{n-1} is an arbitrary test function, the induction assumption implies that

$$\int_{-\infty}^{\infty} f(x_1, \ldots, x_n)\, \varphi_1(x_1)\, dx_1 = 0 \qquad (2)$$

on a set Q of full measure in the space of points $(x_2, \ldots, x_n)$.†

Keeping the point $(x_2, \ldots, x_n)$ of Q fixed, we would like to apply the result of Sec. 1.3.2 and hence conclude that $f(x_1, \ldots, x_n)$ vanishes almost everywhere in R_n. But the situation is complicated by the fact that Q depends here on the function $\varphi_1(x_1)$. Therefore, to complete the proof, we make use of the following fact mentioned at the end of Sec. 1.3.2. In the one-dimensional case, it is sufficient that

$$\int_{-\infty}^{\infty} f(x)\, \varphi(x)\, dx = 0$$

on some countable set of functions $\varphi(x)$ in order to conclude that $f(x) = 0$ almost everywhere. It is just such a countable set of functions that we take in (2). For each of them, there is a set $Q = Q(\varphi)$ of full measure for which (2) holds. Then (2) holds simultaneously for all of these functions on the intersection of the sets $Q = Q(\varphi)$, which is also a set of full measure. But now the result of Sec. 1.3 may be applied at the points of the latter set and this leads to the conclusion that $f(x_1, \ldots, x_n)$ is equal to zero almost everywhere in R_n.

Thus the class of ordinary functions is imbedded in K' in a biunique manner.

1.6.2. Besides the regular functionals, there are still the singular ones not expressible in the form (1).

The following are several examples of generalized functions in R_n which generally speaking do not reduce to ordinary functions:

1°. The *delta-function* $\delta(x)$, which operates through the formula

$$(\delta(x), \varphi(x)) = \varphi(0).$$

Or more generally, the *shifted delta-function*, which operates according to the formula

$$(\delta(x - a), \varphi(x)) = \varphi(a).$$

---

† That is, Q comprises all points of the subspace except for a set of (n-1)-dimensional measure zero. See G. E. Shilov and B. L. Gurevich, *Integral, Measure and Derivative: A Unified Approach*, Prentice-Hall, Inc., 1966 (translated by R. A. Silverman), Chapt. 1, Sec. 1.4.

3*

2°. *A functional of order of singularity not exceeding p* is by definition given by

$$(f, \varphi) = \sum_{|k| \leq p} \int_{R_n} f_k(x) \, D^k \varphi(x) \, dx, \tag{3}$$

where $k = (k_1, k_2, ..., k_n)$ is a multi-index,

$$D^k = \frac{\partial^{k_1 + \cdots + k_n}}{\partial x_1^{k_1} \cdots \partial x_n^{k_n}},$$

all of the $f_k(x)$ are ordinary functions, and the summation extends over all indices k for which $|k| = k_1 + \cdots + k_n \leq p$.

One says that the order of singularity of a functional f is exactly p if f can be written in the form (3) for a given p but cannot be so written for a smaller p.

It will be shown below that the functional $\delta(x)$, as in the case $n = 1$, has order of singularity 1.

3°. *A functional carried on a surface L:*

$$(f, \varphi) = \int_L f(x) \, \varphi(x) \, dl_x,$$

where dl_x is the surface element of L at the point x. A more general functional of this sort is, for example,

$$(f, \varphi) = \int_L \sum_{k=0}^{m} f_k(x) \, D^k \varphi(x) \, dl_x,$$

the $f_k(x)$ all being ordinary functions on the surface L.

4°. *Generalized functions of the argument* (ω, x).

Let $\omega = (\omega_1, ..., \omega_n)$ be a point on the unit sphere Ω in R_n. To each test function $\varphi(x_1, ..., x_n)$, we assign a function

$$\Phi(\xi; \omega) = \int_{(\omega, x) = \xi} \varphi(x) \, dx.$$

The integral on the right-hand side extends over the hyperplane which lies orthogonal to the vector ω at a (directed) distance ξ from the origin.

If a rotation of axes $Ux = x'$ is performed which carries the vector ω into the first basis vector $(1, 0, ..., 0)$, then we can write

$$\Phi(\xi; \omega) = \int_{x_1' = \xi} \varphi_1(x_1', x_2', ..., x_n') \, dx_2' \cdots dx_n',$$

where $\varphi_1(x') = \varphi(x)$.

$\Phi(\xi; \omega)$ clearly has compact support along with $\varphi(x)$ and is infinitely differentiable in ξ for each fixed ω. Thus, $\Phi(\xi; \omega)$ is a test function of the argument ξ. With each generalized function $f \in K_1'$ we can now associate a functional $f_\omega \in K_n'$ depending on the parameter ω through the formula

$$(f_\omega, \varphi(x_1, ..., x_n)) = (f, \Phi(\xi; \omega)).$$

In particular,

$$(\delta_\omega, \varphi) = (\delta(\xi), \Phi(\xi; \omega)) = \int_{(\omega,x)=0} \varphi(x)\, dx = \int_{x_1'=0} \varphi_1(0, x_2', ..., x_n')\, dx_2'...dx_n'$$

and

$$(\delta_\omega^{(q)}, \varphi) = (\delta^{(q)}(\xi), \Phi(\xi; \omega)) = (-1)^q \left(\delta(\xi), \frac{\partial^q \Phi(\xi; \omega)}{\partial \xi^q}\right)$$

$$= (-1)^q \int_{x_1'=0} \frac{\partial^q \varphi_1(0, x_2', ..., x_n')}{\partial x_1'^q}\, dx_2' \cdots dx_n'.$$

The functional f_ω will also be denoted by $f((\omega, x))$.

1.6.3. Instead of considering test and generalized functions throughout R_n, we may confine ourselves to some region $G \subset R_n$. Namely, $K(G)$ is defined as the space of all test functions $\varphi(x)$ in $K(R_n)$ with support *strictly inside* G (i.e., such that the closure of $\{x : \varphi(x) \neq 0\}$ lies inside G). A sequence $\varphi_\nu(x) \in K(G)$ is said to *converge in $K(G)$ to zero* if all $\varphi_\nu(x)$ are carried in the same set lying strictly in G and tend uniformly to zero together with all their derivatives. We shall call a continuous linear functional on $K(G)$ a *generalized function in the region G*. The major portion of the results of this section and those to follow carry over at once to generalized functions in a region.

Problems

1. An ordinary function $f(x)$ is said to be *spherically symmetric* if $f(Ux) = f(x)$ for every rotation U of R_n. Give a definition of a spherically symmetric generalized function.
Ans. $(f, \varphi(Ux)) = (f, \varphi(x))$ for any $\varphi(x) \in K$.

2. A spherically symmetric generalized function f is uniquely determined on all of K by its values on the spherically symmetric test functions.
Hint. For any $\varphi \in K$, the spherical mean $\varphi_m(x)$ is also a function in K

which is the limit of arithmetic means of "rotations" of φ. Hence, $(f, \varphi) = (f, \varphi_m)$.

3. An ordinary function $f(x)$ is said to be *homogeneous of degree p* if for any positive t, $f(tx) = t^p f(x)$. Give a definition of a homogeneous generalized function of degree p.

Ans. $(f, \varphi(tx)) = t^{-p-n}(f, \varphi(x))$ for any $\varphi \in K$.

4. Show that the delta-function $\delta(x)$ is homogeneous and find its degree of homogeneity.

Ans. $p = -n$.

5. Show that a functional carried on a surface (Example 3°) is singular. *Hint.* Produce an argument generalizing the proof that the delta-function is singular (see Sec. 1.3).

6. *The regularization problem.* Let $f(x_1, \ldots, x_n)$ be locally integrable everywhere except in a neighborhood of $x = 0$ where it satisfies the inequality

$$|f(x)| \leq \frac{C}{r^m} .$$

Show that there exists a functional $f \in K'$ which operates on each test function vanishing in a neighborhood of $x = 0$ according to the formula

$$(f, \varphi) = \int_{R_n} f(x) \, \varphi(x) \, dx .$$

Hint. For example,

$$(f, \varphi) = \int_{R_n} f(x) \left\{ \varphi(x) - \left[\sum_{|k| \leq m} \frac{x_1^{k_1} \ldots x_n^{k_n}}{k_1! \ldots k_n!} D^k \varphi(0) \right] h(x) \right\} dx,$$

where $|k| = k_1 + \cdots + k_n$ and $h(x)$ is equal to 1 for $|x| \leq 1$ and to 0 for $|x| > 1$.

7. Show that Prob. 6 cannot be solved for a function $f(x)$ which can be estimated in some positive solid angle by

$$|f(x)| > \frac{A_m}{r_m} \quad (|x| = r < r_0),$$

for any $m = 0, 1, 2, \ldots$ (there exists no regularization for $f(x)$ in K'). *Hint.* Generalize the method of solution of Prob. 5 of Sec. 1.3.

7. Operations on Generalized Functions of Several Variables

1.7.1. We can confine ourselves here to simply listing the operations if we keep in mind that the entire discussion of Sec. 1.4 for the case $n = 1$ carries over at once to the general case.

Thus we have the following operations:

1) Addition and multiplication by real numbers:

$$(\alpha_1 f_1 + \alpha_2 f_2, \varphi) = \alpha_1(f_1, \varphi) + \alpha_2(f_2, \varphi).$$

2) Multiplication by an infinitely differentiable function $\alpha(x)$:

$$(\alpha(x)f, \varphi) = (f, \alpha(x)\varphi).$$

3) Differentiation (with respect to any argument):

$$\left(\frac{\partial f}{\partial x_k}, \varphi\right) = -\left(f, \frac{\partial \varphi}{\partial x_k}\right). \tag{1}$$

The derivative of a functional f of arbitrary order $|k| = k_1 + \cdots + k_n$ may be defined inductively. This yields

$$(D^k f, \varphi) \equiv \left(\frac{\partial^{k_1 + \cdots + k_n} f}{\partial x_1^{k_1} \ldots \partial x_n^{k_n}}, \varphi\right)$$

$$= (-1)^{|k|}\left(f, \frac{\partial^{k_1 + \cdots + k_n} \varphi}{\partial x_1^{k_1} \ldots \partial x_n^{k_n}}\right) \equiv (f, (-1)^{|k|} D^k \varphi).$$

If an ordinary function $f(x)$ is absolutely continuous in x_1 for almost all $x_2, \ldots, x_n$ and has a derivative $\partial f/\partial x_1$ in the usual sense (also an ordinary function of $x_1, x_2, \ldots, x_n$), then the derivative of $f(x)$ with respect to x_1 in the sense of generalized functions leads to the ordinary derivative $\partial f/\partial x_1$.

1.7.2. We now give several examples involving differentiation.

1°. For any generalized function f,

$$\frac{\partial^2 f}{\partial x_j \partial x_k} = \frac{\partial^2 f}{\partial x_k \partial x_j}.$$

In other words, the order of differentiation is immaterial.

To prove this, we observe that the analogous property certainly holds for test functions. Thus,

$$\left(\frac{\partial^2 f}{\partial x_j \, \partial x_k}, \varphi \right) = \left(f, \frac{\partial^2 \varphi}{\partial x_j \, \partial x_k} \right) = \left(f, \frac{\partial^2 \varphi}{\partial x_k \, \partial x_j} \right) = \left(\frac{\partial^2 f}{\partial x_k \, \partial x_j}, \varphi \right),$$

as asserted.

2°. *Weak derivatives.* The (ordinary) function $g(x_1, ..., x_n)$ is called a weak derivative of the (ordinary) function $f(x_1, ..., x_n)$ of order $|k| = k_1 + \cdots + k_n$ in a region G (according to Sobolev and Friedrichs) if

$$(-1)^{|k|} \int_G g(x) \, \varphi(x) \, dx = \int_G f(x) \, D^k\varphi(x) \, dx \qquad (2)$$

for every (infinitely) differentiable function $\varphi(x_1, ..., x_n)$ whose support lies strictly in G. Let us show that $g = D^k f$ in K'. In fact, the definition (2) may be expressed as

$$(-1)^{|k|}(g, \varphi) = (f, D^k\varphi). \qquad (3)$$

At the same time, by the definition of derivative in K', we have

$$(D^k f, \varphi) = (-1)^{|k|}(f, D^k\varphi). \qquad (4)$$

Comparing (3) and (4), we see that $g = D^k f$, as asserted.

Thus, the fact that a functional f has a weak derivative means that its corresponding derivative in K' is an ordinary function.

3°. Let $\theta = \theta(x_1, ..., x_n)$ be a function equal to 1 when all its arguments are positive and 0 otherwise. We wish to determine $\dfrac{\partial^n \theta}{\partial x_1 \, ... \, \partial x_n}$.

We have

$$\left(\frac{\partial^n \theta}{\partial x_1 \, ... \, \partial x_n}, \varphi \right) = (-1)^n \left(\theta, \frac{\partial^n \varphi}{\partial x_1 \, ... \, \partial x_n} \right)$$

$$= (-1)^n \int_0^\infty \cdots \int_0^\infty \frac{\partial^n \varphi}{\partial x_1 \, ... \, \partial x_n} \, dx_1 \, ... \, dx_n$$

$$= \varphi(0, ..., 0) = (\delta, \varphi),$$

and so

$$\frac{\partial^n \theta}{\partial x_1 \, ... \, \partial x_n} = \delta(x).$$

4°. Differentiation of the generalized function $f((\omega, x))$ (Sec. 1.6.2, 4°). Let us show that the familiar chain rule is applicable to f, or

$$\frac{\partial f((\omega, x))}{\partial x_j} = \frac{df((\omega, x))}{d(\omega, x)} \omega_j \tag{5}$$

To this end, we first establish that

$$\int\limits_{(\omega,x)=\xi} \frac{\partial \varphi}{\partial x_j} dx = \omega_j \frac{d}{d\xi} \int\limits_{(\omega,x)=\xi} \varphi(x) dx \tag{6}$$

for every test function $\varphi(x)$. We have

$$\frac{1}{\Delta\xi} \left\{ \int\limits_{(\omega,x)=\xi+\Delta\xi} \varphi(x) dx - \int\limits_{(\omega,x)=\xi} \varphi(x) dx \right\} =$$

$$= \frac{1}{\Delta\xi} \int\limits_{(\omega,x)=\xi} [\varphi(x + \omega\,\Delta\xi) - \varphi(x)] dx =$$

$$= \int\limits_{(\omega,x)=\xi} \frac{\varphi(x + \omega\,\Delta\xi) - \varphi(x)}{\Delta\xi} dx \to \int\limits_{(\omega,x)=\xi} \sum \omega_k \frac{\partial\varphi}{\partial x_k} dx = \int\limits_{(\omega,x)=\xi} \frac{\partial\varphi}{\partial\omega} dx.$$

We decompose the basis vector e_j into components in the directions of the vector ω and a vector θ lying in the plane $(\omega, x) = \xi$. Then

$$\frac{\partial\varphi}{\partial x_j} = \frac{\partial\varphi}{\partial\omega}(\omega, e_j) + \frac{\partial\varphi}{\partial\theta}(e_j, \theta) = \omega_j \frac{\partial\varphi}{\partial\omega} + (e_j, \theta)\frac{\partial\varphi}{\partial\theta}.$$

Hence,

$$\int\limits_{(\omega,x)=\xi} \frac{\partial\varphi}{\partial x_j} dx = \omega_j \int\limits_{(\omega,x)=\xi} \frac{\partial\varphi}{d\omega} dx + (e_j, \theta) \int\limits_{(\omega,x)=\xi} \frac{\partial\varphi}{\partial\theta} dx$$

$$= \omega_j \int\limits_{(\omega,x)=\xi} \frac{\partial\varphi}{\partial\omega} dx = \omega_j \frac{d}{d\xi} \int\limits_{(\omega,x)=\xi} \varphi(x) dx,$$

since the integral of $\partial\varphi/\partial\theta$ is clearly zero. Thus, formula (6) is proved. We now proceed to derive formula (5). We know that

$$(f((\omega, x)), \varphi(x)) = (f(\xi), \Phi(\xi; \omega)),$$

where

$$\Phi(\xi; \omega) = \int\limits_{(\omega,x)=\xi} \varphi(x) dx.$$

Hence, by the basic differentiation rule,

$$\left(\frac{\partial}{\partial x_j}f((\omega, x)), \varphi(x)\right) = -\left(f((\omega, x)), \frac{\partial \varphi}{\partial x_j}\right)$$

$$= -\left(f(\xi), \int\limits_{(\omega,x)=\xi} \frac{\partial \varphi}{\partial x_j}\, dx\right).$$

On the other hand,

$$(f'_\omega(\xi),\ \varphi(x)) = -\left(f(\xi),\ \frac{d}{d\xi}\int\limits_{(\omega,x)=\xi} \varphi(x)dx\right),$$

and so it merely remains to apply formula (6).

5°. A functional of order of singularity not exceeding p is a sum of derivatives of ordinary functions of at most p-th order since

$$(f, \varphi) = \sum_{|k|\leq p}\int_{R_n} f_k(x)\, D^k\varphi(x)\, dx = \sum_{|k|\leq p}(f_k, D^k\varphi) = \sum_{|k|\leq p}(-1)^{|k|}(D^kf_k, \varphi)$$

$$= \left(\sum_{|k|\leq p}(-1)^{|k|}D^kf_k, \varphi\right).$$

The converse statement is also obviously true: A generalized function which is equal to a sum of derivatives of ordinary functions of at most p-th order has order of singularity not exceeding p.

Example 3° shows that $\delta(x)$ has order n at the most. As we shall soon see, its order of singularity is actually 1.

6°. Consider a region G with a piecewise smooth boundary Γ (Fig. 4). Suppose $f(x)$ is a function vanishing exterior to G and having continuous

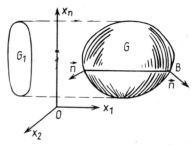

FIG. 4. G_1 is the projection of G on the plane $x_1 = 0$ and $\vec{n}$ is the outward normal to G.

first partials in $G \cup \Gamma$. We shall determine $\partial f / \partial x_1$. According to formula (1),

$$\left(\frac{\partial f}{\partial x_1}, \varphi \right) = -\left(f, \frac{\partial \varphi}{\partial x_1} \right) = -\int_G f(x) \frac{\partial \varphi}{\partial x_1} \, dx_1 \ldots dx_n$$

$$= -\int_{G_1} \left\{ \int_{A(x_2,\ldots,x_n)}^{B(x_2,\ldots,x_n)} f(x) \frac{\partial \varphi}{\partial x_1} \, dx_1 \right\} dx_2 \ldots dx_n$$

$$= -\int_{G_1} \left\{ f(B) \, \varphi(B) - f(A) \, \varphi(A) - \int_A^B \frac{\partial f}{\partial x_1} \varphi \, dx_1 \right\} dx_2 \ldots dx_n$$

$$= \int_G \frac{\partial f}{dx_1} \varphi \, dx - \int_\Gamma f(\xi) \, \varphi(\xi) \cos (\widehat{\xi, n}) \, dl_\xi.$$

We see that the desired derivative is the sum of two functionals, one being an ordinary function equaling $\partial f / \partial x_1$ inside G and zero outside G and the other being a singular functional carried on Γ. This result generalizes formula (4) of Sec. 1.4 to the n-dimensional case.

7°. The classical Green's formula

$$\int_G f(x) \, \Delta \varphi(x) \, dx = \int_G \Delta f(x) \, \varphi(x) \, dx + \int_\Gamma \left(f \frac{\partial \varphi}{\partial n} - \frac{\partial f}{\partial n} \varphi \right) dl$$

(Δ is the Laplacian, G is a bounded region, Γ is its boundary, and $\partial / \partial n$ represents differentiation in the direction of the outward normal) in conjunction with the relation $(\Delta f, \varphi) = (f, \Delta \varphi)$ may be given the following interpretation. Suppose $f(x)$ is an ordinary function with continuous second-order partial derivatives (in the usual sense) in the closure of G and vanishes outside G. Then in K', the Laplacian converts $f(x)$ into a sum of two functionals, one of them being an ordinary function equal to zero outside G and to the (ordinary) $\Delta f(x)$ inside G and the other being a singular functional carried on Γ.

If we write the analogous equation for a function $f(x)$ vanishing inside

G and combine it with the preceding one, we obtain

$$\int_{R_n} f(x)\,\Delta\varphi(x)\,dx = \int_{R_n} \Delta f(x)\,\varphi(x)\,dx$$

$$+ \int_{\Gamma} \left[(f_i - f_e)\frac{\partial\varphi}{\partial n} - \frac{\partial}{\partial n}(f_i - f_e)\,\varphi \right] dl,$$

where f_i and f_e are the boundary values of $f(x)$ on Γ from the inside and outside of G, respectively. Thus, the functional Δf is expressed in terms of the ordinary function Δf and the jumps in f and $\partial f/\partial n$ across the boundary of G.

8°. The Laplacian and spherically symmetric functions.

Let us derive the classical expression for the Laplacian of a twice differentiable function $f(r)$ (depending only on the radial coordinate r). We have

$$\frac{\partial}{\partial x_j} f(r) = f'(r)\frac{\partial r}{\partial x_j} = f'(r)\frac{x_j}{r},$$

$$\frac{\partial^2}{\partial x_j^2} f(r) = f''(r)\frac{x_j^2}{r^2} + f''(r)\frac{r - \dfrac{x_j^2}{r}}{r^2} = f''(r)\frac{x_j^2}{r^2} + f'(r)\frac{r^2 - x_j^2}{r^3},$$

and so

$$\Delta f(r) = \sum_j \frac{\partial^2 f}{\partial x_j^2} = f''(r) + f'(r)\frac{n-1}{r}. \tag{7}$$

Suppose $f(r) = r^p$. Then we obtain

$$\Delta r^p = p(p + n - 2) r^{p-2}. \tag{8}$$

Thus, the Laplacian lowers the power on r by two. The entire calculation remains valid when we differentiate in K' as long as the discussion concerns ordinary functions. Since r^p is integrable in a neighborhood of the origin for $p > -n$, formula (8) remains meaningful in K' for $p > 2-n$. But if $p = 2 - n$, the right-hand side of (8) ceases to have meaning in K' for $n > 2$. It is well-known that r^{2-n} is harmonic everywhere except at the origin, i.e., $\Delta r^{2-n} = 0$ for $x \neq 0$. Let us determine what the expression Δr^{2-n} is in K' assuming that $n > 2$.

By definition,

$$\left(\Delta \frac{1}{r^{n-2}}, \varphi\right) = \left(\frac{1}{r^{n-2}}, \Delta\varphi\right) = \int_{R_n} \frac{\Delta\varphi}{r^{n-2}} \, dx = \lim_{\varepsilon \to 0} \int_{r \geq \varepsilon} \frac{\Delta\varphi}{r^{n-2}} \, dx.$$

To the resulting integral, we apply Green's formula (Example 7°) taking the region G to be a spherical layer $\varepsilon \leq r \leq a$, where a is so large that $\varphi(x)$ vanishes identically outside the sphere $r = a$. This yields

$$\int_{r \geq \varepsilon} \frac{\Delta\varphi}{r^{n-2}} \, dx = \int_{r \geq \varepsilon} \varphi \Delta \frac{1}{r^{n-2}} \, dx - \int_{r=\varepsilon} \frac{\partial\varphi}{\partial r} \frac{1}{r^{n-2}} \, dl + \int_{r=\varepsilon} \varphi \frac{\partial}{\partial r} \frac{1}{r^{n-2}} \, dl,$$

dl being the surface element of the sphere $r = \varepsilon$. We then have

$$\int_{r \geq \varepsilon} \varphi \Delta \frac{1}{r^{n-2}} \, dx = 0$$

since r^{2-n} is harmonic outside the ball $r < \varepsilon$. Moreover,

$$\int_{r=\varepsilon} \frac{\partial\varphi}{\partial r} \frac{1}{r^{n-2}} \, dl = \frac{1}{\varepsilon^{n-2}} \int_{r=\varepsilon} \frac{\partial\varphi}{\partial r} \, dl = O(\varepsilon),$$

$$\int_{r=\varepsilon} \varphi \frac{\partial}{\partial r} \frac{1}{r^{n-2}} \, dl = -\frac{n-2}{\varepsilon^{n-1}} \int_{r=\varepsilon} \varphi \, dl = -(n-2)\,\Omega_n S_\varepsilon[\varphi],$$

where Ω_n is the surface area of a unit sphere in R_n and $S_\varepsilon[\varphi]$ is the mean value of $\varphi(x)$ over a sphere of radius ε. Letting $\varepsilon \to 0$, we find that $S_\varepsilon[\varphi] \to \varphi(0)$ and therefore,

$$\left(\Delta \frac{1}{r^{n-2}}, \varphi\right) = \lim_{\varepsilon \to 0} \int_{r \geq \varepsilon} \frac{\Delta\varphi}{r^{n-2}} \, dx = -(n-2)\,\Omega_n \varphi(0)$$

$$= (-(n-2)\,\Omega_n \delta(x), \varphi(x)).$$

Hence,

$$\Delta \frac{1}{r^{n-2}} = -(n-2)\,\Omega_n \delta(x). \tag{9}$$

In the special case $n = 3$, $\Omega_n = 4\pi$ and so

$$\Delta \frac{1}{r} = -4\pi \delta(x).$$

REMARK. The result obtained shows particularly that the order of singularity of the generalized function $\delta(x_1, \ldots, x_n)$ is equal to 1 independently of n. Indeed, $1/r^{n-2}$ and its first derivatives are ordinary functions and by the foregoing,

$$\Delta \frac{1}{r^{n-2}} = \sum_{j=1}^{n} \frac{\partial}{\partial x_j}\left(\frac{\partial}{\partial x_j}\frac{1}{r^{n-2}}\right) = -(n-2)\,\Omega_n\,\delta(x).$$

Thus, $\delta(x)$ is a sum of first derivatives of ordinary functions.

9°. Let

$$P\left(\frac{\partial}{\partial x}\right) \equiv P\left(\frac{\partial}{\partial x_1}, \ldots, \frac{\partial}{\partial x_n}\right)$$

be a linear partial differential operator. A generalized function $\mathscr{E}(x)$ that satisfies the equation

$$P\left(\frac{\partial}{\partial x}\right)\mathscr{E}(x) = \delta(x)$$

will be called a *fundamental function of the operator* $P\left(\frac{\partial}{\partial x}\right)$.† For example,

$$-\frac{1}{(n-2)\,\Omega_n}\frac{1}{r^{n-2}}$$

is a fundamental function for the Laplacian Δ for $n > 2$ (Example 8°).

Generally speaking, a fundamental function $\mathscr{E}(x)$ is not uniquely determined. Any solution $u(x)$ of the homogeneous equation

$$P\left(\frac{\partial}{\partial x}\right)u(x) = 0$$

may always be added to it.

Let us find the fundamental function $\mathscr{E}_n^{2m}(x)$ for the iterated Laplacian Δ^m in n-dimensional space. Once $\mathscr{E}_n^{2(m-1)}$ is known, $\mathscr{E}_n^{2m}$ may be determined by solving the equation

$$\Delta\mathscr{E}_n^{2m} = \mathscr{E}_n^{2(m-1)}. \tag{10}$$

In view of the spherical symmetry of the Laplacian, it is natural to look

† Also a fundamental solution of the equation $P\left(\frac{\partial}{\partial x}\right)u = 0.$

for $\mathscr{E}_n^{2m}$ as a function of $r = \sqrt{x_1^2 + \cdots + x_n^2}$. Using formula (7), we can write

$$\Delta f(r) = f''(r) + \frac{n-1}{r} f'(r) = h(r). \tag{11}$$

A general solution can be found for (11) by reducing it to first order by means of the substitution $f'(r) = g(r)$. The solution is given by

$$f(r) = \int r^{1-n} \left\{ \int^r \varrho^{n-1} h(\varrho)\, d\varrho \right\} dr.$$

If $h(r) = r^\lambda$ and $\lambda \neq -2$ or $-n$, then

$$f(r) = \frac{r^{\lambda+2}}{(\lambda+2)(\lambda+n)}. \tag{12}$$

This result also follows from (8). If $h(r) = r^{-2}$,

$$f(r) = \frac{1}{n-2} \log r. \tag{13}$$

Thus, the logarithmic function likewise turns up as a fundamental function. Consider yet the case $h(r) = r^\lambda \log r$ with $\lambda \neq -2$ or $-n$. In this instance,

$$f(r) = \int r^{1-n} \left[\frac{r^{\lambda+n}}{\lambda+n} \log r - \frac{r^{\lambda+n}}{(\lambda+n)^2} \right] dr$$

$$= \frac{r^{\lambda+2} \log r}{(\lambda+2)(\lambda+n)} - \frac{(n+2\lambda+2)\, r^{\lambda+2}}{(\lambda+2)^2 (\lambda+n)^2}. \tag{14}$$

Formulas (12)–(14) are sufficient for finding the desired fundamental functions. We know that $\mathscr{E}_n^2 = c_n r^{2-n}$. Thus, solving equation (10) for $m = 2$ $(\lambda = 2 - n)$, we find with the help of formula (12) that

$$\mathscr{E}_n^4 = c_n \frac{r^{4-n}}{2(4-n)} \qquad (n \neq 2, n \neq 4).$$

Then in similar fashion,

$$\mathscr{E}_n^6 = c_n \frac{r^{6-n}}{2\cdot 4\,(4-n)(6-n)} \quad (n \neq 2, n \neq 4, n \neq 6).$$

At each step, the exponent on r is lowered by two and corresponding factors appear in the denominator. If n is odd, the exponent $2m - n$

never becomes equal to -2. Therefore, the following formula is valid for odd n and any m:

$$\mathscr{E}_n^{2m} = c_n \frac{r^{2m-n}}{2 \cdot 4 \cdots (2m-2)(2m-n)\cdots(4-n)} \equiv c_{n,\,2m}r^{2m-n}. \quad (15)$$

If n is even, formula (15) holds only for sufficiently small m, namely, for $2m \le n - 2$. If $2m \ge n$, the successive exponents $2 - n$, $4 - n$, ... will reach the value -2 at some stage after which the functions that occur will contain logarithms according to formulas (13) and (14):

$$\mathscr{E}_n^{n-2} = c_{n,\,n-2}r^{-2},$$

$$\mathscr{E}_n^n = \frac{c_{n,\,n-2}}{n-2} \log r,$$

. . . .

$$\mathscr{E}_n^{2m} = \frac{c_{n,\,n-2}}{n-2} \frac{r^{2m-n}\log r}{2\cdots(2m-n)\,n(n+2)\cdots(2m-2)} \equiv b_{n,\,2m}r^{2m-n}\log r.$$

The terms involving powers of r from r^2 to r^{2m-2} have been omitted here since the Laplacian Δ^m annihilates them. Thus, for $n > 2$, we obtain the formula

$$\mathscr{E}_n^{2m} = \begin{cases} c_{n,\,2m}r^{2m-n}, & \text{if } n \text{ is odd or } n \text{ is even and more than } 2m; \\ b_{n,\,2m}r^{2m-n}\log r, & \text{if } n \text{ is even and at most } 2m. \end{cases} \quad (16)$$

For $n = 2$, it is easy to verify directly just as in Example 8° that $\mathscr{E}_2^2 = -2\pi \log r$. Thus, formula (16) persists for $n = 2$ and $m = 1$. But then it continues to hold also for all $\mathscr{E}_2^{2m}$ since they can be expressed in terms of $\mathscr{E}_2^2$ in the indicated way. Hence, formula (16) is valid for $n = 2$.

1.7.3. If $f \in K'$ is a constant, then $\partial f/\partial x_j$ is obviously zero for $j = 1$, 2, ..., n. The converse statement is also true as we now show.

THEOREM 1. *If $f \in K'$ and $\partial f/\partial x_j = 0$ for $j = 1, 2, ..., n$, then f is a constant.*

We first prove two simple lemmas.

LEMMA 1. *Every test function $\varphi(x_1, ..., x_n)$ can be represented as the limit in K of linear combinations of test functions of the form $\varphi_1(x_1)\,\varphi_2(x_2)$ ... $\varphi_n(x_n)$, where $\varphi_j(x_j)$ is a one-dimensional test function.*
Proof. Let the block $B = \{x: |x_j| \le a_j\}$ contain the support of $\varphi(x)$. By Weierstrass' theorem, there exists a polynomial $P_\nu(x_1, ..., x_n)$ which differs from $\varphi(x)$ by less than $\dfrac{1}{\nu}$ in the block $B' = \{x: |x_j| \le 2a_j\}$, for each

$v = 1, 2, \ldots$ Furthermore, the derivatives of $\varphi(x)$ up to v-th order are also approximated uniformly by the corresponding derivatives of P_v. Now let $e(x_j)$ be a test function equal to 1 for $|x_j| \leq a_j$ and to 0 for $|x_j| \geq 2a_j$. Then as $v \to \infty$, the function $P_v(x_1, \ldots, x_n) e(x_1) \ldots e(x_n)$ tends to $\varphi(x)$ in K, as required.

LEMMA 2. *If* $\dfrac{\partial f}{\partial x_1} \equiv 0$, *then for any test function of the form* $\alpha(x_1)\beta(x_2, \ldots, x_n)$,

$$(f, \alpha\beta) = (f, \alpha_1\beta) \int\limits_{-\infty}^{\infty} \alpha(x_1)\, dx_1,$$

where $\alpha_1(x_1)$ *is a fixed arbitrary test function such that*

$$\int\limits_{-\infty}^{\infty} \alpha_1(x_1)\, dx_1 = 1.$$

Proof. Given $\alpha_1(x_1)$, we can write down the representation

$$\alpha(x_1) = \alpha_1(x_1) \int\limits_{-\infty}^{\infty} \alpha(x_1)\, dx_1 + \alpha_0(x_1),$$

in which $\int\limits_{-\infty}^{\infty} \alpha_0(x_1)\, dx_1 = 0$. Therefore, $\alpha_0(x_1) = \gamma'(x_1)$, where $\gamma(x_1)$ is again a test function (Sec. 1.5.1).

Hence,

$$(f, \alpha\beta) = \left(f, \left[\alpha_1 \int\limits_{-\infty}^{\infty} \alpha(x_1)\, dx_1 + \gamma' \right] \beta \right)$$

$$= (f, \alpha_1\beta) \int\limits_{-\infty}^{\infty} \alpha(x_1)\, dx_1 + (f, \gamma'\beta) = \left(f, \alpha_1\beta \int\limits_{-\infty}^{\infty} \alpha(x_1)\, dx_1 \right),$$

inasmuch as

$$(f, \gamma'\beta) = \left(f, \frac{\partial}{\partial x_1}(\gamma\beta) \right) = -\left(\frac{\partial f}{\partial x_1}, \gamma\beta \right) = 0.$$

The lemma is proved.

We proceed to prove Theorem 1. Consider a test function of the form $\alpha(x_1)\beta(x_2)\ldots \omega(x_n)$. By applying Lemma 2 sequentially, we find that

$$(f, \alpha\beta \ldots \omega) = (f, \alpha_1\beta \ldots \omega) \int\limits_{-\infty}^{\infty} \alpha(x_1)\, dx_1$$

$$= (f, \alpha_1\beta_1 \ldots \omega) \int\limits_{-\infty}^{\infty} \alpha(x_1)\, dx_1 \int\limits_{-\infty}^{\infty} \beta(x_2)\, dx_2 = \cdots$$

$$= (f, \alpha_1\beta_1 \ldots \omega_1) \int\limits_{-\infty}^{\infty} \alpha(x_1)\, dx_1 \int\limits_{-\infty}^{\infty} \beta(x_2)\, dx_2 \ldots \int\limits_{-\infty}^{\infty} \omega(x_n)\, dx_n$$

$$= \int\limits_{R_n} C\alpha\beta \ldots \omega\, dx,$$

where $C = (f, \alpha_1\beta_1 \ldots \omega_1)$. Thus, on test functions of the form $\alpha\beta \ldots \omega$, the functional f is a constant C. The proof is completed by making use of Lemma 1 and the continuity of the functional f.

1.7.4. THEOREM 2. *A generalized function f for which all of its derivatives $\dfrac{\partial f}{\partial x_j} = g_j$ are continuous functions can be represented by*

$$f(x) = f(0) + \int_0^{x_1} g_1(\xi_1, 0, \ldots, 0)\, d\xi_1$$

$$+ \int_0^{x_2} g_2(x_1, \xi_2, 0, \ldots, 0)\, d\xi_2 + \cdots$$

$$+ \int_0^{x_n} g_n(x_1, x_2, \ldots, x_{n-1}, \xi_n)\, d\xi_n. \tag{17}$$

f itself is a continuous and differentiable function having the partial derivatives $\dfrac{\partial f}{\partial x_j} = g_j$ in the usual sense.

The functions $g_1, \ldots, g_n$ are not being assumed to be differentiable in the usual sense. Nevertheless, when considering their derivatives in K', we have

$$\frac{\partial g_j}{dx_k} = \frac{\partial g_k}{\partial x_j} \tag{18}$$

for any i and k since the mixed derivatives $\dfrac{\partial^2 f}{\partial x_k\, \partial x_j}$ and $\dfrac{\partial^2 f}{\partial x_j\, \partial x_k}$ are equal.

To prove the theorem, consider the function

$$F(x) = \int_0^{x_1} g_1(\xi_1, 0, \ldots, 0)\, d\xi_1 + \cdots + \int_0^{x_n} g_n(x_1, x_2, \ldots, x_{n-1}, \xi_n)\, d\xi_n.$$

We aim to show that

$$\frac{\partial F}{\partial x_1} = g_1, \ldots, \frac{\partial F}{\partial x_n} = g_n,$$

from which it will follow by Theorem 1 that F and f differ by a constant:

$$f(x) = C + F(x). \tag{19}$$

Setting $x = 0$ in this, we find that $C = f(0)$ and so (19) becomes our required formula (17).

The fact that $\dfrac{\partial F}{\partial x_n} = g_n$ is trivial. Let us prove that $\dfrac{\partial F}{\partial x_{n-1}} = g_{n-1}$. To this end, we must show that

$$\frac{\partial}{\partial x_{n-1}} \left[\int_0^{x_{n-1}} g_{n-1}(x_1, \ldots, x_{n-2}, \xi_{n-1}, 0)\, d\xi_{n-1} \right.$$

$$\left. + \int_0^{x_n} g_n(x_1, \ldots, x_{n-1}, \xi_n)\, d\xi_n \right] = g_{n-1}(x_1, \ldots, x_n),$$

or equivalently,

$$\frac{\partial f_n}{\partial x_{n-1}} - g_{n-1}(x) = -g_{n-1}(x_1, \ldots, x_{n-1}, 0), \qquad (20)$$

where

$$f_n(x) = \int_0^{x_n} g_n(x_1, \ldots, x_{n-1}, \xi_n)\, d\xi_n. \qquad (21)$$

The integral (21) cannot be differentiated formally (under the integral sign) since g_n need not have an ordinary derivative with respect to x_{n-1}. We first verify that the expression on the left-hand side of (20) is independent of x_n. By the condition (18), we have

$$\frac{\partial}{\partial x_n}\left(\frac{\partial f_n}{\partial x_{n-1}} - g_{n-1} \right) = \frac{\partial^2 f_n}{\partial x_n\, \partial x_{n-1}} - \frac{\partial g_{n-1}}{\partial x_n} = \frac{\partial^2 f_n}{\partial x_{n-1} \partial x_n} - \frac{\partial g_n}{\partial x_{n-1}}$$

$$= \frac{\partial}{\partial x_{n-1}}\left(\frac{\partial f_n}{\partial x_n} - g_n \right) = 0, \qquad (22)$$

since the expression in the last parentheses clearly vanishes.

We now compute the left-hand side of (20). Take a test function $\varphi(x)$ of the form $\alpha(x_1, \ldots, x_{n-1})\,\beta(x_n)$, where α and β are test functions of the indicated arguments. By Lemma 2 above, we may write

$$\left(\frac{\partial f_n}{\partial x_{n-1}} - g_{n-1}, \alpha\beta \right) = C(\alpha) \int_{-\infty}^{\infty} \beta(x_n)\, dx_n.$$

On the other hand,

$$\left(\frac{\partial f_n}{\partial x_{n-1}} - g_{n-1}, \alpha\beta \right) = \left(f_n, -\frac{\partial \alpha}{\partial x_{n-1}}\beta \right) - (g_{n-1}, \alpha\beta)$$

$$= -\int_{-\infty}^{\infty} \left[\int_{R_{n-1}} \left[f_n(x) \frac{\partial \alpha}{\partial x_{n-1}} + g_{n-1}(x) \cdot \alpha \right] dx_1 \ldots dx_{n-1} \right] \beta(x_n)\, dx_n,$$

4*

and therefore,

$$C(\alpha) = - \int_{R_{n-1}} \left[f_n(x) \frac{\partial \alpha}{\partial x_{n-1}} + g_{n-1}(x) \cdot \alpha \right] dx_1 \dots dx_{n-1}.$$

Now $C(\alpha)$ is independent of x_n and so we may set $x_n = 0$. But $f_n(x_1, \dots, x_{n-1}, 0) = 0$, and we can conclude that

$$C(\alpha) = - \int_{R_{n-1}} g_{n-1}(x_1, \dots, x_{n-1}, 0) \, \alpha(x) \, dx_1 \dots dx_{n-1}$$

and

$$\left(\frac{\partial f_n}{\partial x_{n-1}} - g_{n-1}, \alpha\beta \right) = - \int_{R_n} g_{n-1}(x_1, \dots, x_{n-1}, 0) \, \alpha\beta \, dx_1 \dots dx_n.$$

Since α and β are arbitrary, this leads to (20).†

In an entirely similar fashion,

$$\frac{\partial}{\partial x_{n-2}} \left[\int_0^{x_{n-2}} g_{n-2}(x_1, \dots, x_{n-3}, \xi_{n-2}, 0, 0) \, d\xi_{n-2} \right.$$

$$\left. + \int_0^{x_{n-1}} g_{n-1}(x_1, \dots, x_{n-2}, \xi_{n-1}, 0) \, d\xi_{n-1} \right] = g_{n-2}(x_1, \dots, x_{n-1}, 0). \quad (23)$$

Equation (23) enables us to calculate $\dfrac{\partial F}{\partial x_{n-2}}$. Namely,

$$\frac{\partial F}{\partial x_{n-2}} = \frac{\partial}{\partial x_{n-2}} \left[\int_0^{x_{n-2}} g_{n-2}(x_1, \dots, x_{n-3}, \xi_{n-2}, 0, 0) \, d\xi_{n-2} \right.$$

$$+ \int_0^{x_{n-1}} g_{n-1}(x_1, \dots, x_{n-2}, \xi_{n-1}, 0) \, d\xi_{n-1}$$

$$\left. + \int_0^{x_n} g_n(x_1, \dots, x_{n-1}, \xi_n) d\xi_n \right] = g_{n-2}(x_1, \dots, x_{n-1}, 0)$$

$$+ \frac{\partial}{\partial x_{n-2}} \int_0^{x_n} g_n(x_1, \dots, x_{n-1}, \xi_n) \, d\xi_n = g_{n-2}(x),$$

† $\partial f_n / \partial x_j = g_j(x) - g_j(x_1, \dots, x_{n-1}, 0)$ $(j \leq n - 2)$ by the same sort of reasoning.

and so on. As a result, the relations $\dfrac{\partial F}{\partial x_j} = g_j$ are found to hold for any j, as reguired.

Thus, the representation (17) for $f(x)$ has been obtained. It is apparent from it that $f(x)$ is a continuous function and $\dfrac{\partial f}{\partial x_n} = g_n$ in the usual sense.

But since the choice of the order of the variables of integration is arbitrary, equation (17) holds for any other succession of the variables. Therefore on moving the j-th coordinate out to the last place, we can conclude that $\dfrac{\partial f}{\partial x_j} = g_j$ in the usual sense.

Problems

1. Find Δr^λ for $-n < \lambda < -n + 2$.

Ans.

$$(\Delta r^\lambda, \varphi) = \lambda(\lambda + n - 2)\,\Omega_n \int_0^\infty r^{\lambda+n-3}[S_r(\varphi) - \varphi(0)]\,dr,$$

where $S_r[\varphi]$ is the mean value of φ over a sphere of radius r centered at the origin.

Hint. Use Green's formula as in Sec. 1.7.2, 8°, the Taylor expansion of $\varphi(x)$, and the fact that

$$\varepsilon^{\lambda+n-2} = -(\lambda + n - 2) \int_\varepsilon^\infty r^{\lambda+n-3}\,dr.$$

2. Prove that the equation $\dfrac{\partial f}{\partial x_1} = g$ has a solution for every generalized function $g(x_1, ..., x_n)$.

Hint. The functional f is uniquely defined on every $\varphi_0 \in K$ which is the partial of some other test function ψ with respect to x_1. Given any test function φ, one can write

$$\varphi(x_1, ..., x_n) = \varphi_1(x_1) \int_{-\infty}^\infty \varphi(\xi_1, x_2, ..., x_n)\,d\xi_1 + \varphi_0(x_1, ..., x_n),$$

where $\varphi_1(x_1)$ is a fixed test function whose integral is 1 and f is known on φ_0 by the preceding statement. The formula

$$(f_0, \varphi) = (g, \psi), \quad \text{where} \quad \frac{\partial \psi}{\partial x_1} = \varphi_0,$$

furnishes one such solution.

3. Prove that the system of equations $\dfrac{\partial f}{\partial x_j} = g_j \, (j = 1, 2, ..., n)$ has a solution for any functionals g_j that satisfy the compatibility conditions

$$\frac{\partial g_j}{\partial x_k} = \frac{\partial g_k}{\partial x_j}.$$

Hint. Look for a solution of the equation $\dfrac{\partial f}{\partial x_1} = g_1$ in the form $f_0 + f_{n-1}$, where f_{n-1} is an arbitrary functional not depending on x_1. The second equation $\dfrac{\partial f}{\partial x_2} = g_2$ then becomes

$$\frac{\partial f_{n-1}}{\partial x_2} = g_2 - \frac{\partial f_0}{\partial x_2}.$$

The right-hand side of this is independent of x_1 owing to the compatibility condition. One can now find a solution of the form

$$f_{n-1} = f^0_{n-1} + f_{n-2},$$

where f_{n-2} is an arbitrary functional independent of x_1 and x_2, etc.

4. Show that the derivative of a homogeneous generalized function of degree p with respect to x_j is a homogeneous generalized function of degree $p - 1$ (Prob. 3, Sec. 1.6).

5. Show that homogeneous generalized functions of different degrees are linearly independent.

6. Consider the non-linear differential equation in two independent variables,

$$\frac{\partial u}{\partial t} + \frac{\partial \Phi(u)}{\partial x} = 0,$$

where $\Phi(u)$ is a given differentiable function. We wish to look at the solutions $u = u(x, t)$ of this equation in K' which are ordinary piecewise continuous functions. Show that the relation

$$\frac{dx}{dt} = \frac{\Phi(u^+) - \Phi(u^-)}{u^+ - u^-}$$

holds along a discontinuity curve of the solution, u^+ and u^- being the respective right-hand and left-hand limits of u at a discontinuity.

Hint. Use the expression for the derivative of a piecewise smooth function (Sec. 1.7.2, 6°).

7. Find the fundamental function for the operator $\varDelta + k^2$ in three-dimensional space (k is a constant).

Hint. Look for the required function in the form $f(r)$. Use the expression (7) for the Laplacian and then find the classical solutions of the equation

$$f''(r) + \frac{2}{r}f'(r) + k^2 f(r) = 0.$$

Apply the method of Example 8° to find $\varDelta f(r)$ in the generalized sense.
Ans.

$$\mathscr{E}(x) = -\frac{e^{ikr}}{4\pi r} \quad \text{or} \quad -\frac{e^{-ikr}}{4\pi r}.$$

Note. There is an explicit formula (involving Hankel functions) for the fundamental function of any operator $P(\varDelta)$, where P is an arbitrary polynomial and $\varDelta = \dfrac{\partial^2}{\partial x_1^2} + \cdots + \dfrac{\partial^2}{\partial x_n^2}$. See B. P. Paneyakh, *On the existence and uniqueness of the solution of an n-ultraharmonic equation in unbounded space*, Vestnik Moskov. Univ. Ser. I Mat. Mekh., 1959, No. 5, pp. 123–135 (In Russian).

8. We shall say that a generalized function f approaches zero as $|x| \to \infty$ if for every $\varphi(x) \in K$,

$$\lim_{|h| \to \infty} (f, \varphi(x + h)) = 0. \tag{1}$$

Clearly, if $f = f(x)$ is an ordinary function which approaches zero at infinity in the customary sense, then definition (1) will be satisfied. It may also hold for an ordinary function which does not approach xero at infinity in the customary sense $\left(\text{an example is } f(x) = \dfrac{d}{dx}\dfrac{\sin x^2}{x}\right)$. Show that if $f = f(x)$ is a polynomial and condition (1) is satisfied, then $f(x) \equiv 0$.

Hint. If $f(x)$ is a polynomial, the expression $(f, \varphi(x + h))$ is a polynomial in h.

Note. The theorem remains valid under the assumption that $f(x)$ is an entire analytic function whose order is less than 1.

Chapter 2

Special Topics in Generalized Function Theory

From the elementary introduction to generalized functions of Chapter One, we now move on to a deeper and more systematic study of their properties. The aim is to get us to a point where we can actively apply generalized functions to problems in analysis and differential equations.

In the next few sections, we shall obtain expressions for generalized functions in terms of the derivatives of ordinary functions. We shall introduce and study two new operations, the convolution and Fourier transform. A very important role will be played by x_+^λ, as well as other generalized functions.

2.1. Local Properties and the Support of a Generalized Function

2.1.1. We know that it makes no sense to talk about the values of a generalized function at isolated points. However, a well-defined meaning can be attached to the statement "a generalized function f is equal to zero in a region G". It signifies that $(f, \varphi) = 0$ no matter what test function φ we take with support strictly in G.† Thus, $\delta(x_1, \ldots, x_n)$ is equal to zero everywhere in R_n with the origin deleted, a functional carried on a surface (Sec. 1.6.2, 3°) is equal to zero everywhere off the surface, and so on. If $f \equiv f(x)$ is an ordinary function, the condition "f is equal to 0 in a region G" means that $f(x)$ vanishes almost everywhere in G in the customary sense.

Moreover, if $h(x)$ is an infinitely differentiable function vanishing in G and f is any generalized function, then the product hf is a generalized function equal to zero in G. Indeed, if $\varphi \in K$ and its support is strictly inside G, then

$$(hf, \varphi) = (f, h\varphi) = 0$$

since the function $h(x)\, \varphi(x)$ vanishes identically.

† That is, such that the closure of the set $\{x : \varphi(x) \neq 0\}$ lies in G.

If a generalized function f is equal to zero in G, its complement $R_n - G$ is called the *support* of f and is denoted by supp f. One also says that f is *carried in the set* $R_n - G$. Thus, the support of $\delta(x_1, ..., x_n)$ is a single point, the origin; $\delta(x)$ is carried at this point. A generalized function is said to have *compact support* if its support is a bounded set.

If a test function $\varphi(x)$ *vanishes on the support of a functional* f *and in a neighborhood of it, then* $(f, \varphi) = 0$ since in this case φ is carried inside the region where f is equal to zero. Hence, it follows that *any change made in a test function* φ *outside a neighborhood of the support of a generalized function* f *has no effect on the value of* (f, φ). Actually, such a change is equivalent to adding to φ another test function ψ equal to zero in a neighborhood of the support of the functional f. For then $(f, \psi) = 0$ and hence $(f, \varphi + \psi) = (f, \varphi)$.

This last fact may be used at times to broaden the domain of definition of a functional f by the simple expedient of including in it (besides the test functions) any function $\hat{\varphi}(x)$ coinciding with some test function $\varphi(x)$ in a neighborhood of the support of f. Namely, we can define

$$(f, \hat{\varphi}) = (f, \varphi).$$

By the above assertion, the expression on the right-hand side remains unchanged when we replace φ by another test function φ_1 also coinciding with φ in a neighborhood of the support of f. The expression $(f, \hat{\varphi})$ is therefore determined just by the function $\hat{\varphi}$.

Thus, the expressions $(f, 1)$, (f, x), ..., (f, ψ), where $\psi(x)$ is *any* infinitely differentiable function (of arbitrary growth at infinity), all have a well-defined meaning when the functional f has compact support.

Moreover, if $\psi_\nu(x)$ is a sequence of functions which converges uniformly to 0 in some neighborhood of the support of the functional f together with all its derivatives, then $(f, \psi_\nu) \to 0$ independently of the behavior of the sequence ψ_ν outside this neighborhood.

To see this, multiply ψ_ν by a fixed function $h(x)$ equal to 0 outside the indicated neighborhood and to 1 in a smaller neighborhood of the support of f. Then $(f, \psi_\nu) = (f, h\psi_\nu)$, while at the same time the product $h\psi_\nu \to 0$ in K. Hence, it follows that $(f, h\psi_\nu) \to 0$.

Thus, a functional having compact support possesses considerably stronger continuity properties than an arbitrary functional on K.

2.1.2. *If a generalized function f is equal to* 0 *in a region G, then each of its derivatives* $\dfrac{\partial f}{\partial x_k}$ *is equal to zero in G.* In fact, if $(f, \varphi) = 0$ for every test function φ carried strictly in G, then for each such function

$$\left(\frac{\partial f}{\partial x_k}, \varphi \right) = - \left(f, \frac{\partial \varphi}{\partial x_k} \right) = 0.$$

Let us return to the example involving the determination of $\varDelta \dfrac{1}{r^{n-2}}$ (Sec. 1.7.2, 8°). The function $\dfrac{1}{r^{n-2}}$ is infinitely differentiable everywhere except at the origin and $\varDelta \dfrac{1}{r^{n-2}} = 0$ in the customary sense. Therefore, we may state beforehand that the generalized function $\varDelta \dfrac{1}{r^{n-2}}$ is equal to zero everywhere outside the origin (and hence is carried at a single point, the origin itself). It will be recalled that the relation $\varDelta \dfrac{1}{r^{n-2}} = c\, \delta(x)$ was derived in Sec. 1.7.2, 8°.

If a generalized function f is carried in a set E, then all of its derivatives are carried in the same set.

By hypothesis, the generalized function f is equal to 0 in the complement of E. Thus, by what we just proved, all the derivatives of f are also zero in the complement of E.

Two generalized functions f and g are defined to be *equal in a region G* if their difference $f - g$ is equal to zero in G. In particular, a generalized function f is *regular in G* if it is equal to some ordinary function $g(x)$ in this region. Thus, $\delta(x)$ is regular (and equal to 0) everywhere outside the origin.

As a consequence, we obtain the following:

If two generalized functions f and g are equal in G, their derivatives $\dfrac{\partial f}{\partial x_k}$ *and* $\dfrac{\partial g}{\partial x_k}$ *are also equal in this region.*

In particular, if a generalized function f is a constant in G, then $\dfrac{\partial f}{\partial x_k}$ is equal to zero in $G (k = 1, 2, ..., n)$. Conversely, *if all first order partials of a generalized function are equal to zero in a (connected) region G, then f is constant in G.*

If G is a block $B \subset R_n$, the proof given in Sec. 1.7.3 can be repeated verbatim with R_n being replaced by B. The general result follows from the fact that a connected set is the union of a chained family of blocks.

Every generalized function in R_n is of course also a generalized function in any region $G \subset R_n$. The converse is false: a generalized function defined in G cannot be extended in general to all of R_n. A slightly weaker fact holds. Namely, if f is a given generalized function in G, then for any subregion G' lying strictly in G, there is a generalized function $f_1 \in K'(R_n)$ which is equal to f in G'. One merely must take a test function $a(x)$ with support strictly in G and equal to 1 on G' and then set $(f_1, \varphi) = (f, a\varphi)$.

2.1.3. In the following subsections, we shall examine some further important properties of test functions. In particular, the method of "partitioning unity" (Sec. 2.1.4) will play a large role in many of our subsequent considerations.

LEMMA 1. *Let F be a bounded closed set and G a region containing F. Then there exists a test function h which is 1 on F, 0 outside G and varies between 0 and 1 elsewhere.*

Proof. Let V_δ be a δ-neighborhood of F, i.e., the union of open balls of radius δ with centers at the points of F. The number δ is chosen small enough so that $V_{2\delta} \subset G$.

Consider the function

$$\psi(x; \delta) = \begin{cases} C_\delta e^{-\frac{\delta^2}{\delta^2 - r^2}} & \text{for} \quad r = |x| < \delta, \\ 0 & \text{for} \quad r = |x| > \delta. \end{cases} \tag{1}$$

Here

$$C_\delta^{-1} = \int\limits_{|x| \leq \delta} e^{-\frac{\delta^2}{\delta^2 - r^2}} dx$$

and thus $\int\limits_{R_n} \psi(x; \delta) \, dx = 1$.

We assert that the function

$$h(x) = \int\limits_{V_\delta} \psi(x - \xi; \delta) \, d\xi \tag{2}$$

satisfies the conditions. In fact, for any x in F, the integration in (2) is always over a ball of radius δ with center at x and consequently yields the value 1. For any x not in G, the integrand in (2) is zero on V_δ and the integral has the value 0. For all remaining points, the value of $h(x)$ evidently lies between 0 and 1. Finally, $h(x)$ is infinitely differentiable together with $\psi(x; \delta)$, and so the lemma is proved.

As a corollary, we can conclude that *if $f(x)$ is an arbitrary infinitely differentiable function, F is a bounded closed set, and G is a region containing F, then there is a test function $\varphi(x)$ coinciding with $f(x)$ on F and equaling 0 outside G.* It clearly suffices to take $\varphi(x) = f(x)\,h(x)$, where $h(x)$ is the above test function equaling 1 on F and 0 outside G.

2.1.4. Let the bounded open regions $U_1, ..., U_m, ...$ be a countable cover of R_n which is *locally finite*. By this, we mean that each point x has a neighborhood $W(x)$ which is covered by just a finite number of sets of the family $\{U_m\}$ and is disjoint from the remaining U_m. We wish to construct infinitely differentiable functions $e_1(x), ..., e_m(x), ...$ such that

a) $0 \leqq e_m(x) \leqq 1, \quad m = 1, 2, ...;$

b) $e_m(x) = 0$ outside $U_m, \quad (m = 1, 2, ...);$

c) $e_1(x) + e_2(x) + \cdots + e_m(x) + \cdots = 1.$

Owing to b), the left-hand side of the last identity involves only a finite number of non-zero terms for each x. The set of functions $\{e_m(x)\}$ is called a *partition of unity*, or more precisely, a *partition of unity subordinate to the cover* $\{U_m\}$.

We shall first construct another locally finite cover $V_1, ..., V_m, ...$ of R_n such that the closure $\overline{V}_m$ of V_m is contained in U_m ($m = 1, 2, ...$). To this end, we observe that the complement of $U_2 \cup \cdots \cup U_m \cup \cdots$ is a closed set F_1 covered by U_1. As V_1, we choose any region containing F_1 and contained with its closure in U_1.

Suppose now that $V_1, ..., V_{m-1}$ have already been determined so that $\overline{V}_k \subset U_k$ ($k = 1, 2, ..., m - 1$) and $V_1, ..., V_{m-1}, U_m, U_{m+1}, ...$ is a locally finite cover of R_n. The complement of $V_1 \cup V_2 \cup \cdots \cup V_{m-1} \cup U_{m+1} \cup \cdots$ is a closed set F_m completely covered by U_m. As V_m, we may take any open region containing F_m and contained in U_m together with its closure, etc.

Since $\overline{V}_m$ is a bounded set, by the result of the preceding section, there exists an infinitely differentiable function $h_m(x)$ lying between 0 and 1 for all x and equal to 1 on $\overline{V}_m$ and to 0 outside U_m. We now let

$$h(x) = \sum_{m=1}^{\infty} h_m(x). \tag{3}$$

This function exists for all x and is automatically at least 1. Since only a finite number of terms in the sum (3) are different from zero in any neighborhood $W(x)$, $h(x)$ is infinitely differentiable.

Finally, the functions

$$e_m(x) = \frac{h_m(x)}{h(x)} \quad (m = 1, 2, ...)$$

clearly satisfy the above three requirements.

2.1.5. Lemma 2. *Let $G_1, G_2, ...$ be an arbitrary family of regions. Then every test function $\varphi(x)$ with support strictly in $G = \bigcup_\nu G_\nu$ may be represented by a finite sum of $\varphi_k \in K$:*

$$\varphi = \varphi_1 + \cdots + \varphi_p, \tag{4}$$

with the support of φ_k strictly in G_k $(k = 1, 2, ..., p)$.

First of all, we can confine ourselves right away to the case where the family $G_1, G_2, ...$ is *finite*. In fact, in the general case, the closure of $\Phi = \{x: \varphi(x) \neq 0\}$ is a bounded set for which $\{G_\nu\}$ is an open cover. A finite subcover can always be extracted from it and then all of the remaining regions discarded.

Thus, G may be assumed to be the union of a finite family of regions $G_1, ..., G_p$.

Such a family together with the region V complementary to $\overline{\Phi}$ forms a cover of R_n. If unbounded, each of the G_k or V may be replaced by a (countable) union of bounded regions which collectively comprise a locally finite cover of R_n (Sec. 2.1.4). Consider a partition of unity $\{e_m(x)\}$ subordinate to the regions constructed. The function $\varphi(x)$ can be represented in the form

$$\varphi = \varphi e_1 + \varphi e_2 + \cdots = \varphi_1 + \varphi_2 + \cdots$$

The sum on the right-hand side is finite since only a finite number of the sets in the cover intersect the support of φ. In particular, the terms corresponding to the subsets of V are absent because φ vanishes on them. On combining the terms corresponding to the subsets of $G_1, ..., G_p$, we obtain our required expansion.

We mention that if φ_ν is a sequence of test functions converging to 0 in K, then each component in the expansion (4) for φ_ν, if obtained in the indicated way, likewise converges to zero in K.

2.1.6. THEOREM. *If a generalized function f is equal to zero in each of the regions $G_1, G_2, ..., G_\nu, ...,$ then it is equal to zero in their union $G = \bigcup_\nu G_\nu$.*
Proof. Let φ be a given test function with support in G. Employing Lemma 2 above, we can construct the expansion

$$\varphi = \varphi_1 + \cdots + \varphi_p,$$

where the test function φ_k is carried strictly in G_k. Applying f to this expansion and using the condition that $f = 0$ in G_k, we obtain

$$(f, \varphi) = (f, \varphi_1) + \cdots + (f, \varphi_p) = 0,$$

as asserted.

COROLLARY. *A generalized function which is equal to zero in a neigborhood of each point of R_n is the null generalized function, or in other words, $(f,\varphi) = 0$ for every $\varphi \in K$.*

Problems

1. Given that f is a generalized function such that $g(x)f = 0$ in a neighborhood $U(x_0)$ for each x_0, where $g(x)$ is a non-vanishing infinitely differentiable function (depending on the neighborhood $U(x_0)$). Prove that $f = 0$.

Hint. $\dfrac{1}{g(x)}$ is also infinitely differentiable in the neighborhood $U(x_0)$.

2. The condition of Prob. 1 is fulfilled everywhere except at the points of a closed set Φ. Show that f is carried in Φ.

Hint. Use Theorem 1.

3. Construct a test function $\varphi(x)$ with support in the ball $|x| \leq a$ such that 1 can be represented as a series of translates of $\varphi(x)$:

$$1 \equiv \sum_{\nu=0}^\infty \varphi(x + \xi_\nu). \tag{1}$$

Hint. It is enough to consider the case $n = 1$ since the general case may be disposed of by multiplying the equations (1) obtained for each of the coordinates. For $n = 1$, take any infinitely differentiable function $\varphi(x)$ equal to 1 for $0 \leq x \leq a/3$ and to 0 for $x > 2a/3$. Then complete its definition for $x < 0$ by means of the formula $\varphi(-x) = 1 - \varphi(x + a)$.

4. Given that $f(x)$ is defined and locally integrable for $x \neq 0$ and has at most a power singularity in the neighborhood of the origin:

$$|f(x)| \leq \frac{C}{|x|^m} \quad (|x| \leq a).$$

Show that there is an $f \in K'_n$ coinciding with $f(x)$ for $x \neq 0$.

Hint. See Prob. 6 of Sec. 1.6.

2.2. Convergence in the Space of Generalized Functions

2.2.1. A sequence of generalized functions $f_1, \ldots, f_\nu, \ldots$ *converges in K' to a generalized function f* if

$$(f, \varphi) = \lim_{\nu \to \infty} (f_\nu, \varphi)$$

for every $\varphi \in K$. It is apparent from the definition that f is uniquely determined by the sequence f_ν.

Similarly, a series of generalized functions $g_1 + g_2 + \cdots + g_\nu + \cdots$ *converges to a generalized function f* if the sequence of partial sums of the series,

$$f_1 = g_1, \ldots, f_\nu = g_1 + \cdots + g_\nu, \ldots$$

converges to f in the above sense.

For instance, if $f_1 = f_1(x), \ldots, f_\nu = f_\nu(x), \ldots$ is a sequence of ordinary functions converging to $f(x)$ almost everywhere and possessing an ordinary (i.e., locally integrable) majorant $F(x)$ so that

$$|f_\nu(x)| \leq F(x), \quad \nu = 1, 2, \ldots,$$

then it converges to f in the sense of generalized functions. More generally if the sequence $f_\nu(x)$ converges to $f(x)$ in the L_1 metric so that

$$\int_B |f_\nu(x) - f(x)| \, dx \to 0$$

for each bounded block, then it also converges to f in K'. For we have,

$$(f, \varphi) - (f_\nu, \varphi) = (f - f_\nu, \varphi) = \int_B (f - f_\nu) \varphi \, dx,$$

where B is a block outside of which $\varphi(x)$ vanishes. Hence,

$$|(f, \varphi) - (f_\nu, \varphi)| \leq \max |\varphi(x)| \int_B |f(x) - f_\nu(x)| \, dx \to 0.$$

Every generalized function f is a sum of generalized functions each having compact support. To see this, take an arbitrary locally finite cover of R_n (Sec. 2.1.4) and a partition of unity $\{e_m(x)\}$ subordinate to it. The multiplication of the series of test functions

$$1 \equiv e_1(x) + \cdots + e_m(x) + \cdots$$

by f yields

$$f = e_1 f + \cdots + e_m f + \cdots = f_1 + \cdots + f_m + \cdots, \qquad (1)$$

an expansion in generalized functions having compact support. The series (1) is obviously convergent since for each $\varphi \in K$ the expansion

$$(f, \varphi) = (f_1, \varphi) + \cdots + (f_m, \varphi) + \cdots$$

contains just a finite number of non-zero terms.

The limiting operation is linear. Stated otherwise, if $f_\nu \to f$ in K', $g_\nu \to g$ in K', and α and β are real numbers, then

$$\alpha f_\nu + \beta g_\nu \to \alpha f + \beta g \quad \text{in} \quad K'.$$

Similarly, if $\alpha(x)$ is infinitely differentiable and $f_\nu \to f$, then $\alpha(x) f_\nu \to \alpha(x) f$. Indeed, for every $\varphi \in K$,

$$(\alpha(x) f_\nu, \varphi) = (f_\nu, \alpha\varphi) \to (f, \alpha\varphi) = (\alpha(x) f, \varphi).$$

When the space K is complex, the convergence of the sequence f_ν to f implies that $\bar{f}_\nu$ converges to $\bar{f}$ since

$$(\bar{f}_\nu, \varphi) = \overline{(f_\nu, \bar{\varphi})} \to \overline{(f, \bar{\varphi})} = (\bar{f}, \varphi).$$

Somewhat less usual is the fact that *the convergence of f_ν to f always implies the convergence of* $\dfrac{\partial f_\nu}{\partial x_k}$ *to* $\dfrac{\partial f}{\partial x_k}$. As we know, to prove the analogous fact in classical analysis, we need additional assumptions. For generalized functions, it is a simple matter, namely,

$$\left(\frac{\partial f_\nu}{\partial x_k}, \varphi \right) = \left(f_\nu, -\frac{\partial \varphi}{\partial x_k} \right) \to \left(f, -\frac{\partial \varphi}{\partial x_k} \right) = \left(\frac{\partial f}{\partial x_k}, \varphi \right).$$

EXAMPLE 1°. The sequence $f_\nu(x) = \dfrac{1}{\nu} \sin \nu x \; (-\infty < x < \infty)$ converges uniformly to 0 as $\nu \to \infty$ and therefore converges to 0 in K'. The sequence of derivatives $f'_v(x) = \cos vx$ does not converge to zero in the usual sense but does so as a sequence of generalized functions. We have

$$(\cos vx, \varphi) = \int\limits_{-\infty}^{\infty} \cos vx \, \varphi(x) \, dx = -\frac{1}{v} \int\limits_{-\infty}^{\infty} \sin vx \, \varphi'(x) \, dx \to 0.$$

The sequence of higher derivatives

$$f''_v = -v \sin vx, \quad f'''_v = -v^2 \cos vx, \ldots$$

also converge to zero in K'.

2.2.2. A sequence of ordinary functions $f_v(x)$ will be called *delta-defining* if it converges in K' to the singular function $\delta(x)$. Such sequences often occur in problems relating to differential equations. The following lemma gives some general sufficient conditions for a sequence $f_v(x)$ to be delta-defining.

LEMMA. *If for any block B_0, the quantities*

$$\left| \int\limits_{B} f_v(x) \, dx \right|, \quad B \subset B_0,$$

are bounded by a constant depending on neither B nor v (and thus only on B_0) and if

$$\lim_{v \to \infty} \int\limits_{B} f_v(x) \, dx = \begin{cases} 0 & \text{when the origin is exterior to } B, \\ 1 & \text{when the origin is interior to } B, \end{cases}$$

then

$$\lim_{v \to \infty} f_v(x) = \delta(x) \quad \text{in } K'.$$

Proof. Consider the sequence of integrals

$$F_v(x) = \int\limits_{-1}^{x_1} \cdots \int\limits_{-1}^{x_n} f_v(\xi) \, d\xi_1 d\xi_2 \ldots d\xi_n.$$

The sequence is uniformly bounded in each bounded block. Furthermore, it converges to zero if at least one of the $x_1, x_2, \ldots, x_n$ is negative and converges to 1 if all x_k are positive. Hence, it follows that

$$\lim_{v \to \infty} F_v(x) = \theta(x_1, \ldots, x_n) \quad \text{in K'}.$$

Differentiating this limit relation (cf. Sec. 1.7.2, 3°), we obtain

$$\lim_{\nu \to \infty} f_\nu(x) = \lim_{\nu \to \infty} \frac{\partial^n F_\nu(x)}{\partial x_1 \ldots \partial x_n} = \frac{\partial^n \theta(x)}{\partial x_1 \ldots \partial x_n} = \delta(x) \quad \text{(in } K'\text{)},$$

as asserted.

EXAMPLE 2°. Let $n = 1$ and

$$f_\varepsilon(x) = \frac{1}{\pi} \frac{\varepsilon}{x^2 + \varepsilon^2}, \quad \varepsilon \to 0$$

(see Fig. 5). We have

$$\int_a^b f(x)\, dx = \frac{1}{\pi} \left[\arctan \frac{b}{\varepsilon} - \arctan \frac{a}{\varepsilon} \right],$$

and it easily follows from this that the hypotheses of the lemma are

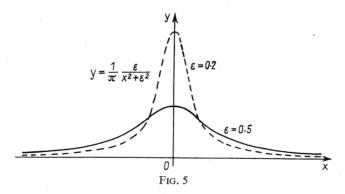

FIG. 5

satisfied. Applying it, we conclude that

$$\lim_{\varepsilon \to 0} f_\varepsilon(x) = \delta(x).$$

EXAMPLE 3°. Let $n = 1$, $c > 0$, and

$$f_t(x) = \frac{1}{c\sqrt{\pi t}} e^{-x^2/c^2 t} \quad (t \to 0+).$$

Making the substitution $y = x/c\sqrt{t}$, we find that

$$\int_a^b f_t(x)\, dx = \frac{1}{\sqrt{\pi}} \int_{a/c\sqrt{t}}^{b/c\sqrt{t}} e^{-y^2} dy.$$

If a and b have different signs, the resulting integral has the limit 1. But if a and b have the same sign, the integral tends to zero. Applying the lemma, we conclude that

$$\lim_{t \to 0} f_t(x) = \delta(x).$$

It should be pointed out that a delta-defining sequence need not consist of positive terms alone (see Prob. 1).

2.2.3. Of great importance is the fact that K' is complete under the convergence defined in it:

THEOREM. *Let $f_1, f_2, \ldots$ be a sequence of continuous linear functionals in K' such that for each $\varphi \in K$ the sequence (f_ν, φ) is convergent ($\nu \to \infty$). Then the functional on K determined by the relation*

$$(f, \varphi) = \lim_{\nu \to \infty} (f_\nu, \varphi)$$

is also linear and continuous.

The proof is based on a lemma which is important in its own right.

LEMMA. *If a sequence of functionals $f_\nu \in K'$ ($\nu = 1, 2, \ldots$) is such that the sequence (f_ν, φ) has a limit (f, φ) for each $\varphi \in K$, then for every sequence of test functions $\varphi_\nu \to \varphi$ in K, the limit of (f_ν, φ_ν) also exists as $\nu \to \infty$ and is equal to (f, φ).*

Assuming the lemma has been established, we shall use it to infer the completeness of K'.

The linearity of the functional $f = \lim_{\nu \to \infty} f_\nu$ is easy to prove since

$$(f, \alpha_1 \varphi_1 + \alpha_2 \varphi_2) = \lim_{\nu \to \infty} (f_\nu, \alpha_1 \varphi_1 + \alpha_2 \varphi_2) = \lim_{\nu \to \infty} [\alpha_1 (f_\nu, \varphi_1) + \alpha_2 (f_\nu, \varphi_2)]$$

$$= \alpha_1 \lim_{\nu \to \infty} (f_\nu, \varphi_1) + \alpha_2 \lim_{\nu \to \infty} (f_\nu, \varphi_2)$$

$$= \alpha_1 (f, \varphi_1) + \alpha_2 (f, \varphi_2).$$

Now consider the question of its continuity. Let $\varphi_\nu \to 0$ in K. It is necessary to show that $(f, \varphi_\nu) \to 0$. Assume the contrary. Then going to a subsequence, if necessary, we may assume $|(f, \varphi_\nu)| > 2a$ for all $\nu = 1, 2, \cdots$, where a is a positive constant. Since $(f, \varphi_\nu) = \lim_{\mu \to \infty} (f_\mu, \varphi_\nu)$, to each $\nu = 1, 2, \ldots$, there is a $\mu = \mu(\nu)$ such that $|(f_\mu, \varphi_\nu)| > a$. Again choosing a subsequence, we may consider $\mu(\nu) \equiv \nu$. Thus, $|(f_\nu, \varphi_\nu)| > a$ for $\nu = 1, 2, \ldots$ On the other hand, by the lemma the limit of (f_ν, φ_ν) exists

5*

as $v \to \infty$ and equals $(f, 0) = 0$. The resulting contradiction shows that f is indeed continuous, as asserted.

It remains to prove the lemma. There is no loss of generality in taking $\varphi(x) \equiv 0$. Otherwise, we would replace φ_v by $\varphi_v - \varphi$. We have to prove that $(f_v, \varphi_v) \to 0$.

Assume the contrary. Then going, if necessary, to a subsequence, we may assume that $|(f_v, \varphi_v)| \geq C > 0$. It will be recalled that the convergence of a sequence φ_v to zero in K means that all φ_v vanish outside a bounded region and φ_v converges to zero uniformly in R_n together with its derivatives of any order. Once more choosing a subsequence, we may assume that

$$|D^l \varphi_v(x)| \leq \frac{1}{4^v} \quad (|l| \leq v = 0, 1, 2, \ldots).$$

Let $\psi_v = 2^v \varphi_v$. The ψ_v satisfy the inequalities

$$|D^l \psi_v(x)| \leq \frac{1}{2^v} \quad (|l| \leq v = 0, 1, 2, \ldots),$$

thus showing that ψ_v also converges to zero in K. Moreover, every series of the form

$$\sum_{(k)} \psi_{v_k}(x)$$

converges in K. At the same time, the quantity $|(f_v, \psi_v)| = 2^v|(f_v, \varphi_v)| \geq 2^v C$ and therefore tends to infinity.

We now proceed to construct two subsequences f_{v_k} and ψ_{v_k} as follows.

We first choose f_{v_1} and ψ_{v_1} so that $|(f_{v_1}, \psi_{v_1})| > 1$. Suppose f_{v_j} and ψ_{v_j} have been determined for $j = 1, 2, \ldots, k-1$. We then take ψ_{v_k} to be a term in the sequence ψ_v of such high index that

$$|(f_{v_j}, \psi_{v_k})| < \frac{1}{2^{k-j}} \quad (j = 1, \ldots, k-1),$$

and after that we determine f_{v_k} so that

$$|(f_{v_k}, \psi_{v_k})| > \sum_{j=1}^{k-1} |(f_{v_k}, \psi_{v_j})| + k. \tag{2}$$

The first thing is possible because the sequence ψ_v converges to zero in K and therefore for each generalized function f_0, we have $(f_0, \psi_v) \to 0$.

The second is possible because $|(f_v, \psi_v)| \to \infty$ and $(f_v, \psi_{v_j}) \to (f, \psi_{v_j})$, $j \leqq k - 1$.

The ψ_{v_k} and f_{v_k} may be so determined ad infinitum. We now let

$$\psi = \sum_{k=1}^{\infty} \psi_{v_k}.$$

By construction, the series on the right-hand side is convergent in K and therefore ψ is in K. Further,

$$(f_{v_k}, \psi) = \sum_{j=1}^{k-1} (f_{v_k}, \psi_{v_j}) + (f_{v_k}, \psi_{v_k}) + \sum_{j=k+1}^{\infty} (f_{v_k}, \psi_{v_j}).$$

But by (2) and the fact that

$$\sum_{j=k+1}^{\infty} |(f_{v_k}, \psi_{v_j})| < \sum_{j=k+1}^{\infty} \frac{1}{2^{j-k}} = 1,$$

we have

$$|(f_{v_k}, \psi)| > k - 1.$$

In other words, as $k \to \infty$, $|(f_{v_k}, \psi)| \to \infty$. But this contradicts the relation $\lim_{v \to \infty} (f_v, \psi) = (f, \psi)$. The lemma is proved.†

2.2.4. Suppose that with each value of a real or complex parameter λ varying over a region Λ there is associated a generalized function f_λ. In accordance with the definition of Sec. 2.2.1, f is *the limit of f_λ as $\lambda \to \lambda_0$*, if (f_λ, φ) approaches (f, φ) for every φ. The function f_λ is said to be *continuous in λ in the region Λ* if for each λ_0 in Λ, $f_{\lambda_0} = \lim_{\lambda \to \lambda_0} f_\lambda$.

Let us examine the important question of whether *a generalized function f_λ can be extended continuously in its parameter λ*. Imagine f_λ to be defined and continuous on a set Λ. Suppose that λ_0 is an accumulation point of Λ at which f_λ is initially undefined. The question raised is this. It is possible to define f_λ at λ_0 so as to obtain a generalized function continuous on $\Lambda \cup \{\lambda_0\}$?

An obvious necessary condition for this to be possible is that each of the numerical functions (f_λ, φ) be extendable continuously to λ_0. *The condition is also sufficient.* Indeed, if the limit of (f_λ, φ) exists for every φ in K and every sequence $\lambda_v \to \lambda_0$, with $\lambda_v \in \Lambda$, then by the completeness of K', there is an $f = f_{\lambda_0}$ in K' which is the limit of f_{λ_v}. By standard reason-

† The proof is due to M. L. Brodskii.

ing, it can be shown that the limit is independent of the choice of sequence $\lambda_v \to \lambda_0$.

Moreover, if f_λ is a generalized function depending continuously on λ, then its spatial derivatives are also continuous in λ. For, as we have seen in Sec. 2.2.1, the fact that $f_{\lambda_v} \to f_{\lambda_0}$ implies that

$$\frac{\partial}{\partial x_k} f_{\lambda_v} \to \frac{\partial}{\partial x_k} f_{\lambda_0}.$$

A generalized function continuous in a parameter may be integrated with respect to the parameter. Suppose, for example, that f_λ is continuous in λ along a rectifiable curve Γ. We partition Γ into m parts at the points $\lambda_0, \lambda_1, ..., \lambda_m$. Then choosing an arbitrary point λ'_j in each interval $(\lambda_{j-1}, \lambda_j)$, we form the Riemann type sum

$$s_m = \sum_{j=1}^{m} f'_{\lambda_j} \Delta\lambda_j.$$

Since the function (f_λ, φ) is continuous, the expression

$$(s_m, \varphi) = \sum_{j=1}^{m} (f'_{\lambda_j}, \varphi) \Delta\lambda_j$$

has a limit as $\max |\Delta\lambda_j| \to 0$ independent of the partition of Γ and the choice of intermediary points λ'_j for each φ. The limit equals the integral of (f_λ, φ) and defines a continuous linear functional on K. It is called *the integral of the generalized function f_λ along Γ* and is denoted in the customary way by

$$\int_\Gamma f_\lambda \, d\lambda \quad \text{or} \quad \int_\Gamma f_\lambda(x) \, d\lambda.$$

Because differentiation is a continuous operation, we have

$$\frac{\partial}{\partial x_k} \int_\Gamma f_\lambda \, d\lambda = \int_\Gamma \frac{\partial f_\lambda}{\partial x_k} \, d\lambda.$$

Thus, integration of a generalized function with respect to a parameter commutes with spatial differentiation.

Of course, integration may be performed not only along a curve but also over a region of any dimensionality.

2.2.5. A generalized function g is called the *derivative* of a generalized function f_λ with respect to λ at $\lambda = \lambda_0$ if

$$g = \lim_{\lambda \to \lambda_0} \frac{f_\lambda - f_{\lambda_0}}{\lambda - \lambda_0}.$$

A necessary and sufficient condition for $\partial f_\lambda / \partial \lambda$ to exist at $\lambda = \lambda_0$ is that each of the functions (f_λ, φ) be differentiable in λ at $\lambda = \lambda_0$. The necessity is trivial. The proof of the sufficiency goes as follows. By hypothesis, for each φ in K and every sequence $\lambda_\nu \to \lambda_0$, the limit of

$$\frac{(f_\lambda, \varphi) - (f_{\lambda_0}, \varphi)}{\lambda - \lambda_0} = \left(\frac{f_\lambda - f_{\lambda_0}}{\lambda - \lambda_0}, \varphi \right)$$

exists. But then, as was pointed out above, the generalized function $(f_\lambda - f_{\lambda_0})/(\lambda - \lambda_0)$, defined except for $\lambda = \lambda_0$, can be extended continuously to $\lambda = \lambda_0$. In other words, there exists a generalized function which is the limit of $(f_\lambda - f_{\lambda_0})/(\lambda - \lambda_0)$ as $\lambda \to \lambda_0$, as asserted.

If f_λ has a derivative with respect to λ for every $\lambda \in \Lambda$, then f_λ is said to be *differentiable in Λ*.

Higher order derivatives with respect to a parameter and differentiability are defined similarly.

If f_λ is differentiable in Λ, then it is easy to verify that all of its first-order spatial derivatives are differentiable with respect to λ and that

$$\frac{\partial}{\partial \lambda} \left(\frac{\partial}{\partial x_k} f_\lambda \right) = \frac{\partial}{\partial x_k} \left(\frac{\partial}{\partial \lambda} f_\lambda \right). \tag{3}$$

In fact, for any $\varphi \in K$, the function

$$\left(\frac{\partial}{\partial x_k} f_\lambda, \varphi \right) = \left(f_\lambda, -\frac{\partial \varphi}{\partial x_k} \right)$$

is differentiable with respect to λ and the derivative

$$\frac{\partial}{\partial \lambda} \left(f_\lambda, -\frac{\partial \varphi}{\partial x_k} \right) = \left(\frac{\partial f_\lambda}{\partial \lambda}, -\frac{\partial \varphi}{\partial x_k} \right) = \left(\frac{\partial}{\partial x_k} \frac{\partial f_\lambda}{\partial \lambda}, \varphi \right).$$

This means that the functional $\partial f_\lambda / \partial x_k$ has a derivative with respect to λ and so (3) holds, as asserted.

2.2.6. Let λ be a complex parameter varying over a domain Λ. A generalized function f_λ is defined to be an *analytic function of λ in Λ* if f_λ is differentiable throughout Λ. In that event, all of the functions (f_λ, φ) are ordinary analytic functions of λ in the domain Λ. Conversely, *if f_λ is a generalized function such that every (f_λ, φ) is an analytic function of λ in Λ, then f_λ is also an analytic function of λ. Furthermore, all of its derivatives $\partial f_\lambda / \partial \lambda$, $\partial^2 f_\lambda / \partial \lambda^2$, ... exist everywhere in Λ, and f_λ has a Taylor expansion*

$$f_\lambda = f_{\lambda_0} + (\lambda - \lambda_0)\frac{\partial f_{\lambda_0}}{\partial \lambda} + \frac{1}{2}(\lambda - \lambda_0)^2 \frac{\partial^2 f_{\lambda_0}}{\partial \lambda^2} + \cdots \tag{4}$$

in the neighborhood of each λ_0 in Λ. Indeed, $\partial f_\lambda / \partial \lambda$ exists since by assumption the functions (f_λ, φ) each have a derivative with respect to λ at $\lambda = \lambda_0$. But all of the higher derivatives, $\partial^2 f_\lambda / \partial \lambda^2$, ..., also exist for the same reason. Now for each $\varphi \in K$, the ordinary analytic function (f_λ, φ) has the Taylor expansion

$$(f_\lambda, \varphi) = (f_{\lambda_0}, \varphi) + (\lambda - \lambda_0)\frac{\partial}{\partial \lambda}(f_\lambda, \varphi)\bigg|_{\lambda = \lambda_0} + \cdots$$

$$= (f_{\lambda_0}, \varphi) + (\lambda - \lambda_0)\left(\frac{\partial f_{\lambda_0}}{\partial \lambda}, \varphi\right) + \cdots$$

$$= \left(f_{\lambda_0} + (\lambda - \lambda_0)\frac{\partial f_{\lambda_0}}{d\lambda} + \cdots, \varphi\right),$$

and this implies the validity of (4).

Two analytic functions f_λ and g_λ defined in a domain Λ and coinciding on a set of values of λ having an accumulation point inside Λ coincide for all $\lambda \in \Lambda$. This is a consequence of the classical uniqueness theorem for analytic functions which implies that the expressions (f_λ, φ) and (g_λ, φ) have to be equal throughout Λ for every $\varphi \in K$.

This property is the basis for the important method of analytic continuation of a functional f_λ in the parameter λ. Suppose that f_λ is analytic in a domain Λ and that all of the functions (f_λ, φ) can be analytically continued to a larger domain Λ_1. Then for each $\lambda \in \Lambda_1$, the quantities (f_λ, φ) also define a continuous linear functional on K. For we know that the analytic continuation to a point λ of Λ_1 can always be accomplished by a finite number of Taylor series expansions. But each Taylor series

$$(f_\lambda, \varphi) = (f_{\lambda_0}, \varphi) + (\lambda - \lambda_0)\left(\frac{\partial f_{\lambda_0}}{\partial \lambda}, \varphi\right) + \cdots$$

is convergent for every $\varphi \in K$ and has a radius of convergence dependent only on the configurations of Λ and Λ_1 and not on φ. Thus, the series represents a continuous linear functional within its circle of convergence, as was to be proved.

The spatial derivatives of an analytic generalized function f_λ are also clearly analytic generalized functions of λ.

If $f_{t,\lambda}$ is a functional which is continuous in a parameter t in a bounded closed region T and analytic in λ in a domain Λ (for any t), then

$$\int_T f_{t,\lambda}\, dt = g_\lambda$$

remains analytic in λ throughout Λ. This follows from the fact that for any test function φ, the integral

$$\int_T (f_{t,\lambda}, \varphi)\, dt$$

of an analytic function $(f_{t,\lambda}, \varphi)$ is again an analytic function of λ in Λ.†

If $f_{t,\lambda}$ can be continued analytically from a domain Λ to a domain Λ_1 for each t, then its integral g_λ can also be analytically continued to Λ_1. Moreover, the analytic continuation of the integral (the function g_λ) coincides with the integral of the analytic continuation (i.e., the integral of f_λ) by the uniqueness of an analytic continuation.

2.2.7. Consider a generalized function of the argument (ω, x) (Sec. 1.6.2, 4°). Let us recall how it is constructed. If ω is a fixed unit vector, we can associate with each test function $\varphi(x) = \varphi(x_1, ..., x_n)$ the function

$$\Phi(\xi; \omega) = \int_{(\omega, x)\,=\,\xi} \varphi(x)\, dx.$$

The latter is a test function of the argument ξ, $-\infty < \xi < \infty$.

With every generalized function $f(\xi) \in K_1'$, we can then associate a functional $f_\omega \in K_n'$ such that

$$(f_\omega, \varphi) = (f(\xi), \Phi(\xi; \omega)).$$

We shall study f_ω here in its dependence on the parameter ω, with ω varying over the unit sphere Ω. $\Phi(\xi; \omega)$ can be considered a test function depending on the parameter ω.

† See E. T. Copson, *Theory of Functions of a Complex Variable*, Oxford University Press, 1935, Sec. 5.5.

$\Phi(\xi; \omega)$ is carried in a fixed closed interval of the ξ-axis (which is independent of ω). As the integral of $\varphi(x_1, \ldots, x_n)$ over $(\omega, x) = \xi$, it is a continuous and infinitely differentiable function of ω (owing to the uniform continuity of $\varphi(x_1, \ldots, x_n)$ and each of its derivatives).

The functional f_ω on K_n is therefore also continuous and infinitely differentiable in ω. In particular, *it can be integrated with respect to ω over Ω.*

Let us evaluate the integral when $f(\xi)$ is an ordinary function. Let $\Omega_n(\varrho)$ and $S_n(\varrho)$ denote the surface area and volume of a ball of radius ϱ in R_n. These quantities have the trivial relationship

$$\Omega_n(\varrho) = \frac{dS_n(\varrho)}{d\varrho}.$$

For conciseness, denote $\Omega_n(1)$ by Ω_n and $S_n(1)$ by S_n. For dimensional reasons, $\Omega_n(\varrho) = \varrho^{n-1}\Omega_n(1) = \varrho^{n-1}\Omega_n$ and $S_n(\varrho) = \varrho^n S_n(1) = \varrho^n S_n$. Therefore,

$$\Omega_n = n\varrho^{n-1}S_n(\varrho) \mid_{\varrho=1} = nS_n.$$

A little later on we shall find explicit expressions for Ω_n and S_n. We wish to evaluate

$$\int_{\Omega_n} f((\omega, x)) \, d\omega. \tag{5}$$

Instead of (5), we first evaluate the integral over the ball $S_n(\varrho)$ of radius ϱ given by

$$\int_{S_n(\varrho)} f((x, y)) \, dy.$$

As a function of $y, f((x, y))$ is constant in planes orthogonal to the vector x. The area of a section of $S_n(\varrho)$ by any such plane lying at a distance $h \leq \varrho$ from the origin is

$$(\varrho^2 - h^2)^{\frac{n-1}{2}} S_{n-1}.$$

Therefore

$$\int_{S_n(\varrho)} f((x, y)) \, dy = S_{n-1} \int_{-\varrho}^{\varrho} f(rh) (\varrho^2 - h^2)^{\frac{n-1}{2}} \, dh \quad (r = |x|).$$

To obtain the integral over Ω_n, we must differentiate the resulting relation with respect to ϱ and then set $\varrho = 1$. We have

$$
\int_{\Omega_n} f((\omega, x))\, d\omega = S_{n-1}(n-1) \int_{-1}^{1} f(rh)\,(1-h^2)^{\frac{n-3}{2}}\, dh
$$

$$
= \frac{\Omega_{n-1}}{r^{n-2}} \int_{-r}^{r} f(\xi)\,(r^2 - \xi^2)^{\frac{n-3}{2}}\, d\xi. \tag{6}
$$

In particular, if $f(\xi) \equiv 1$, then (5) yields the surface area Ω_n of the unit sphere in R_n and so

$$
\Omega_n = \Omega_{n-1} \int_{-1}^{1} (1-h^2)^{\frac{n-3}{2}}\, dh = \Omega_{n-1}\, \frac{\Gamma\left(\dfrac{n-1}{2}\right)\Gamma\left(\dfrac{1}{2}\right)}{\Gamma\left(\dfrac{n}{2}\right)}. \tag{7}
$$

This formula easily leads to an expression for Ω_n for any n.

Let $\Omega_n \Gamma\left(\dfrac{n}{2}\right) = A_n$. From (7), we have

$$
A_n = \sqrt{\pi}\, A_{n-1}.
$$

Since $A_2 = 2\pi$, we find for any $n > 1$ that

$$
A_n = 2\pi(\sqrt{\pi})^{n-2} = 2\pi^{\frac{n}{2}}.
$$

Hence

$$
\Omega_n = 2\frac{\pi^{\frac{n}{2}}}{\Gamma\left(\dfrac{n}{2}\right)}, \qquad S_n = \frac{2}{n}\,\frac{\pi^{\frac{n}{2}}}{\Gamma\left(\dfrac{n}{2}\right)}. \tag{8}
$$

Problems

1. Prove that as $\nu \to \infty$,

$$
\frac{1}{\pi}\,\frac{\sin \nu x}{x} \to \delta(x) \quad \text{in } K_1'.
$$

2. Find a sequence of ordinary functions $g_\nu(x)$ converging (in K') to the generalized function $\delta^{(k)}(x)$.
Hint. Set $g_\nu(x) = f_\nu^{(k)}(x)$, where $f_\nu(x)$ is a delta-defining sequence of infinitely differentiable functions.

3. Multiplication by a function $\beta(x)$ which is not infinitely differentiable cannot be "reasonably" defined for all of K'. "Reasonably" means the following: The multiplication of ordinary functions by $\beta(x)$ should coincide with ordinary multiplication; for generalized functions, it should always be a continuous operation on K', i.e., one such that $f_\nu \to f$ in K' implies that $\beta f_\nu \to \beta f$.

Hint. Let the k-th derivative of $\beta(x)$ have a discontinuity at $x = 0$. Construct a sequence $f_\nu(x) \to \delta^{(k)}(x)$ in such a way that $(\beta(x) f_\nu(x), \varphi(x))$ has no limit for certain φ.

4. If the series

$$\sum_{m=0}^{\infty} a_m \, \delta^{(m)}(x)$$

is convergent in K_1', then all of the coefficients a_m vanish from some point onward.

Hint. Use a test function $\varphi(x)$ for which $|\varphi^{(m)}(0)| > |a_m^{-1}|$.

5. Let $\varphi_\lambda(x)$ be a test function differentiable with respect to λ in K and f_λ a generalized function differentiable with respect to λ in K'. Then $(f_\lambda, \varphi_\lambda)$ is a differentiable function of λ and

$$\frac{d}{d\lambda}(f_\lambda, \varphi_\lambda) = \left(\frac{df_\lambda}{d\lambda}, \varphi_\lambda \right) + \left(f_\lambda, \frac{d\varphi_\lambda}{d\lambda} \right).$$

Hint. Use the lemma of Sec. 2.2.3.

2.3. The Structure of Generalized Functions

2.3.1. Let $f_1(x), f_2(x), \ldots, f_\nu(x), \ldots$ be a sequence of ordinary functions vanishing in any given block $B = \{x : \alpha_j \leq x_j \leq \beta_j\}$ beginning with some index $\nu(B)$. Let $P_\nu(D)$ be an arbitrary linear partial differential operator with constant coefficients. Then the series

$$\sum_{\nu=1}^{\infty} P_\nu(D) f_\nu(x) \tag{1}$$

converges in K'. Indeed, for any test function φ, the series

$$\sum_{\nu=1}^{\infty} (P_\nu(D) f_\nu(x), \varphi(x)) = \sum_{\nu=1}^{\infty} (f_\nu(x), P_\nu(-D) \varphi(x)) \tag{2}$$

contains just a finite number of terms (since the support of $\varphi(x)$ lies in some block B). Therefore, (2) is defined for any test function φ. It is easy to verify that the sum determines a continuous linear functional on K.

In this section, we shall show that *every generalized function on K has the form* (1). If we take only a finite number of functions $f_1(x), ..., f_p(x)$ which moreover have compact support, then

$$\sum_{v=1}^{p} P_v(D) \, f_v(x) \tag{3}$$

is clearly a generalized function with compact support. We shall show that *every generalized function with compact support can be represented by* (3). Finally, we shall obtain the general form of any f in K' carried at a single point. If the point is the origin, f is given by

$$f = \sum_{|k| \leq p} a_k D^k \, \delta(x), \quad D^k = \frac{\partial^{|k|}}{\partial x_n^{k_1} ... \partial x_n^{k_n}}. \tag{4}$$

2.3.2. All of these results rest on the following property of generalized functions:

LEMMA 1. *To each generalized function f and to each bounded region G, there is a number $N = N(f, G)$ such that for any φ in K with support in G,*

$$|(f, \varphi)| \leq C \sum_{|k| \leq N} \max_{x \in G} |D^k \varphi(x)| \tag{5}$$

with C a constant depending on f and G.

Proof. Suppose that for some f and G the required N does not exist. This means that to each $N = 1, 2, ...,$ there is a test function $\varphi = \varphi_N(x)$ with support in G such that

$$|(f, \varphi_N)| \geq N \sum_{|k| \leq N} \max_{x \in G} |D^k \varphi_N(x)|. \tag{6}$$

If we replace $\varphi_N(x)$ by a multiple of it $\psi_N = C_N \varphi_N$, we do not disturb the inequality (6). Accordingly,

$$|(f, \psi_N)| \geq N \sum_{|k| \leq N} \max |D^k \psi_N(x)| \tag{7}$$

for any C_N.

Choose C_N so that

$$\sum_{|k| \leq N} \max_{x \in G} |D^k \psi_N(x)| = \frac{1}{N} \quad (N = 1, 2, ...).$$

The sequence $\psi_1(x), ..., \psi_N(x), ...$ converges to 0 in K. The $\psi_N(x)$ are all carried in the bounded region G and converge uniformly to 0 as $N \to \infty$

as do their derivatives of any fixed order. Since f is a continuous functional, $(f, \psi_N) \to 0$. On the other hand, (7) implies that $|(f, \psi_N)| > 1$. The resulting contradiction proves the validity of our assertion.

The inequality (5) can be extended in a certain way (if G has a sufficiently simple configuration). To this end, we apply the Schwarz inequality. For any $x \in G$, we have

$$|D^k \varphi(x)| = \left| \int_{a_1}^{x_1} \cdots \int_{a_n}^{x_n} \frac{\partial^n D^k \varphi(\xi)}{\partial x_1 \ldots \partial x_n} \, d\xi \right|$$

$$\leqq C_1 \sqrt{\int_G \left(\frac{\partial^n D^k \varphi(\xi)}{\partial x_1 \ldots \partial x_n} \right)^2 d\xi} \leqq C_1 \sqrt{\sum_{|k| \leq N+n} \int_G (D^k \varphi(x))^2 \, dx}.$$

where $(a_1, \ldots, a_0)$ is a point (possibly outside G) at which $\varphi(x)$ and all its derivatives vanish. Combining this result with (5), we finally obtain

$$|(f, \varphi)| \leqq C \sum_{|k| \leqq N} \max_{x \in G} |D^k \varphi(x)| \leqq C_2 \sqrt{\sum_{|k| \leq N+n} \int_G (D^k \varphi(x))^2 \, dx}. \quad (8)$$

2.3.3. Consider the Hilbert space $\mathscr{H}_m$ of test functions carried in G with inner product

$$(\varphi, \psi)_m = \sum_{|k| \leqq m} \int_G D^k \varphi(x) \, D^k \psi(x) \, dx$$

and corresponding norm

$$\|\varphi\|_m^2 = \sum_{|k| \leqq m} \int_G (D^k \varphi(x))^2 \, dx.$$

It follows from (8) that the functional f is bounded in the norm of $\mathscr{H}_m$ with $m = N + n$ and can therefore be extended as a bounded functional to the completion of $\mathscr{H}_m$. Let us obtain the completion $\overline{\mathscr{H}}_m$ of $\mathscr{H}_m$. If $\varphi_1, \ldots, \varphi_\nu, \ldots$ is a Cauchy sequence in $\mathscr{H}_m$, then

$$\|\varphi_\nu - \varphi_\mu\|_m^2 = \sum_{|k| \leqq m} \int_G \left(D^k (\varphi_\nu(x) - \varphi_\mu(x)) \right)^2 dx$$

$$\geqq \int_G (D^k \varphi_\nu(x) - D^k \varphi_\mu(x))^2 \, dx$$

for any fixed k, $|k| \leqq m$. This means that each sequence $D^k \varphi_\nu(x)$ is a Cauchy sequence under convergence in $L_2(G)$ (the usual space of square integrable functions in G) and hence has an $L_2(G)$ limit $\varphi^{[k]}(x)$. The superscript $[k]$ indicates that $\varphi^{[k]}(x)$ originated from k-th order derivatives of $\varphi_\nu(x)$ and does not generally speaking mean that $\varphi^{[k]}(x)$ is a k-th order derivative of $\varphi(x)$. Thus, associated with each element of $\overline{\mathscr{H}}_m$ is a family $\{\varphi^{[k]}(x)\}$, $|k| \leq m$, of square integrable functions in G. Moreover, by continuity of the inner product,

$$(\{\varphi^{[k]}(x)\}, \{\psi^{[k]}(x)\})_m = \sum_{|k| \leq m} \int_G \varphi^{[k]}(x)\, \psi^{[k]}(x)\, dx.$$

We now apply the Riesz representation theorem for continuous linear functionals in a complete Hilbert space. As we know, the theorem asserts that every such functional is an inner product with a fixed element of the space.† In particular, corresponding to our functional f is a certain family of square integrable functions $\{f^{[k]}(x)\}$, $|k| \leq m$, such that for any element $\{\varphi^{[k]}(x)\} \in \overline{\mathscr{H}}_m$,

$$(f, \varphi) = \sum_{|k| \leq m} \int_G f^{[k]}(x)\, \varphi^{[k]}(x)\, dx.$$

If $\varphi(x)$ is one of the initial infinitely differentiable functions, we have

$$(f, \varphi) = \sum_{|k| \leq m} \int_G f_k(x)\, D^k \varphi(x) = \sum_{|k| \leq m} (f_k(x), D^k \varphi(x)),$$

where the $f_k(x) = f^{[k]}(x)$ are ordinary (square integrable) functions in G. Thus, we have obtained the following theorem.

THEOREM 1. *To every generalized function f and bounded region G, there exists an m and ordinary functions $f_k(x)$ ($|k| \leq m$), such that for any $\varphi \in K$ with support in G,*

$$(f, \varphi) = \sum_{|k| \leq m} (f_k(x),\ D^k \varphi(x)).$$

Using the definition of derivative, we can write this as

$$(f, \varphi) = \left(\sum_{|k| \leq m} (-1)^{|k|} D^k f_k,\ \varphi \right)$$

or

$$f = \sum_{|k| \leq m} D^k g_k,$$

where $g_k = (-1)^{|k|} f_k$ is once more an ordinary function for all k. Thus,
    Each generalized function in a bounded region G is a finite sum of derivatives of ordinary functions with support in G.

† See K. Yosida, *Functional Analysis*, Academic Press, Inc., New York, 1965, Chapt. 3, Sec. 6.

The functions g_k may be considered to be bounded in G, to be continuous, and to even have a given number of continuous derivatives. This may be achieved by replacing them by certain m-fold indefinite integrals.

Recalling the definition of order of singularity (Sec. 1.6.2, 2°), we can formulate the result as follows:

A generalized function in a bounded region G has finite order of singularity.

2.3.4. Let us now clarify the way in which a generalized function f operates on an *arbitrary* $\varphi(x) \in K$. To this end, consider a locally finite cover of R_n consisting of a countable family of bounded regions $G_1, \ldots, G_p, \ldots$. By Theorem 1, to each G_p, there is an $m = m(p)$ and a family of ordinary functions $f_{p1}, \ldots, f_{p,m(p)}$ carried in G_p such that

$$(f, \varphi) = \sum_{|k| \leq m(p)} (f_{pk}, D^k \varphi)$$

for each test function $\varphi(x)$ carried strictly in G_p.

Now according to Sec. 2.1.4, unity has an expansion of the form

$$1 \equiv e_1(x) + \cdots + e_p(x) + \cdots$$

in infinitely differentiable functions $e_p(x)$ each vanishing respectively outside G_p. Let $\varphi(x)$ be an arbitrary test function. Then we can write $\varphi = e_1 \varphi + e_2 \varphi + \cdots + e_p \varphi + \cdots$ and hence

$$
\begin{aligned}
(f, \varphi) &= \sum_p (f, e_p \varphi) = \sum_p \sum_{|k| \leq m(p)} (f_{pk}, D^k e_p \varphi) \\
&= \sum_p \sum_{|k| \leq m(p)} \left(f_{pk}, \sum_{|j| \leq |k|} \binom{k}{j} D^{k-j} e_p D^j \varphi \right) \\
&= \sum_p \sum_{|k| \leq m(p)} \left(\sum_{|j| \leq |k|} \binom{k}{j} f_{pk} D^{k-j} e_p, D^j \varphi \right) \\
&= \sum_p \sum_{|k| \leq m(p)} \left(\sum_{|j| \leq |k|} (-D)^j \binom{k}{j} (f_{pk} D^{k-j} e_p), \varphi \right) \\
&= \sum_p \sum_{|j| \leq m(p)} (D^j g_{jp}, \varphi).
\end{aligned}
\tag{9}
$$

Here

$$g_{jp} = (-1)^{|j|} \sum_{|j| \leq |k| \leq m(p)} \binom{k}{j} f_{pk} D^{k-j} e_p$$

is once more an ordinary function with support in G_p and $\binom{k}{j}$ symbolizes the product of binomial coefficients $\binom{k_1}{j_1} \binom{k_2}{j_2} \ldots \binom{k_n}{j_n}$. As p increases, the support of g_{jp} recedes from the origin. Therefore, for each x, the series

$$g_j(x) = \sum_p g_{jp}(x)$$

(summed for fixed j over those p for which $|j| \leq m(p)$)) has only a finite number of non-zero terms. Thus, $g_j(x)$ exists for every x. As $|j|$ increases, the support of $g_j(x)$ moves further and further away from the origin. By changing the order of summation in (9), we can bring (9) into the form

$$(f, \varphi) = \sum_j (D^j g_j(x), \varphi(x)).$$

Hence,

$$f = \sum_j D^j g_j(x). \tag{10}$$

The series (10) is infinite. However, in each finite region, only a finite number of its terms are different from zero.

Formula (10) furnishes the general form of a functional on K. In other words, every generalized function is expressible in the form (10). On the other hand, as we pointed out at the outset of this section, if the $g_j(x)$ are ordinary functions whose supports recede indefinitely from the origin, then (10) converges and defines a certain generalized function.

2.3.5. Suppose now that f is a generalized function with compact support. In this case, $G_1, G_2, ..., G_p, ...$ can be taken so that f is zero in each of them beginning with G_2. Therefore all f_{pk} can be made equal to zero for $p \geq 2$ and the sum (9) reduces to

$$(f, \varphi) = \sum_{|j| \leq m(1)} (D^j g_{j1}, \varphi).$$

Thus, f has the representation

$$f = \sum_{|j| \leq m} D^j g_j(x), \quad m = m(1), \tag{11}$$

where each $g_j(x)$ is an ordinary function vanishing outside G_1. Since G_1 can be made to fit the support of f arbitrarily closely, we have the following important addendum to formula (11): *the $g_j(x)$ may always be chosen so that the support of each is contained in an arbitrarily small neighborhood of the support of f.*

As a corollary, we have

6   Shilov

A functional with compact support has finite order of singularity through-out the entire space.

If the support of f is a closed bounded region G_0, the $g_k(x)$ in (11) may be taken so that their supports lie in G_0 itself (rather than a neighborhood of it). This results, which rests on an extension theorem for smooth functions, makes up the content of Prob. 2.

2.3.6. Consider finally a functional f carried at a single point, for definiteness, the origin. Each block B containing the origin in its interior is a support of f. Take any such B and keep it fixed. We can then find a representation for f of the form (11), where each $g_j(x)$ is an ordinary function with support in B, and we can write

$$(f, \varphi) = \sum_{|j| \leq m} (g_j(x), D^j \varphi). \tag{12}$$

We wish to show that *the functional f is equal to zero on each $\varphi(x)$ in K vanishing at the origin together with all its derivatives up to m-th order.* It suffices to establish that there exists a sequence of test functions $\varphi_\nu(x)$ each equal to $\varphi(x)$ (in some neighborhood of the origin) which converges to 0 uniformly in B together with its derivatives up to m-th order. For, suppose that such a sequence $\varphi_\nu(x)$ has been determined. Then, firstly, we have $(f, \varphi) = (f, \varphi_\nu)$ for each ν, since f is carried at the origin. And secondly, by (12),

$$(f, \varphi_\nu) = \sum_{|j| \leq m} \int_B g_j(x) D^j \varphi_\nu(x) \, dx \to 0,$$

from which it follows that $(f, \varphi) = 0$, as required.

Now let φ be any test function. It can be expressed as†

$$\varphi(x) = \sum_{|j| \leq m} D^j \varphi(0) \frac{x^j}{j!} + h_{m+1}(x), \tag{13}$$

where $h_{m+1}(x)$ vanishes at the origin together with its derivatives up to order m. The terms on the right-hand side are not test functions. We therefore multiply (13) by a fixed test function $e(x)$ equaling 1 in a neighborhood of the origin. This yields

$$\varphi(x) \, e(x) = \sum_{|j| \leq m} D^j \varphi(0) \frac{x^j \, e(x)}{j!} + h_{m+1}(x) \, e(x), \tag{14}$$

† Here x^j stands for the product $x_1^{j_1} x_2^{j_2} \cdots x_n^{j_n}$ and $j! = j_1! \, j_2! \ldots j_n!$

thus furnishing an expansion for φe in terms of test functions. Apply f to both sides of (14). Since $\varphi(x)$ coincides with $\varphi(x) e(x)$ in a neighborhood of the origin, $(f, \varphi e) = (f, \varphi)$ and so

$$(f, \varphi) = (f, \varphi e) = \sum_{|j| \leq m} D^j \varphi(0) \left(f, \frac{x^j e(x)}{j!} \right) + (f, h_{m+1} e).$$

We know that $D^j \varphi(0) = ((-1)^{|j|} D^j \delta, \varphi)$. Since the test function $h_{m+1} e$ vanishes at the origin with all its derivatives up to m-th order, we have $(f, h_{m+1} e) = 0$. Finally, let $a_j = (-1)^{|j|}(f, x^j e / j!)$. Putting this all together, we find that for any φ in K,

$$(f, \varphi) = \sum_{|j| \leq m} (a_j D^j \delta, \varphi)$$

and hence

$$f = \sum_{|j| \leq m} a_j D^j \delta. \tag{15}$$

Every generalized function carried at the origin can be expressed in the form (15). A generalized function carried at another point x_0 naturally has the form

$$f = \sum_{|j| \leq m} a_j D^j \delta(x - x_0).$$

We still have to determine for every $\varphi(x)$ in K vanishing at $x = 0$ together with its derivatives up to m-th order, a sequence of test functions $\varphi_\nu(x)$ each equaling $\varphi(x)$ in a neighborhood of the origin and converging to 0 uniformly in B with its derivatives up to m-th order. To this end, consider a fixed test function $h(x)$ equaling 1 in a certain spherical neighborhood of the origin, say, of radius ϱ and vanishing outside a larger spherical neighborhood, say, of radius 2ϱ.

Then the sequence of functions

$$\varphi_\nu(x) = \varphi(x) h(\nu x) \quad (\nu = 1, 2, \ldots) \tag{16}$$

satisfies the required condition.

First, each of the $\varphi_\nu(x)$ defined by (16) actually equals $\varphi(x)$ in some neighborhood of the origin.

Let us now estimate the derivatives of $\varphi_\nu(x)$. By Leibniz's formula,

$$D^k \varphi_\nu(x) = \sum_{|j| \leq |k|} \binom{k}{j} D^{k-j} \varphi(x) D^j h(\nu x).$$

6*

Denote by M a constant exceeding all values of $|D^k h(x)|$, $|k| \leq m$. Clearly, $|D^j h(\nu x)| \leq \nu^{|j|} M$. The derivatives of $\varphi(x)$ have to be estimated just for $|x| \leq 2\varrho/\nu$ since $h(\nu x)$ vanishes outside this ball. In this neighborhood, we have $\max |D^m \varphi(x)| = o(1)$ as $\nu \to \infty$. Integration then yields $\max |D^{m-1}\varphi| = o(1/\nu), \ldots, \max |D^{m-j}\varphi| = o(1/\nu^{|j|})$. Hence,

$$\max_B |D^k \varphi_\nu(x)| \leq \sum_{|j| \leq |k|} \binom{k}{j} \max |D^{k-j}\varphi(x)| \max |D^j h(\nu x)|$$

$$\leq C\, o\left(\frac{1}{\nu^{|j|}}\right) \nu^{|j|} M = o(1).$$

In other words, the sequence $D^k \varphi_\nu(x)$, $|k| \leq m$, converges to zero uniformly as $\nu \to \infty$, which completes the proof.

2.3.7. As an application of the preceding theorems, we shall consider the integral over the n-dimensional unit sphere Ω given by

$$\int_\Omega f((\omega, x))\, d\omega, \tag{17}$$

where $f(\xi)$ is a generalized function of a single independent variable ξ (Sec. 2.2.7).

We are interested in the smoothness properties of (17) as a function of x. However, we shall examine this question for the more general integral

$$\int_\Omega f((\omega, x))\, g(\omega)\, d\omega, \tag{18}$$

where $g(\omega)$ is an infinitely differentiable function over Ω.

If $f(\xi)$ is itself infinitely differentiable for all ξ, then the integral (18) is clearly an infinitely differentiable function of x.

Consider the case where $f(\xi)$ is ordinary and infinitely differentiable except at $\xi = 0$. Then (18) is an infinitely differentiable function except at $x = 0$.

To see this, choose a coordinate system on the sphere (depending on x) so that the first coordinate is $\xi = (\omega, x)$. Denote the remaining set of coordinates by ω'. Then $d\omega = J(\xi, \omega', x)\, d\omega'$. The Jacobian $J(\xi, \omega', x)$ is infinitely differentiable in ξ, ω', and x providing that $x \neq 0$ and $\xi \neq |x|$. There is no loss of generality in assuming that $f(\xi)$ vanishes in a neighborhood of the points $\xi = \pm|x|$ since otherwise we could always write $f(\xi) = f_0(\xi) + f_1(\xi)$, with $f_0(\xi)$ satisfying the required condition and $f_1(\xi)$ infinitely differentiable everywhere.

We carry out the integration in (18) first with respect to ω'. Set

$$\int J(\xi, \omega', x) \, g(\omega) \, d\omega' = G(\xi, x).$$

$G(\xi, x)$ is also infinitely differentiable with respect to ξ and x for $x \neq 0$ and $\xi \neq \pm|x|$. We then have

$$\int_\Omega f((\omega, x)) \, g(\omega) \, d\omega = \int_{-|x|}^{|x|} f(\xi) \, G(\xi, x) \, d\xi. \qquad (19)$$

Since $f(\xi)$ vanishes in a neighborhood of $\pm|x|$, the integral (19) is infinitely differentiable with respect to x for $x \neq 0$, as asserted.

Finally, suppose that $f(\xi)$ is a generalized function equal except for $\xi = 0$ to an ordinary infinitely differentiable function. $f(\xi)$ may be assumed to have compact support (if necessary, we can isolate an infinitely differentiable term). According to Sec. 2.3.5, it can be expressed as

$$f(\xi) = \sum_{k=0}^{m} f_k^{(k)}(\xi),$$

where each $f_k(\xi)$ is an ordinary function. If $f_k(\xi)$ is replaced by a suitable primitive of itself, it is possible to write $f(\xi)$ in the form

$$f(\xi) = \frac{d^{2m}}{d\xi^{2m}} F(\xi),$$

where $F(\xi)$ is again an ordinary function. For $\xi \neq 0$, $F(\xi)$ is the solution of an ordinary differential equation whose right-hand side is an ordinary function. By Sec. 1.5.4, it is the classical solution of the equation and in particular is infinitely differentiable together with $f(\xi)$. By what was proved above,

$$\Phi(x) = \int_\Omega F((\omega, x)) \, g(\omega) \, d\omega$$

is infinitely differentiable in x for $x \neq 0$. Moreover, by Sec. 2.2.4 and Sec. 1.7.2, 4°, we have

$$\Delta^m \Phi(x) = \int_\Omega \Delta^m F((\omega, x)) \, g(\omega) \, d\omega = \int_\Omega f(\xi) \, (\Sigma \omega_j^2)^m \, g(\omega) \, d\omega$$

$$= \int_\Omega f((\omega, x) \, g(\omega) \, d\omega.$$

Therefore, the function (18) is infinitely differentiable for $x \neq 0$ together with $\Phi(x)$, as asserted.

Such specific expressions as

$$\int_\Omega \delta^{(m)}((\omega, x))\, d\omega$$

are infinitely differentiable functions of x for $x \neq 0$. We shall evaluate them further on in Sec. 2.4.

Problems

1. A functional f carried in a closed bounded set F is equal to zero on every test function $\varphi(x)$ vanishing over F together with all its derivatives.

Hint. Generalize the procedure of Secs. 2.3.6. Instead of $h(\nu x)$, consider

$$e(x) * h(\nu x) = \int h(\nu\xi)\, e(x - \xi)\, d\xi,$$

where $e(x)$ equals 1 in some neighborhood of F and vanishes outside some larger neighborhood.

2. According to a theorem of Whitney,† every $\varphi(x_1, ..., x_n)$ in K with continuous derivatives up to m-th order over a closed bounded region $\overline{G}$ can be extended to all of R_n provided

$$\max_{R_n} |D^k\varphi(x)| \leq C_m(G) \max_{x \in G} |D^k\varphi(x)| \quad (|k| \leq m).$$

(Whitney actually employs a slightly more general class of sets.) Using this theorem, show that a functional f with support in $\overline{G}$ has the representation

$$f = \sum_{|j| \leq m} D^j f_j(x),$$

where each $f_j(x)$ is an ordinary function with support in $\overline{G}$.

Hint. Generalize the method of Secs. 2.3.2-3.

3. Given that $f(x)$ satisfies a linear differential equation $P(D)f(x) \neq 0$ for $x \neq 0$ and behaves like

$$|f(x)| \leq \frac{c}{|x|^p}$$

in a neighborhood of $x = 0$. Let $\mathscr{E}(x)$ be the fundamental function for $P(D)$ (Sec. 1.7.2, 9°). Show that $f(x)$ has the representation

$$f(x) = R(D)\,\mathscr{E}(x) + f_0(x),$$

---

† A proof may be found in L. Hörmander's article, *On the division of distributions by polynomials*, Ark. Mat., 3, (1958), pp. 555–568.

(for $x \neq 0$), where $P(D) f_0(x) = 0$ everywhere and $R(D)$ is a certain new differential operator.

Hint. There exists a generalized function f which is equal to $f(x)$ for $x \neq 0$ (Prob. 4, Sec. 2.1). The expression $P(D) f$ is equal to zero everywhere outside the origin and by Sec. 2.3.6 has the form $R(D) \delta(x)$. Consider $P(D) [f - R(D) \mathscr{E}(x)]$.

4. Let S' denote the set of all generalized functions f in K' which are representable by finite sums of the form

$$f = \sum_{|j| \leq m} D^j f_j(x),$$

with $f_j(x)$ an ordinary function, $|j| \leq m$, having no more than power growth at infinity. Show that each $f \in S'$ may be defined on the space S of infinitely differentiable functions φ which satisfy the conditions

$$\|\varphi\|_{kq}^2 \equiv \int_{R_n} (x^k D^q \varphi(x))^2 \, dx < \infty \quad (|k|, |q| = 0, 1, 2, \ldots)$$

so that the quantity (f, φ) is continuous on S in the following sense. If $\|\varphi_\nu\|_{kq} \to 0$ as $\nu \to \infty$ for any fixed k and q, then $(f, \varphi_\nu) \to 0$.
Hint. Take

$$(f, \varphi) = \sum_{|j| \leq m} (-1)^{|j|} \int_{R_n} f_j(x) \, D^j \varphi(x) \, dx. \tag{1}$$

5. Show that a homogeneous generalized function f of degree λ in K' (Sec. 1.6, Prob. 3) belongs to S'.

Hint. If f operates on a test function φ with support in $1 < |x| < 4$ like the m-th derivative of an ordinary function $f(x)$ bounded by M, then it operates on a test function ψ_p with support in $2^p < |x| < 2^{p+2}$ like the m-th derivative of a function bounded by $2^{p(\lambda+m)} M$. Every φ in K can be represented by a sum of functions such as ψ_p.

6. Show that formula (1) of Prob. 4 furnishes the general form of a continuous linear functional on S. As before, the $f_j(x)$ are ordinary functions increasing no faster than a power of $|x|$ at infinity.
Hint. Show that there are k and q for which f is continuous in the norm $\|\varphi\|_{kq}$. Then repeat the argument of Sec. 2.3.3 appropriate to this case.

7. A generalized function f in K' is known to satisfy the inequality

$$|(f, \varphi(x + h))| \leq C \|\varphi\|_m (1 + |h|)^p$$

for $\varphi(x)$ in K carried in $|x| \leq 1$, where C, p, and m are constants and $\|\varphi\|_m$ has the same meaning as in Sec. 2.3.3.

Show that f can be extended from K to S as a continuous linear functional.

Hint. Consider $\psi(x) \in K$ carried in the ball $|x| \leq 1$ such that its translates sum to 1 (Sec. 2.1, Prob. 3):

$$1 = \sum_{\nu=0}^{\infty} \psi(x + h_\nu).$$

Every $\varphi \in S$ can be represented in the form

$$\varphi(x) = \sum_{\nu=0}^{\infty} \varphi(x)\, \psi(x + h_\nu) = \sum_{\nu=0}^{\infty} \varphi_\nu(x + h_\nu),$$

where $\varphi_\nu(x) = \varphi(x - h_\nu)\, \psi(x)$ is carried in $|x| \leq 1$. Show that $\|\varphi_\nu\|_m$ approaches 0 faster than any negative power of $|h_\nu|$ as $\nu \to \infty$. Define

$$(f, \varphi) = \sum_{\nu=0}^{\infty} (f, \varphi_\nu(x + h_\nu)).$$

8. A generalized function $f \in K'$ is said to have power growth at infinity if the function of h,

$$(f, \varphi(x + h)) \tag{1}$$

increases no faster than some power of $|h|$ as $|h| \to \infty$ for any given $\varphi \in K$. Show that any such $f \in S'$.

Hint. First show that for all φ with fixed support, the powers of $|h|$ with which the growth of (1) can be estimated are bounded. To do this, assume the contrary. Then using the fact that the specified powers are unaffected when φ is multiplied by a constant, construct a test function φ of the form $\sum_\nu \lambda_\nu \varphi_\nu(x)$ (see Sec. 1.2, Prob. 2) for which the function (1) increases faster than any power of $|h|$. Then having the inequality

$$|(f, \varphi(x + h))| \leq C_\varphi (1 + |h|)^q$$

with q fixed, show that C_φ may be chosen to be $C\|\varphi\|_m$ for some fixed m (the same method!). The problem is thereby reduced to Prob. 7 (V. P. Palamodov).

9. Let D_m denote the linear space of functions $\varphi(x)$ with compact support having continuous derivatives up to m-th order. Convergence in D_m is defined by the rule: $\varphi_\nu(x) \to 0$ in D_m if all $\varphi_\nu(x)$ vanish outside the same bounded set and the sequence $\varphi_\nu(x)$ converges uniformly to 0 together with its derivatives up to m-th order. Show that K is dense in D_m.

Hint. Use the idea of Sec. 2.1.3.

10. Show that every continuous linear functional f in K' with order of singularity $s(f)$ not exceeding m can be extended from K to all of D_m as a continuous linear functional. But if $s(f) \geq m + 1$, such an extension is generally speaking impossible.

Hint. Apply the definition for a functional with $s(f) \leq m$. The functional $1/x$, which has order of singularity 1 (Sec. 1.3, Prob. 3), cannot be extended from K to D_0 (the space of continuous functions with compact support).

11. Determine the general form of a continuous linear functional on D_m.

Hint. For $m = 0$, use Riesz's theorem† to obtain the representation

$$(f, \varphi) = \int_{R_n} \varphi \, d\mu,$$

where μ is a Stieltjes measure of bounded variation (locally).

In general, D_m is isomorphic to the space of m-th order derivatives of functions $\varphi \in D_m$. This new space is a closed subspace and a direct sum of subspaces of D_0. Use the Hahn-Banach extension theorem for linear functionals (see K. Yosida, *ibid.*, Chapt. 4).

Ans.

$$(f, \varphi) = \sum_{|k| \leq m} \int_{R_n} \frac{\partial^m \varphi(x)}{\partial x_1^{k_1} \dots \partial x_n^{k_n}} \, d\mu_k(x),$$

where the $\mu_k(x)$ are Stieltjes measures of bounded variation.

2.4. Special Generalized Functions

In various applications, a large role is played by the specific generalized functions x_+^λ, x_-^λ, r^λ and others related to them. Their properties will be examined in this section.

2.4.1. Consider the function x_λ^+ equal to x^λ for $x > 0$ and to 0 for $x < 0$.

---

† See F. Riesz and B. Sz. Nagy, *Functional Analysis*, Ungar. Publishing Company New York, 1955, p. 110.

We want to construct and study its corresponding generalized function. For $\operatorname{Re} \lambda > -1$, the function determines a regular functional

$$(x_+^\lambda, \varphi) = \int_0^\infty x^\lambda \varphi(x)\, dx. \tag{1}$$

The function (1) is evidently analytic in λ since it has a derivative with respect to λ given by

$$\int_0^\infty x^\lambda \log x \varphi(x)\, dx.$$

This means that the functional x_λ^+ is analytic in λ for $\operatorname{Re} \lambda > -1$ (cf. Sec. 2.2.6). We would like to continue it analytically to the entire λ-plane. We write the right-hand side of (1) in the form

$$\int_0^1 x^\lambda [\varphi(x) - \varphi(0)]\, dx + \int_1^\infty x^\lambda \varphi(x)\, dx + \frac{\varphi(0)}{\lambda + 1}. \tag{2}$$

The first term is defined for $\operatorname{Re} \lambda > -2$, the second for any λ, and the third for $\lambda \neq -1$. Hence, it furnishes an analytical continuation of the function (1) to the domain $\operatorname{Re} \lambda > -2, \lambda \neq -1$.

The functional x_+^λ can be continued analytically to the domain $\operatorname{Re} \lambda > -n - 1, \lambda \neq -1, -2, ..., -n$, in similar fashion to obtain

$$\int_0^\infty x^\lambda \varphi(x)\, dx$$

$$= \int_0^1 x^\lambda \left[\varphi(x) - \varphi(0) - x\varphi'(0) - \cdots - \frac{x^{n-1}}{(n-1)!} \varphi^{(n-1)}(0) \right] dx$$

$$+ \int_1^\infty x^\lambda \varphi(x)\, dx + \sum_{k=1}^n \frac{\varphi^{(k-1)}(0)}{(k-1)!\,(\lambda + k)}. \tag{3}$$

The right-hand side of this provides a regularization of the integral on the left-hand side. By the same token, the generalized function x_+^λ is defined for all $\lambda \neq -1, -2, ...$

Formula (3) can be converted into a simpler form in the strip $-n-1 < \operatorname{Re} \lambda < -n$. There we have

$$(x_+^\lambda, \varphi) = \int_0^\infty x^\lambda \left[\varphi(x) - \varphi(0) - x\varphi'(0) - \cdots - \frac{x^{n-1}}{(n-1)!} \varphi^{(n-1)}(0) \right] dx,$$

$$\tag{4}$$

owing to the fact that for $1 \leqq k \leqq n$,

$$\int_1^\infty x^{\lambda+k-1}\, dx = -\frac{1}{\lambda+k}\,.$$
(5)

Formula (3) shows that as a function of λ, (x_+^λ, φ) has first-order poles at $\lambda = -1, -2, -3, \ldots$, its residue at $\lambda = -k$ being $\varphi^{(k-1)}(0)/(k-1)!$ Since $\varphi^{(k-1)}(0) = (-1)^{k-1}\,(\delta^{(k-1)}(x), \varphi(x))$, *the functional x_+^λ has a first-order pole at $\lambda = -k$ with residue* $\dfrac{(-1)^{k-1}}{(k-1)!}\, \delta^{(k-1)}(x)$.

2.4.2. Consider now the function x_-^λ equal to $|x|^\lambda$ for $x < 0$ and to 0 for $x > 0$. For $\text{Re }\lambda > -1$, it defines a regular functional

$$(x_-^\lambda, \varphi) = \int_{-\infty}^0 |x|^\lambda \varphi(x)\, dx\,.$$
(6)

The functional can be continued into the half-plane $\text{Re }\lambda \leqq -1$ similarly to x_+^λ. For that purpose, it is easiest to replace x by $-x$ and represent (x_-^λ, φ) in the form

$$(x_-^\lambda, \varphi(x)) = \int_0^\infty x^\lambda \varphi(-x)\, dx = (x_+^\lambda, \varphi(-x)).$$

This allows us to carry over at once all of the results obtained for x_+^λ to x_-^λ if we merely replace $\varphi(x)$ by $\varphi(-x)$ in corresponding formulas. The quantities $\varphi^{(j)}(0)$ also have to be replaced by $(-1)^j \varphi^{(j)}(0)$.

We see particularly that the generalized function x_-^λ, like x_+^λ, exists and is analytic in the entire λ-plane except at the points $\lambda = -1, -2, \ldots$ At $\lambda = -k$, it has a simple pole with residue $\delta^{(k-1)}(x)/(k-1)!$

The quantity (x_-^λ, φ) can be determined in the strip $-n-1 < \text{Re }\lambda < -n$ through the formula

$$(x_-^\lambda, \varphi(x)) = (x_+^\lambda, \varphi)\,(-x))$$

$$= \int_0^\infty x^\lambda \left[\varphi(-x) - \varphi(0) + x\varphi'(0) - \cdots - \frac{(-1)^{n-1}x^{n-1}}{(n-1)!}\, \varphi^{(n-1)}(0) \right] dx.$$
(7)

2.4.3. A generalized function f is called *odd* if

$$(f(x), \varphi(x)) = -(f(x), \varphi(-x))$$

and *even* if

$$(f(x), \varphi(x)) = (f(x), \varphi(-x)).$$

Using the two generalized functions derived above, we can compose
the following even and odd combinations:

$$|x|^\lambda = x_+^\lambda + x_-^\lambda, \tag{8}$$

$$|x|^\lambda \operatorname{sgn} x = x_+^\lambda - x_-^\lambda. \tag{9}$$

Let us examine the singularities of $|x|^\lambda$ and $|x|^\lambda \operatorname{sgn} x$. Since the gene-
ralized functions x_+^λ and x_-^λ each have a pole at $\lambda = -k$ with the respec-
tive residues $(-1)^{k-1} \delta^{(k-1)}(x)/(k-1)!$ and $\delta^{(k-1)}(x)/(k-1)!$, $|x|^\lambda$ has
poles just at $\lambda = -1, -3, \ldots, -2m-1, \ldots$ The residue of $|x|^\lambda$ at
$\lambda = -2m-1$ is $2\delta^{(2m)}(x)/(2m)!$. At $\lambda = -2m$ ($m = 1, 2, \ldots$), $|x|^\lambda$ has
a definite value. For such λ, we shall naturally write x^{-2m} instead of
$|x|^{-2m}$.

Similarly, the generalized function $|x|^\lambda \operatorname{sgn} x$ has poles at $\lambda = -2, -4,$
$\ldots, -2m, \ldots$ with residues $-2\delta^{(2m-1)}(x)/(2m-1)!$. At $\lambda = -2m-1$
($m = 0, 1, \ldots$), $|x|^\lambda \operatorname{sgn} x$ has a definite value and we write x^{-2m-1}
instead of $|x|^{-2m-1} \operatorname{sgn} x$. Thus, the generalized functions x^{-n} have been
defined for all $n = 1, 2, \ldots$

Let us obtain explicit expressions for $|x|^\lambda$ and $|x|^\lambda \operatorname{sgn} x$. To this end,
we make use of (4) and (7). In the strip $-n-1 < \operatorname{Re} \lambda < -n$,

$$(x_+^\lambda, \varphi)$$

$$= \int_0^\infty x^\lambda \left[\varphi(x) - \varphi(0) - x\varphi'(0) - \cdots - \frac{x^{n-1}}{(n-1)!} \varphi^{(n-1)}(0) \right] dx,$$

$$(x_-^\lambda, \varphi)$$

$$= \int_0^\infty x^\lambda \left[\varphi(-x) - \varphi(0) + x\varphi'(0) - \cdots - \frac{(-1)^{n-1}x^{n-1}}{(n-1)!} \varphi^{(n-1)}(0) \right] dx.$$

If we substitute $2m$ for n in these and add and subtract, we find that

$$(|x|^\lambda, \varphi) = \int_0^\infty x^\lambda \left\{ \varphi(x) + \varphi(-x) \right.$$

$$\left. -2 \left[\varphi(0) + \frac{x^2}{2!} \varphi''(0) + \cdots + \frac{x^{2m-2}}{(2m-2)!} \varphi^{(2m-2)}(0) \right] \right\} dx \tag{10}$$

and

$$(|x|^\lambda \operatorname{sgn} x, \varphi) = \int_0^\infty x^\lambda \bigg\{ \varphi(x) - \varphi(-x)$$

$$-2\bigg[x\varphi'(0) + \frac{x^3}{3!} \varphi'''(0) + \cdots + \frac{x^{2m-1}}{(2m-1)!} \varphi^{(2m-1)}(0) \bigg] \bigg\} dx. \quad (11)$$

The first expansion converges for $-2m - 1 < \operatorname{Re} \lambda < -2m + 1$ and the second for $-2m - 2 < \operatorname{Re} \lambda < -2m$. Thus,

$$(x^{-2m}, \varphi) = \int_0^\infty x^{-2m} \bigg\{ \varphi(x) + \varphi(-x)$$

$$-2\bigg[\varphi(0) + \frac{x^2}{2!} \varphi''(0) + \cdots + \frac{x^{2m-2}}{(2m-2)!} \varphi^{(2m-2)}(0) \bigg] \bigg\} dx, \quad (12)$$

$$(x^{-2m-1}, \varphi) = \int_0^\infty x^{-2m-1} \bigg\{ \varphi(x) - \varphi(-x)$$

$$-2\bigg[x\varphi'(0) + \frac{x^3}{3!} \varphi'''(0) + \cdots + \frac{x^{2m-1}}{(2m-1)!} \varphi^{(2m-1)}(0) \bigg] \bigg\} dx. \quad (13)$$

For example,

$$(x^{-2}, \varphi) = \int_0^\infty \frac{\varphi(x) + \varphi(-x) - 2\varphi(0)}{x^2} dx, \quad (14)$$

$$(x^{-1}, \varphi) = \int_0^\infty \frac{\varphi(x) - \varphi(-x)}{x} dx; \quad (15)$$

The last expression is equal to the Cauchy principal value of the integral of $\varphi(x)/x$ (see Sec. 1.3.3, 2°).

2.4.4. (a) *Multiplication by a function.* Let us verify that

$$x^m x_+^\lambda = x_+^{m+\lambda} \quad (\lambda \neq -1, -2, \ldots) \quad (16)$$

(*m* a positive integer). The formula needs to be proved since the left and right-hand sides have independent meanings. The proof is a simple matter. The left and right-hand sides are analytic in λ and coincide for

Re $\lambda > -1$. Therefore, they coincide over their full region of analyticity, i.e., at least for all $\lambda \neq -1, -2, ...$ But the right-hand side is also analytic for $\lambda = -1, ..., -m$, and this means that the analyticity of the left-hand side is also maintained. In particular, we have

$$\lim_{\lambda \to -k} x^m x_+^\lambda = x_+^{m-k} \quad (k = 1, 2, ..., m).$$

Thus, formula (16) is valid for $\lambda = -1, ..., -m$ if the left-hand side is interpreted as an appropriate limit. Similarly,

$$x^m x_-^\lambda = x_-^{m+\lambda} (-1)^m \qquad (\lambda \neq -m-1, -m-2, ...), \quad (17)$$

$$x^m |x|^\lambda = |x|^{m+\lambda} (\operatorname{sgn} x)^m \qquad (\lambda \neq -m-1, -m-2, ...), \quad (18)$$

$$x^m |x|^\lambda \operatorname{sgn} x = |x|^{m+\lambda} (\operatorname{sgn} x)^{m+1} \quad (\lambda \neq -m-1, -m-2, ...). \quad (19)$$

If k and m are any positive integers, then

$$x^k x^{-m} = x^{k-m}. \tag{20}$$

Finally, let $f(x)$ be an infinitely differentiable function having an m-fold zero at $x = 0$. Then $f(x) = x^m g(x)$, $g(0) \neq 0$, and we can write

$$f(x) x_+^\lambda = g(x) x^m x_+^\lambda = g(x) x_+^{m+\lambda};$$

$$f(x) x^{-k} = g(x) x^{m-k}. \tag{21}$$

(b) Differentiation. For Re $\lambda > 0$, we have the obvious relation $dx_+^\lambda/dx = \lambda x_+^{\lambda-1}$. That is to say, $(x_+^\lambda, \varphi'(x)) = -(\lambda x_+^{\lambda-1}, \varphi(x))$. Both sides of this last relation can be analytically continued throughout the λ-plane (with the points $0, -1, ...$ excluded). By the uniqueness property, it therefore holds throughout the λ-plane. Thus,

$$\frac{dx_+^\lambda}{dx} = \lambda x_+^{\lambda-1} \qquad (\lambda \neq 0, -1, -2, ...). \tag{22}$$

Similarly,

$$\frac{dx_-^\lambda}{dx} = -\lambda x_-^{\lambda-1} \quad (\lambda \neq 0, -1, -2, ...). \tag{23}$$

For $\lambda = 0, x_+^\lambda = \theta(x)$ and $x_-^\lambda = \theta(-x)$ and so differentiation is performed according to our previously derived formulas (Sec. 1.4.3)

$$\theta'(x) = \delta(x), \quad \theta'(-x) = -\delta(x).$$

Further,

$$\frac{d}{dx}|x|^\lambda = \frac{d}{dx}(x_+^\lambda + x_-^\lambda) = \lambda x_+^{\lambda-1} - \lambda x_-^{\lambda-1} = \lambda |x|^{\lambda-1}\,\mathrm{sgn}\,x, \quad (24)$$

$$\frac{d}{dx}|x|^\lambda\,\mathrm{sgn}\,x = \frac{d}{dx}(x_+^\lambda - x_-^\lambda) = \lambda x_+^{\lambda-1} + \lambda x_-^{\lambda-1} = \lambda |x|^{\lambda-1}$$

$$(\lambda \neq 0, -1, \dots). \quad (25)$$

The right and left-hand sides of (24) admit analytic continuation to negative even values of λ and the right and left-hand sides of (25), to negative odd values of λ.

In particular, for $\lambda = -n$, we obtain

$$\frac{d}{dx}(x^{-n}) = -nx^{-n-1} \quad (n \neq 0). \quad (26)$$

On inverting formulas (22)-(26), we arrive at expressions for the indefinite integrals of our generalized functions. Thus for $m \neq 1$, (26) yields

$$\int x^{-m}\,dx = \frac{x^{-m+1}}{-m+1} + C. \quad (27)$$

If $m = 1$, (27) does not allow one to determine the corresponding indefinite integral. However, we know from Sec. 1.4.3 that $\log|x|$ is the appropriate indefinite integral and we can write

$$\int x^{-1}\,dx = \log|x| + C. \quad (28)$$

2.4.5. Let $r = \sqrt{x_1^2 + \cdots + x_n^2}$. We wish to consider the functional r^λ defined by

$$(r^\lambda, \varphi) = \int_{R_n} r^\lambda \varphi(x)\,dx. \quad (29)$$

The formula is meaningful for $\mathrm{Re}\,\lambda > -n$. Since formal differentiation yields

$$\frac{d}{d\lambda}(r^\lambda, \varphi) = \int_{R_n} r^\lambda \log r\varphi(x)\,dx,$$

r^λ represents an analytic function of λ for $\mathrm{Re}\,\lambda > -n$. For $\mathrm{Re}\,\lambda \leq -n$, r^λ is not locally integrable. We use analytic continuation to define the functional r^λ.

By introducing spherical coordinates in the integral in (29), we can write it in the form

$$(r^\lambda, \varphi) = \int_0^\infty r^\lambda \left\{ \int_{\Omega_n(r)} \varphi(x)\, d\Omega_n(r) \right\} dr,$$

where $d\Omega_n(r)$ is the surface element of a sphere of radius r. The inner integral can be expressed as

$$\int_{\Omega_n(r)} \varphi(x)\, d\Omega_n(r) = \Omega_n r^{n-1} S_r[\varphi].$$

Here, Ω_n is the surface area of an n-dimensional unit sphere and $S_r[\varphi]$ is the mean value of $\varphi(x)$ over a sphere of radius r. Thus we arrive at the formula

$$(r^\lambda, \varphi) = \Omega_n \int_0^\infty r^{\lambda+n-1} S_r[\varphi]\, dr. \tag{30}$$

We now establish some properties of the function $S_r[\varphi]$ (which is defined for $r \geqq 0$).

$S_r[\varphi]$ has compact support. It is infinitely differentiable, and all its odd order derivatives vanish at $r = 0$.

Since $\varphi(x)$ vanishes for r sufficiently large, so does its mean value $S_r[\varphi]$. Thus, $S_r[\varphi]$ has compact support.

It is likewise obvious that $S_r[\varphi]$ is infinitely differentiable for $r > 0$.

To see that the derivatives of $S_r[\varphi]$ all exist at $r = 0$ also, we use Taylor's theorem with remainder to expand $\varphi(x)$ through terms of order r^{2m+1}. Denoting the remainder term by R_{2m+1}, we then have

$$S_r[\varphi] = \frac{1}{\Omega_n r^{n-1}} \int_{\Omega_n(r)} \left[\varphi(0) + \sum \frac{\partial \varphi(0)}{dx_j} x_j + \frac{1}{2!} \sum \frac{\partial^2 \varphi(0)}{\partial x_i\, \partial x_j} x_i x_j \right.$$

$$\left. + \frac{1}{3!} \sum \frac{\partial^3 \varphi(0)}{\partial x_i\, \partial x_j\, \partial x_k} x_i x_j x_k + \cdots + R_{2m+1} \right] d\Omega_n(r).$$

Clearly, each term in the integrand containing an odd number of factors x_l (other than the remainder term) drops out after integration. Terms containing an even number of, say, $2k$ factors x_l yield a term of the form $a_k r^{2k}$ after integration and simplification. Thus we obtain

$$S_r[\varphi] = \varphi(0) + a_1 r^2 + a_2 r^4 + \cdots + a_m r^{2m} + O(r^{2m-1}). \tag{31}$$

This expression shows that $S_r[\varphi]$ has $2m$ derivatives at $r = 0$ and that the odd order derivatives vanish. Since m may be chosen arbitrarily, $S_r[\varphi]$ is infinitely differentiable at $r = 0$ and all its odd order derivatives vanish there.

Hence, $S_r[\varphi]$ may be interpreted as an even test function of the variable r. The integral (30) is then the result of applying the functional $\Omega_n \xi_+^\mu$ $(\mu = \lambda + n - 1)$ to $S_\xi[\varphi]$. But we very well know that ξ_+^μ, which is analytic for Re $\mu > -1$ (or for Re $\lambda > -n$), can be analytically continued to the entire μ-plane except for the points $\mu = -1, -2, \ldots$ ($\lambda = -n$, $-n + 1, \ldots$) at which it has first order poles. The residue of the function $(\xi_+^\mu, S_\xi[\varphi])$ at the pole $\mu = -m$ ($\lambda = -n - m + 1$) is equal to

$$\frac{1}{(m-1)!}((-1)^{m-1}\delta^{(m-1)}(\xi), S_\xi[\varphi]) = \frac{1}{(m-1)!}\frac{d^{m-1}}{d\xi^{m-1}}S_\xi[\varphi]\bigg|_{\xi=0}.$$

But since the odd derivatives of $S_\xi[\varphi]$ vanish at $\xi = 0$, there are actually no poles for even values of m. All that is left is a sequence of poles at $m = 1, 3, 5, \ldots$, or equivalently, at $\lambda = -n, -n - 2, -n - 4, \ldots$ Thus, the residue of the function (r^λ, φ) at $\lambda = -n - 2k$ ($k = 0, 1, \ldots$) is given by

$$\Omega_n\frac{(\delta^{(2k)}(\xi), S_\xi[\varphi])}{(2k)!} = \frac{\Omega_n}{(2k)!}\frac{d^{2k}}{d\xi^{2k}}S_\xi[\varphi]\bigg|_{\xi=0}. \tag{32}$$

In particular, (r_λ, φ) has a first order pole at $\lambda = -n$ with residue $\Omega_n S_0[\varphi] = \Omega_n \varphi(0)$. This means that the generalized function r^λ has a first order pole at $\lambda = -n$ with residue $\Omega_n \delta(x)$.

2.4.6. Let $\omega = (\omega_1, \ldots, \omega_n)$ be a point of the n-dimensional unit sphere Ω, and let $(\omega, x) = \omega_1 x_1 + \cdots + \omega_n x_n$. Consider the following integral:

$$\int_\Omega |(\omega, x)|^\lambda d\omega.$$

It exists as a proper integral for Re $\lambda > 0$ (and as an improper one for Re $\lambda > -1$). It represents a certain function of $x = (x_1, \ldots, x_n)$ which we temporarily label $G(x, \lambda)$. The function is spherically symmetric in x. For if h is any rotation of R_n, then

$$G(hx, \lambda) = \int_\Omega |(\omega, hx)|^\lambda d\omega = \int_\Omega |(h'\omega, x)|^\lambda d\omega = \int_\Omega |(\omega', x)|^\lambda d\omega',$$

where $\omega' = h'\omega$ and $h' = h^{-1}$ is the rotation inverse to h.

7   Shilov

Thus, $G(x, \lambda)$ is a function of r and λ. Denote it now by $F(r, \lambda)$. Substituting αx for x, where α is a positive number, we find that

$$F(\alpha r, \lambda) = \int_{\Omega} |(\omega, \alpha x)|^{\lambda} \, d\omega = \alpha^{\lambda} \int_{\Omega} |(\omega, x)|^{\lambda} \, d\omega = \alpha^{\lambda} F(r, \lambda).$$

Setting $r = 1$ in this, we obtain

$$F(\alpha, \lambda) = \alpha^{\lambda} F(1, \lambda) = C(\lambda) \, \alpha^{\lambda}.$$

Thus, $F(r, \lambda)$ is proportional to r^{λ}. It remains to determine the coefficient $C(\lambda)$. We have $C(\lambda) = F(1, \lambda) = G(e, \lambda)$, where e is any unit vector. Take $e = (0, 0, ..., 0, 1)$. We then conclude that

$$C(\lambda) = \int_{\Omega} |\omega_n|^{\lambda} \, d\omega = 2 \int_0^{\frac{\pi}{2}} \cos^{\lambda} \theta_{n-1} \sin^{n-2} \theta_{n-1} \, d\theta_{n-1} \Omega_{n-1},$$

where θ_{n-1} is the angle between e and ω, and Ω_{n-1} is the surface area of an $(n-1)$-dimensional unit sphere. As a result,

$$C(\lambda) = 2\Omega_{n-1} \int_0^{\frac{\pi}{2}} \cos^{\lambda} \theta_{n-1} \sin^{n-2} \theta_{n-1} \, d\theta_{n-1} = 2\pi^{\frac{n-1}{2}} \frac{\Gamma\left(\dfrac{\lambda + 1}{2}\right)}{\Gamma\left(\dfrac{\lambda + n}{2}\right)}, \quad (33)$$

since by (8) of Sec. 2.2.7 and the classical formula for the Beta function,[†] we have respectively,

$$\Omega_{n-1} = \frac{2\pi^{\frac{n-1}{2}}}{\Gamma\left(\dfrac{n-1}{2}\right)}$$

and

$$\int_0^{\frac{\pi}{2}} \sin^p \theta \cos^q \theta \, d\theta = \frac{1}{2} B\left(\frac{p+1}{2}, \frac{q+1}{2}\right).$$

† See E. T. Whittaker and G. N. Watson, *Modern Analysis*, Cambridge Univ. Press. 4th ed., Chapt. 12, Secs. 12.41–42.

We can write the resulting formula as follows:

$$\int_\Omega \frac{|(\omega, x)|^\lambda}{\Gamma\left(\frac{\lambda + 1}{2}\right)} \, d\omega = c_n \frac{r^\lambda}{\Gamma\left(\frac{\lambda + n}{2}\right)}, \quad c_n = 2\pi^{\frac{n-1}{2}}. \tag{34}$$

Regarding the right and left-hand sides as functionals on K, we shall continue them analytically to the entire λ-plane.

As we know, r^λ can be analytically continued to the entire λ-plane with simple poles at $\lambda = -n, -n - 2, -n - 4, \dots$ The Gamma function $\Gamma(\zeta)$ is known to have simple poles at $0, -1, -2, \dots$ so that $\Gamma\left(\frac{\lambda + n}{2}\right)$ has simple poles at $\lambda = -n, -n - 2, -n - 4, \dots$ Thus, the numerator and denominator of $r^\lambda / \Gamma\left(\frac{\lambda + n}{2}\right)$ each has the exact same simple poles. The quotient therefore is itself an entire function of λ. Let us find its value at $\lambda = -n$. It is equal to the quotient of the residues of the numerator and denominator. The residue of the numerator we know to be $\Omega_n \delta(x)$. The residue of the denominator is easily obtained using the functional equation for the Gamma function,

$$\Gamma(\zeta + 1) = \zeta\Gamma(\zeta). \tag{35}$$

Namely, setting $\zeta = (\lambda + n)/2$, we have

$$\Gamma\left(\frac{\lambda + n}{2}\right) = \frac{2}{\lambda + n}\Gamma\left(\frac{\lambda + n + 2}{2}\right).$$

As $\lambda \to -n$,

$$\Gamma\left(\frac{\lambda + n}{2}\right) = \frac{2}{\lambda + n}\Gamma(1) + O(1).$$

Hence, the residue of $\Gamma\left(\frac{\lambda + n}{2}\right)$ at $\lambda = -n$ is $2\Gamma(1) = 2$. As a result,

$$\left.\frac{r^\lambda}{\Gamma\left(\frac{\lambda + n}{2}\right)}\right|_{\lambda = -n} = \frac{\Omega_n}{2}\delta(x). \tag{36}$$

Let us now examine the analytic continuation of the left-hand side of (34). The analytic continuation of the functional $|\xi|^\lambda$ has simple poles

7*

at $\lambda = -1, -3, \ldots$ However $\Gamma\left(\dfrac{\lambda + 1}{2}\right)$ also has these very same simple poles. Let us determine the value of

$$\left.\frac{|\xi|^\lambda}{\Gamma\left(\dfrac{\lambda + 1}{2}\right)}\right|_{\lambda = -2m-1}$$

The residue of $|\xi|^\lambda$ at $\lambda = -2m - 1$ is $2\delta^{(2m)}(\xi)/(2m)!$ (Sec. 2.4.3.) The residue of $\Gamma\left(\dfrac{\lambda + 1}{2}\right)$ at this point can be determined from the functional equation (35). Namely,

$$\Gamma\left(\frac{\lambda + 1}{2}\right) = \frac{2}{\lambda + 1}\Gamma\left(\frac{\lambda + 3}{2}\right) = \frac{2}{\lambda + 1}\frac{2}{\lambda + 3}\Gamma\left(\frac{\lambda + 5}{2}\right) = \cdots$$

$$= \frac{2^{m+1}\Gamma\left(\dfrac{\lambda + 2m + 3}{2}\right)}{(\lambda + 1)\cdots(\lambda + 2m + 1)}.$$

Hence,

$$\mathrm{Res}_{\lambda = -2m-1}\Gamma\left(\frac{\lambda + 1}{2}\right) = \frac{2(-1)^m}{m!}.$$

As a result,

$$\left.\frac{|\xi|^\lambda}{\Gamma\left(\dfrac{\lambda + 1}{2}\right)}\right|_{\lambda = -2m-1} = (-1)^m \frac{m!}{(2m)!}\delta^{(2m)}(\xi).$$

At $\lambda = -2m$, the numerator and denominator have no singularities. Thus, depending on whether n is even or odd, we obtain

$$\delta(x) = \begin{cases} \dfrac{(-1)^{\frac{n-1}{2}}}{2(2\pi)^{n-1}} \displaystyle\int_\Omega \delta^{(n-1)}((\omega, x))\, d\omega & (n \text{ odd}), \\[4ex] \dfrac{(-1)^{\frac{n}{2}}(n-1)!}{(2\pi)^n} \displaystyle\int_\Omega (\omega, x)^{-n}\, d\omega & (n \text{ even}). \end{cases} \tag{37}$$

These formulas are the so-called *plane-wave expansions* of the delta-function. When applied to a test function φ, they furnish the solution to the *Radon problem* which is to determine the value of $\varphi(x)$ at $x = 0$ knowing its integrals over the hyperplanes $(\omega, x) = \xi$.

Problems

1. *Canonical regularization.* Each quotient $g(x) = f(x)/P(x)$ of an infinitely differentiable function $f(x)$ and polynomial $P(x)$ defines a functional g in K_1' with the following correspondences:

a) to $g_1(x) + g_2(x)$, the sum of g_1 and g_2;

b) to the product of $g(x)$ and an infinitely differentiable function $\alpha(x)$,

the product $\alpha(x) g$ in K';

c) to the usual derivative $g'(x)$, the derivative g' in K';

d) to $g(x) = f(x)$ (i.e., $P(x) \equiv 1$), the functional

$$(f, \varphi) = \int_{-\infty}^{\infty} f(x) \, \varphi(x) \, dx;$$

e) to even $g(x)$, an even functional g and to odd $g(x)$, an odd functional.

Hint. $g(x)$ is uniquely representable by

$$g(x) = \sum_{j,k} \frac{A_{jk}}{(x - x_j)^k} + h(x),$$

where $h(x)$ is infinitely differentiable and the A_{jk} are constants. Assign to $g(x)$ the functional

$$g = \sum_{j,k} A_{jk}(x - x_j)^{-k} + h.$$

2. Show that the conditions a)–e) uniquely determine a canonical regularization.

Hint. Let the functional g correspond to $1/x$. Then to the function $x \dfrac{1}{x} \equiv 1$, there corresponds the functional $xg = 1$. Hence, $g = \dfrac{1}{x} + g_0$, where g_0 is carried at the point $x = 0$. Using the result of Sec. 2.3.6, show that $g_0 = C\delta(x)$. Since g has to be an odd functional, $C = 0$. Proceed now making use of c).

3. Show that

$$\frac{x_+^\lambda}{\Gamma(\lambda + 1)} \qquad (\text{Re } \lambda > -1)$$

can be continued to the entire λ-plane as an entire function and find its values at $\lambda = -1, -2, \ldots$

Ans.

$$\left. \frac{x_+^\lambda}{\Gamma(\lambda + 1)} \right|_{\lambda = -n} = \delta^{(n-1)}(x) \quad (n = 1, 2, \ldots).$$

4. Derive the formula

$$\frac{d}{dx} \frac{x_+^\lambda}{\Gamma(\lambda + 1)} = \frac{x_+^{\lambda - 1}}{\Gamma(\lambda)}.$$

Hint. Establish it for Re $\lambda > 0$ and then use analytic continuation.

5. Let the generalized functions x_+^{-n} and x_-^{-n} be defined by the formulas

$$(x_+^{-n}, \varphi) = \int_0^\infty x^{-n} \left[\varphi(x) - \varphi(0) - \cdots - \frac{x^{n-1}}{(n-1)!} \varphi^{(n-1)}(0) \, \theta(1 - x) \right] dx,$$

$$(x_-^{-n}, \varphi) = \int_0^\infty x^{-n} \left[\varphi(-x) - \varphi(0) - \right.$$

$$\left. \cdots - (-1)^{n-1} \frac{x^{n-1}}{(n-1)!} \varphi^{(n-1)}(0) \, \theta(1 - x) \right] dx,$$

where $\theta(1 - x)$ equals 0 for $x > 1$ and 1 for $x < 1$.

Show that x_+^{-n} is carried on the half-line $x \geq 0$ and equals the ordinary function $1/x^n$ in any closed interval [a, b], with $0 < a < b$; x_-^{-n} is carried on the half-line $x \leq 0$ and equals $1/x^n$ in any interval [d, c], with $c < d < 0$. Then establish the formulas

$$x_+^{-2m} + x_-^{-2m} = x^{-2m},$$

$$x_+^{-2m-1} - x_-^{-2m-1} = x^{-2m-1},$$

$$xx_+^{-n} = x_+^{-n+1},$$

$$xx_-^{-n} = -x_-^{-n+1},$$

$$(x_+^{-n})' = -nx_+^{-n-1} + \frac{(-1)^n}{n!} \delta^{(n)}(x),$$

$$(x_-^{-n})' = nx_-^{-n-1} - \frac{(-1)^n}{n!} \delta^{(n)}(x).$$

6. Show that x_+^λ and x_-^λ are homogeneous functions of degree λ (for any $\lambda \neq -1, -2, \dots$).

Hint. Both sides of the relation $(x_+^\lambda, \varphi(tx)) = t^{-\lambda-1}(x_+^\lambda, \varphi(x))$ can be analytically continued from the half-plane $\mathrm{Re}\,\lambda > -1$ to the entire λ-plane except for the singular points $\lambda = -1, -2, \dots$

7. Show that a generalized homogeneous function f in K_1' of degree λ satisfies the equation

$$xf' = \lambda f.$$

Hint. Differentiate the relation $(f, \varphi(tx)) = t^{-\lambda-1}(f, \varphi(x))$ with respect to t and then set $t = 1$.

8. Show that the $k + m$ linearly independent solutions of the equation

$$x^k y^{(m)}(x) = 0$$

found in Prob. 1, Sec. 1.5 form a complete set of linearly independent solutions.

Hint. For $x > 0$, the solution $y(x)$ is a polynomial $P(x_+)$ and for $x < 0$, another polynomial $Q(x_-)$. The difference $y - [P(x_+) + Q(x_-)]$ is carried at 0 and by Sec. 2.3.6 can be represented by

$$\sum_{j=0}^{p} a_j \, \delta^{(j)}(x).$$

Substitute

$$y = P(x_+) + Q(x_-) + \sum_{j=0}^{p} a_j \delta^{(j)}(x)$$

in the equation and find conditions on the coefficients.

9. Show that the general solution of $xy' = \lambda y$ is given by

$$C_1 x_+^\lambda + C_2 x_-^\lambda \qquad \text{for } \lambda \neq 1, -2, \dots,$$
$$C_1 x^{-n} + C_2 \, \delta^{(n-1)}(x) \quad \text{for } \lambda = -n.$$

Hint. $y(x)$ is equal to $C_1 x_+^\lambda$ for $x > 0$ and to $C_2 x_-^\lambda$ for $x < 0$ (for $\lambda = -n$, they are not solutions, but this does not matter). The difference $y - (C_1 x_+^\lambda + C_2 x_-^\lambda)$ is carried at $x = 0$ and by Sec. 2.3.6 has the form

$$\sum_{k=0}^{p} a_k \, \delta^{(k)}(x).$$

Hence,

$$y = C_1 x_+^\lambda + C_2 x_-^\lambda + \sum_{k=0}^{p} a_k \delta^{(k)}(x).$$

Substitute this in the equation and find conditions on the coefficients.

REMARK. The solution to Prob. 9 furnishes a description of all homogeneous generalized functions in K_1' (see Prob. 7). In particular, the functions x_+^λ and x_-^λ of Prob. 5 are not homogeneous.

10. Show that the value of $\dfrac{r^\lambda}{\Gamma\left(\dfrac{\lambda + n}{2}\right)}$ at $\lambda = -n - 2k$ is

$$\frac{\Omega_n}{n(n + 2) \cdots (n + 2k - 2)} \frac{(-1)^k}{2^{k+1} k!} \Delta^k \delta(x).$$

Hint. Iterate (8) of Sec. 1.7 to obtain

$$r^\lambda = \frac{\Delta^k r^{\lambda + 2k}}{(\lambda + 2)(\lambda + 4) \cdots (\lambda + 2k)(\lambda + n)(\lambda + n + 2) \cdots (\lambda + n + 2k - 2)}.$$

Then to compute the residue at $\lambda = -n - 2k$, apply the final result of Sec. 2.4.5.

NOTE. Comparing this result with (32), deduce that

$$\frac{d^{2k}}{dr^{2k}} S_r[\varphi]\bigg|_{r=0} = S_0^{(2k)}[\varphi] = \frac{\Omega_n}{n(n + 2) \cdots (n + 2k - 2)} \frac{(2k)!}{2^k k!} \Delta^k \varphi(0).$$

This formula makes it possible to write down the following expression for $S_r[\varphi]$ in powers of r:

$$S_r[\varphi] = \varphi(0) + S_0''[\varphi] \frac{r^2}{2!} + \cdots + S_0^{(2m)}[\varphi] \frac{r^{2m}}{(2m)!} + o(r^{2m})$$

$$= \Omega_n \sum_{k=0}^{m} \frac{\Delta^k \varphi(0)}{2^k k! \, n(n + 2) \cdots (n + 2k - 2)} + o(r^{2m})$$

(Pizetti's formula).

11. Evaluate

$$I(x) = \int_\Omega f((\omega, x)) \, d\omega$$

when $f(\xi)$ is equal to a) $\delta^{(k)}(\xi)$; b) ξ^{-k}; c) $|\xi|^\lambda \log |\xi|$.

Hint. Utilize formula (34) and Prob. 10. In c), differentiate with respect to λ.

Ans.

a) $I(x) = \begin{cases} 0 & \text{for } k \text{ odd}; \\ C\Delta^{\frac{k-n+1}{2}} \delta(x) & \text{for } k \text{ even}, n \text{ odd}, k + 1 \geq n; \\ Cr^{-k-1} & \text{for remaining values of } k \text{ and } n. \end{cases}$

b) $I(x) = \begin{cases} 0 & \text{for } k \text{ odd}; \\ C\Delta^{\frac{k-n}{2}} \delta(x) & \text{for } k \geq n; k \text{ even}, n \text{ odd}, \\ Cr^{-k} & \text{for remaining values of } k \text{ and } n. \end{cases}$

c) $I(x) = C_1 r^\lambda \log r + C_2 r^\lambda$.

12. Show that a spherically symmetric homogeneous function f in K_n' of degree λ must be Cr^λ ($\lambda \neq -n, -n-2, \ldots$).

Hint. Suppose first that $\operatorname{Re} \lambda > -n$. With each f in K_n' associate a g in K_1' as follows. Let $\varphi(\xi)$ be an even function in K_1. Then $\varphi(r)$ belongs to K_n. Define $(g, \varphi(\xi)) = (f, \varphi(r))$. For odd φ in K_1, put $(g, \varphi) = 0$. g is an even homogeneous functional of degree $n + \lambda - 1$ and is of the form $C|x|^{n+\lambda-1}$ (by Probs. 7 and 9). Thus, for spherically symmetric $\varphi(r)$,

$$(f, \varphi(r)) = C \int_0^\infty r^{n+\lambda-1} \varphi(r)\, dr = C_1 \int_{R_n} r^\lambda \varphi(r)\, dx.$$

By Prob. 2, Sec. 1.6, $(f, \varphi) = C_1(r^\lambda, \varphi)$ for any φ in K_n. If $\operatorname{Re} \lambda \leq -n$, then for some integer m, $\operatorname{Re} \lambda + 2m > -n$. Therefore, $r^{2m}f = Cr^{2m+\lambda}$. Hence, for $x \neq 0$, f is equal to Cr^λ. If $\lambda \neq -n, -n-2, \ldots$, the functional $f - Cr^\lambda$ is carried at 0 and is therefore a linear combination of the δ-function and its derivatives. By Probs. 6 and 7 of Sec. 1.7, $f - Cr^\lambda = 0$.

Note. For $\lambda = -n, -n-2, \ldots$, see Sec. 2.8, Prob. 4.

13. Let $f(x) = f(x_1, \ldots, x_n)$ be a non-negative homogeneous function of degree 1. Study $f^\lambda(x)$ as an analytic function of λ. It defines a functional on K_n for $\operatorname{Re} \lambda > -n$. Determine its analytic continuation to the half-plane $\operatorname{Re} \lambda > -n - 1$ and find its residue at $\lambda = -n$.

Hint. Let S be the unit ball and Ω the unit sphere in R_n. Then for $\operatorname{Re} \lambda > -n$,

$$(f^\lambda, \varphi) = \int_S f^\lambda(x)[\varphi(x) - \varphi(0)]\, dx + \int_{R_n - S} f^\lambda(x)\, \varphi(x)\, dx + \varphi(0) \int_S f^\lambda(x)\, dx.$$

Since $f^{\lambda}(x)$ is homogeneous of degree λ,

$$\lambda f^{\lambda}(x) = \sum_{k=1}^{n} x_k \frac{\partial f^{\lambda}(x)}{\partial x_k}$$

and

$$\int_S f^{\lambda}(x)\,dx = \frac{1}{\lambda} \sum_{k=1}^{n} \int_S x_k \frac{\partial f^{\lambda}(x)}{\partial x_k}\,dx$$

$$= \frac{1}{\lambda}\left\{ \int_S \sum_{k=1}^{n} \frac{\partial [x_k f^{\lambda}(x)]}{\partial x_k}\,dx - \sum_{k=1}^{n} \int_S f^{\lambda}(x)\,dx \right\}.$$

Hence, by the Divergence theorem,

$$\left(1 + \frac{n}{\lambda}\right) \int_S f^{\lambda}(x)\,dx = \frac{1}{\lambda} \int_S \sum_{k=1}^{n} \frac{\partial [x_k f^{\lambda}(x)]}{\partial x_k}\,dx = \frac{1}{\lambda} \int_{\Omega} f^{\lambda}(x)\,d\omega.$$

Thus,

$$(f^{\lambda}, \varphi) = \int_S f^{\lambda}(x)\,[\varphi(x) - \varphi(0)]\,dx$$

$$+ \int_{R_n - S} f^{\lambda}(x)\,\varphi(x)\,dx + \frac{\varphi(0)}{n + \lambda} \int_{\Omega} f^{\lambda}(x)\,d\omega.$$

All three terms have analytic continuations to the half-plane $\mathrm{Re}\ \lambda > -n - 1$. The residue of f^{λ} at $\lambda = -n$ is

$$\left(\int_{\Omega} f^{-n}(x)\,d\omega \right) \delta(x).$$

14. Let $f(x) = f(x, ..., x_n)$ be a homogeneous function of degree $-n + 1$. It is locally integrable and defines a functional f through the formula

$$(f, \varphi) = \int_{R_n} f(x)\,\varphi(x)\,dx.$$

Show that the derivatives of f are given by the expression

$$\left(\frac{\partial f}{\partial x_k}, \varphi\right) = \int_{|x| \leq 1} \frac{\partial f}{\partial x_k}\,[\varphi(x) - \varphi(0)]\,dx$$

$$+ \int_{|x| \geq 1} \frac{\partial f}{\partial x_k}\,\varphi(x)\,dx + \varphi(0) \int_{|x| = 1} f(x)\,x_k\,d\omega.$$

Hint. Transform the integrals in

$$\left(\frac{\partial f}{\partial x_k}, \varphi\right) = -\left(f, \frac{\partial \varphi}{\partial x_k}\right) = -\int_{R_n} f\frac{\partial \varphi}{\partial x_k} dx = -\int_{|x|\leq 1} f\frac{\partial[\varphi(x) - \varphi(0)]}{\partial x_k} dx$$

$$-\int_{|x|\geq 1} f\frac{\partial \varphi}{\partial x_k} dx = -\int_{|x|\leq 1}\frac{\partial}{\partial x_k}\{f[\varphi(x) - \varphi(0)]\} dx$$

$$+\int_{|x|\leq 1}\frac{\partial f}{\partial x_k}[\varphi(x) - \varphi(0)] dx - \int_{|x|\geq 1}\frac{\partial(f\varphi)}{\partial x_k} dx + \int_{x|\geq 1}\frac{\partial f}{\partial x_k}\varphi \, dx$$

by applying the Divergence theorem.

15. Verify by direct differentiation that

$$\Delta\left(\frac{1}{r^{n-2}}\right) = c_n \delta(x).$$

Hint. The first derivatives can be computed in a straightforward manner, the second with the help of Prob. 14.

16. Show that a homogeneous function $f(x)$ of degree $-n$ satisfies the relation

$$\sum_{k=1}^{n}\frac{\partial}{\partial x_k}(x_k f(x)) = c\delta(x),$$

and find c.

Hint. Use Prob. 14.

Ans.
$$c = \int_{|x|=1} f(x) \, d\omega.$$

17. Is it possible to assign each $g(x) = f(x)/r^m$ a functional $g \in K_n'$ so as to satisfy the conditions

1) $\alpha_1 g_1(x) + \alpha_2 g_2(x) \to \alpha_1 g_1 + \alpha_2 g_2$ (α_1 and α_2 numbers);

2) $\alpha(x) g(x) \to \alpha(x) g$ ($\alpha(x)$ an infinitely differentiable function);

3) $\dfrac{\partial g(x)}{\partial x_j} \to \dfrac{\partial g}{\partial x_j}$ (in K_n);

4) If $g(x)$ is an ordinary function, then

$$(g, \varphi) = \int_{R_n} g(x) \varphi(x) \, dx?$$

Ans. No. For instance, for $g(x) = 1/r^n$, we would have to have

$$\sum \frac{\partial}{\partial x_k}(x_k g) = \sum x_n \frac{\partial g}{\partial x_k} + ng(x) = 0$$

by Euler's theorem, whereas by Prob. 16,

$$\sum \frac{\partial}{\partial x_k}(x_k g) = \delta(x) \int_{|x|=1} g(\xi)\,d\xi \neq 0. \quad \text{(V. V. Grushin)}.$$

For $n > 2$, the same fact can be inferred immediately from the equation $\Delta(r^{2-n}) = c_n \delta(x)$.

Note. The result of Prob. 17 shows that for several variables, canonical regularization over the class of functions of the form $f(x)/r^m$ is impossible (see Prob. 1).

2.5. Convolutions of Generalized Functions

2.5.1. Let $f(x)$ and $g(x)$ be ordinary functions of x, $-\infty < x < \infty$. We form the expression

$$(f * g)(x) = \int_{-\infty}^{\infty} f(\xi)\,g(x - \xi)\,d\xi, \tag{1}$$

the *convolution* of $f(x)$ and $g(x)$.

The integral does not always exist. But it is known for instance that if f and g both belong to $L_1(-\infty, \infty)$, then $f * g$ will exist almost everywhere for $-\infty < x < \infty$ and be an $L_1(-\infty, \infty)$ function as well.† We shall examine another case here where the convolution exists. Namely, one of the functions f or g will be assumed to have compact support.

Suppose that $f(x)$ has compact support and therefore vanishes outside $[a, b]$. We wish to show that $f * g$ exists for almost all x and is an ordinary function. To this end, choose any interval $[c, d]$ and consider the double integral

$$\int_c^d \int_a^b |f(\xi)||g(x - \xi)|\,d\xi\,dx. \tag{2}$$

---

† See S. Bochner, *Lectures on Fourier Integrals*, Princeton University Press, New Jersey, 1959, Chapt. 3, Sec. 13.

By Fubini's theorem, the iterated integral can be evaluated in either order. Consider first

$$\int_a^b |f(\xi)| \left\{ \int_c^d |g(x - \xi)| \, dx \right\} d\xi. \tag{3}$$

The function in the braces is bounded for $a \leqq \xi \leqq b$ since

$$\int_c^d |g(x - \xi)| \, dx = \int_{c-\xi}^{d-\xi} |g(x)| \, dx = G(d - \xi) - G(c - \xi),$$

where G is the indefinite integral of $|g(x)|$. This proves the existence of the integral (3) and with it that of (2). Reversing the order of integration, we can deduce the existence of

$$\int_a^b |f(\xi) \, g(x - \xi)| \, d\xi \tag{4}$$

almost everywhere and with it that of the integral (1). But the function (4) is summable (Lebesgue integrable) with respect to x in $c \leqq x \leqq d$, therefore implying the summability of the function (1) in this interval.

Thus, when f has compact support, $f * g$ is summable in each finite interval and hence is an ordinary function.

Now let $g(x)$ be a function with support in $[c, d]$. Then the integral (1) also actually extends over a finite interval; we have

$$f * g = \int_{x-d}^{x-c} f(\xi) g(x - \xi) \, d\xi. \tag{5}$$

By the foregoing discussion, the convolution

$$g * f = \int_c^d g(\eta) \, f(x - \eta) \, d\eta \tag{6}$$

exists. If we make the substitution $\xi = x - \eta$ in the integral, we obtain

$$g * f = \int_{x-d}^{x-c} g(x - \xi) f(\xi) \, d\xi = \int_{x-d}^{x-c} f(\xi) \, g(x - \xi) \, d\xi = f * g,$$

proving the existence of the integral (5) and its coincidence with the integral (6).

Thus, $f * g$ exists and represents an ordinary function if at least one of the functions f or g has compact support.

Finally, let us characterize the support of $f * g$ knowing the supports of f and g. If $(f * g)(x) \neq 0$, this means that supp $f(\xi)$ and supp $g(x - \xi)$ $= x - $ supp $g(\xi)$ intersect at some point, say, ξ_0. Since $\xi_0 \in x - $ supp $g(\xi)$ we have $x \in \xi_0 + $ supp $g \subset$ supp $f + $ supp g (the arithmetic sum!).†
Thus

$$\text{supp } (f * g) \subset \text{supp} f + \text{supp } g .$$

2.5.2. We shall now carry over the definition of convolution to generalized functions. In order to find the necessary rule, we apply (1) to a test function $\varphi(x)$ and without worrying about the existence of integrals we reverse the order of integration. This yields

$$(f * g, \varphi) = \int_{-\infty}^{\infty} \left\{ \int_{-\infty}^{\infty} f(\xi) g(x - \xi) \, d\xi \right\} \varphi(x) \, dx$$

$$= \int_{-\infty}^{\infty} f(\xi) \left\{ \int_{-\infty}^{\infty} g(x - \xi) \varphi(x) \, dx \right\} d\xi \qquad (7)$$

$$= \int_{-\infty}^{\infty} f(\xi) \left\{ \int_{-\infty}^{\infty} g(x) \varphi(x + \xi) \, dx \right\} d\xi .$$

Equation (7) makes the following definitions natural:‡

DEFINITION 1. For any generalized function $g \in K'$, the function of ξ given by

$$g * \varphi = (g(x), \varphi(x + \xi)) \qquad (8)$$

is called the *convolution of g and φ*.

DEFINITION 2. For any generalized functions f and g, the functional $f * g$ operating through

$$(f * g, \varphi) = \left(f(\xi), (g * \varphi)(\xi) \right) \qquad (9)$$

is called the *convolution of f and g*.

---

† The arithmetic sum (difference) of two sets A and B comprises all ξ of the form $\alpha + \beta$ $(\alpha - \beta)$ with $\alpha \in A$ and $\beta \in B$.
‡ More precisely, natural for real functions. For the case of complex-valued functions, the definitions are appropriate only to functionals of the first of the four types mentioned in Sec. 1.3.4.
The definitions for the remaining three types are as follows:

$$(g * \varphi) = (g, \varphi(x + \xi))_1 = (\bar{g}, \varphi(x + \xi))_2 = (g, \bar{\varphi}(x + \xi))_3 = (\bar{g}, \bar{\varphi}(x + \xi))_4;$$
$$(f * g, \varphi) = (f, g * \varphi)_1 = (\bar{f}, g * \varphi)_2 = (f, \overline{g * \varphi})_3 = (\bar{f}, g * \varphi)_4.$$

Of course, just as for ordinary functions, the convolution does not always exist. But the foregoing discussion shows in any event that if f and g are ordinary functions and one of them has compact support, then Definition 2 leads to the usual convolution of f and g.

Suppose now that f is an arbitrary functional and $g = \delta(x)$. Then for each $\varphi \in K$,

$$g * \varphi = (\delta(x), \varphi(x + \xi)) = \varphi(\xi)$$

and

$$(f * g, \varphi) = (f(\xi), \varphi(\xi)).$$

In other words, $f * \delta = f$. Similarly, for arbitrary g,

$$(\delta * g, \varphi) = (\delta(\xi), (g(x), \varphi(x + \xi))) = (g(x), \varphi(x)),$$

i.e., $\delta * g = g$. Thus, a convolution with the delta-function always exists, and the functional δ plays the same role for convolution as unity does for multiplication. If $g = \delta'(x)$, then

$$g * \varphi = (\delta', \varphi(x + \xi)) = -\varphi'(\xi),$$

$$(f * g, \varphi) = (f(\xi), - \varphi'(\xi)) = (f'(\xi), \varphi(\xi)),$$

so that $f * \delta' = f'$. Similarly, $\delta' * g = g'$, and so the convolution of any functional with δ' results in its derivative.

2.5.3. We continue to examine the convolution of functionals f and g in the general case.

One would naturally expect the convolution of two generalized functions to exist if one of them has compact support.

In this subsection, we consider the convolution of a functional g and a test function φ. Since $\varphi(x + \xi)$ is also a test function, the expression $\psi(\xi) = (g(x), \varphi(x + \xi))$ is well-defined for each ξ.

It is easy to verify that $\psi(\xi) = (g(x), \varphi(x + \xi))$ is a continuous function having continuous derivatives of all orders. In fact, let $\xi_\nu \to \xi_0$. Then $\varphi(x + \xi_\nu)$ approaches $\varphi(x + \xi_0)$ uniformly in x together with all its derivatives and each $\varphi(x + \xi_\nu)$ has support in the same fixed interval. Thus $\varphi(x + \xi_\nu) \to \varphi(x + \xi_0)$ in K. Since g is a continuous functional,

$$\psi(\xi_\nu) = (g(x), \varphi(x + \xi_\nu)) \to (g(x), \varphi(x + \xi_0)) = \psi(\xi_0).$$

This proves the continuity of $\psi(\xi)$. Now consider the difference quotient

$$\frac{\varphi(x + \xi_\nu) - \varphi(x + \xi_0)}{\xi_\nu - \xi_0}.$$

It has the uniform limit $\varphi'(x + \xi_0)$ as $\xi_\nu \to \xi_0$ together with all derivatives. Therefore,

$$\frac{\psi(\xi_\nu) - \psi(\xi_0)}{\xi_\nu - \xi_0} = \left(g, \frac{\varphi(x + \xi_\nu) - \varphi(x + \xi_0)}{\xi_\nu - \xi_0}\right) \to (g, \varphi'(x * \xi)),$$

so that $\psi(\xi)$ has the derivative

$$\psi'(\xi) = (g(x), \varphi'(x + \xi)) = g * \varphi'.$$

On the other hand, since

$$(g(x), \varphi'(x + \xi)) = -(g'(x), \varphi(x + \xi)) = -g' * \varphi,$$

we obtain the string of equalities

$$(g * \varphi)' = g * \varphi' = -g' * \varphi. \qquad (10)$$

Arguing inductively, we see that $\psi(\xi)$ has derivatives of all orders and that

$$(g * \varphi)^{(q)} = g * \varphi^{(q)} = (-1)^q g^{(q)} * \varphi.$$

In general, the function $\psi(\xi)$ does not have compact support and is therefore not a test function. We point out one important case where $\psi(\xi)$ does have compact support. This is where the generalized function g has compact support. We can then even describe the support of $\psi(\xi)$. For instance, suppose $g(x)$ is carried in the interval Δ_g of the x-axis and the test function $\varphi(x)$ in the interval Δ_φ. Then $\varphi(x + \xi)$ will be carried in $\Delta_\varphi - \xi$. If the convolution $\psi = g * \varphi$ is to be different from zero, the intervals Δ_g and $\Delta_\varphi - \xi$ must intersect. Let $\eta \in \Delta_g \cap (\Delta_\varphi - \xi)$ be such a common point. Since $\eta \in \Delta_\varphi - \xi$, we have $\eta + \xi \in \Delta_\varphi$ and $\xi \in \Delta_\varphi - \eta$. But since $\eta \in \Delta_\varphi$, ξ belongs to the arithmetic difference $\Delta_\varphi - \Delta_g$. Thus,

$$\operatorname{supp}(g * \varphi) \subset \operatorname{supp}\varphi - \operatorname{supp}g.$$

But this difference is obviously a subset of some bounded interval Δ. $\psi(\xi)$ vanishes outside Δ and so has compact support.

We now show that taking the convolution with a generalized function g having compact support is a continuous linear operation on K. In other words, if $\varphi = \alpha_1\varphi_1 + \alpha_2\varphi_2$, then $g * \varphi = \alpha_1(g * \varphi_1) + \alpha_2(g * \varphi_2)$, and if $\varphi_\nu \to 0$ in K, then $g * \varphi_\nu \to 0$ in K. The first assertion is obvious:

$$g * \varphi = (g(x), \varphi(x + \xi)) = (g, \alpha_1\varphi_1(x + \xi) + \alpha_2\varphi_2(x + \xi))$$
$$= \alpha_1(g(x), \varphi_1(x + \xi)) + \alpha_2(g(x), \varphi_2(x + \xi))$$
$$= \alpha_1 g * \varphi_1 + \alpha_2 g * \varphi_2.$$

The verification of the second goes as follows.

Let φ_ν be a given sequence that converges to 0 in K. By assumption, the support of each φ_ν lies in a fixed interval $\varDelta$. But then the support of each $g * \varphi_\nu$ lies in the fixed interval $\varDelta - \varDelta_g$, according to what we proved above. We wish to show that the sequence $(g * \varphi_\nu)(\xi)$ converges uniformly to 0. We know that $\varphi_\nu(x)$ and all its derivatives converge uniformly to 0 in $\varDelta$. Suppose that $\psi_\nu(\xi) = (g * \varphi_\nu)(\xi)$ does not converge uniformly to zero. Then to some $\varepsilon > 0$, there is a sequence of points ξ_ν such that

$$|(g * \varphi_\nu)(\xi_\nu)| = |(g(x), \varphi_\nu(x + \xi_\nu))| > \varepsilon. \tag{11}$$

The ξ_ν form a bounded sequence (they belong to $\varDelta - \varDelta_g$), which we may suppose converges to some ξ_0. But then the function $\varphi_\nu(x + \xi_\nu)$ approaches 0 in K together with $\varphi_\nu(x)$. Hence,

$$(g(x), \varphi_\nu(x + \xi_\nu)) \to 0,$$

and this contradicts (11). Thus, $\psi_\nu(\xi) = (g * \varphi_\nu)(\xi)$ approaches 0 uniformly. Similarly, $\psi'_\nu(\xi) = g * \varphi'_\nu$, $\psi''_\nu(\xi) = g * \varphi''_\nu$, ... approach zero uniformly and so $\psi_\nu(\xi) \to 0$ in K, as asserted.

2.5.4. Still assuming that the functional g has compact support, we consider $f * g$ according to Definition 2. Since $(g * \varphi)(\xi)$ is in K whenever φ is, the expression

$$(f * g, \varphi) = (f(\xi), (g * \varphi)(\xi))$$

is well-defined. It clearly represents a linear functional of φ, since by the foregoing results,

$$(f * g, \alpha_1\varphi_1 + \alpha_2\varphi_2) = (f, g * (\alpha_1\varphi_1 + \alpha_2\varphi_2))$$
$$= (f, \alpha_1 g * \varphi_1 + \alpha_2 g * \varphi_2)$$
$$= \alpha_1(f * g, \varphi_1) + \alpha_2(f * g, \varphi_2)$$

for any φ_1 and $\varphi_2 \in K$ and any real numbers α_1 and α_2.

8   Shilov

It moreover is a continuous functional. If $\varphi_\nu \to 0$ in K, then so does $g * \varphi_\nu \to 0$ in K, and by the continuity of f, $(f * g, \varphi_\nu) = (f, g * \varphi_\nu) \to 0$. Thus,

If g is a functional with compact support, then $f * g$ is a continuous linear functional on K.

We now consider the case where g is arbitrary and f has compact support. Although infinitely differentiable, $(g * \varphi)(\xi)$ is not a test function. The operation $g * \varphi$ is clearly still linear. It is also continuous in the following sense. *If $\varphi_\nu \to 0$ in K, then $\psi_\nu = g * \varphi_\nu$ approaches zero uniformly in any interval Δ together with all its derivatives.* To see this, we note that the determination of the value of $\psi(\xi) = (g(x), \varphi(x + \xi))$ for given ξ does not require complete information about the functional $g(x)$. One needs to know the functional merely on the interval in which $\varphi(x + \xi)$ is carried, i.e., on the support of $\varphi(x)$ translated to the left by the amount ξ. In particular, the values of $\psi_\nu(\xi)$ are determined by the properties of g just on the set $\Delta_{\varphi_\nu} - \xi$, where Δ_{φ_ν} is the interval in which the support of φ_ν lies. Therefore, the values of $\psi_\nu(\xi)$ in Δ are determined by the properties of g in the arithmetic difference $\Delta_{\phi_\nu} - \Delta \subset \overline{\Delta} - \Delta$, where $\overline{\Delta}$ is the common support of all φ_ν. Multiplication of $g(x)$ by a test function $e(x)$ equaling 1 in $\overline{\Delta} - \Delta$ leaves all of the $\psi_\nu(\xi)$ unchanged on Δ. At the same time, eg now has compact support. Applying the preceding theorem, we obtain our required result.

We now apply f to $g * \varphi$. The expression $(f, g * \varphi)$ is meaningful since f has compact support. It defines *a continuous linear functional on K.* The linearity follows from the linearity of $g * \varphi$. To prove continuity, suppose that $\varphi_\nu \to 0$ in K. Then $g * \varphi_\nu$ converges uniformly to 0 together with all its derivatives over any interval of the line, in particular, in a neighborhood of the support of f. But then, as we know, $(f, g * \varphi_\nu) \to 0$.

2.5.5. Thus the convolution is properly defined in two cases: when the functional in the first position has compact support or when the functional in the second position has compact support.

In this section, we shall find an explicit expression for $f * g$ in terms of convolutions of ordinary functions. This expression will show in particular that $f * g = g * f$ when f has compact support. As we know, an arbitrary functional g may be written in the form

$$g = \sum_{k=0}^{\infty} g_k^{(k)}(x),$$

where the $g_k(x)$ are ordinary functions whose supports recede from the origin with increasing k. Hence,

$$(g * \varphi)(\xi) = (g(x), \varphi(x + \xi)) = \left(\sum_{k=0}^{\infty} g_k^{(k)}(x), \varphi(x + \xi) \right)$$

$$= \sum_{k=0}^{\infty} (g_k(x), (-1)^k \varphi^{(k)}(x + \xi))$$

$$= \sum_{k=0}^{\infty} (-1)^k \int_{a-\xi}^{b-\xi} g_k(x) \varphi^{(k)}(x + \xi)\, dx$$

$$= \sum_{k=0}^{\infty} (-1)^k \int_{a}^{b} g_k(x - \xi)\, \varphi^{(k)}(x)\, dx, \tag{12}$$

where $[a, b]$ contains the support of $\varphi(x)$. The sum actually includes just a finite number of terms (depending on ξ) since for each ξ the $g_k(x - \xi)$ become equal to zero on $[a, b]$ beginning with a certain index.

From the next to the last relation in (12), it is apparent that the derivatives of the infinitely differentiable function $(g * \varphi)(\xi)$ are given by

$$(g * \varphi)^{(m)}(\xi) = \sum_{k=0}^{\infty} (-1)^k \int_{a-\xi}^{b-\xi} g_k(x)\, \varphi^{(k+m)}(x + \xi)\, dx$$

$$= \sum_{k=0}^{\infty} (-1)^k \int_{a}^{b} g_k(x - \xi)\, \varphi^{(k+m)}(x)\, dx. \tag{13}$$

Consider now $f * g$ assuming that $f(\xi)$ has compact support. We know that such a functional has the form

$$f(\xi) = \sum_{m=0}^{p} f_m^{(m)}(\xi),$$

where each $f_m(\xi)$ has compact support and is carried in, say, $[c, d]$. We then have

$$(f * g, \varphi) = (f, g * \varphi) = \left(\sum_{m=0}^{p} f_m^{(m)}(\xi), (g * \varphi)(\xi) \right)$$

$$= \sum_{m=0}^{p} (-1)^m (f_m(\xi), (g * \varphi)^{(m)}(\xi))$$

$$= \sum_{m=0}^{p} (-1)^m \int_{c}^{d} f_m(\xi)\, (g * \varphi)^{(m)}(\xi)\, d\xi.$$

8*

Applying (12), we find further that

$$(f * g, \varphi) = \sum_{m=0}^{p} (-1)^m \int_c^d f_m(\xi) \left\{ \sum_{k=0}^{p} (-1)^k \int_a^b g_k(x - \xi) \varphi^{(k+m)}(x) \, dx \right\} d\xi.$$

The amount of terms in the k summation may be considered fixed and N in number since ξ varies over a bounded interval. Therefore, the order of summation and integration can be reversed to yield

$$(f * g, \varphi) = \sum_{m=0}^{p} \sum_{k=0}^{N} (-1)^{k+m} \int_c^d f_m(\xi) \left\{ \int_a^b g_k(x - \xi) \varphi^{(k+m)}(x) \, dx \right\} d\xi$$

$$= \sum_{m=0}^{p} \sum_{k=0}^{N} (-1)^{k+m} \int_a^b \left\{ \int_c^d f_m(\xi) g_k(x - \xi) \, d\xi \right\} \varphi^{(k+m)}(x) \, dx$$

$$= \sum_{m=0}^{p} \sum_{k=0}^{N} (-1)^{k+m} ((f_m * g_k)(x), \varphi^{(k+m)}(x)),$$

with $f_m * g_k$ the convolution of ordinary functions. But this result can be written as

$$(f * g, \varphi) = \left(\sum_{m=0}^{p} \sum_{k=0}^{N} (f_m * g_k)^{(k+m)}, \varphi \right),$$

or even as

$$(f * g, \varphi) = \left(\sum_{m=0}^{p} \sum_{k=0}^{\infty} (f_m * g_k)^{(k+m)}, \varphi \right),$$

since by construction the terms in the k summation beginning with the $(N + 1)$-st annihilate $\varphi(x)$. Thus, the required formula expressing $f * g$ in terms of ordinary functions is

$$f * g = \sum_{m=0}^{p} \sum_{k=0}^{\infty} (f_m * g_k)^{(k+m)}. \tag{14}$$

A similar expression results for $f * g$ if f is assumed to be arbitrary and g to have compact support. In this instance, formula (12) has a finite number of terms at the outset, say $p + 1$. In addition, $g * \varphi$ not only is infinitely differentiable but also has compact support, being carried say in $[c, d]$. On the other hand, f is now represented by an infinite sum

$$f = \sum_{k=0}^{\infty} f_k^{(k)}(x)$$

of ordinary functions $f_k(x)$ whose supports recede from the origin with increasing k. Therefore, when it is applied to a function with compact support, only a finite number of terms, say N, come into play. We have

$$(f * g, \varphi) = (f, g * \varphi) = \left(\sum_{k=0}^{N} f_k^{(k)}(\xi), (g * \varphi)(\xi) \right)$$

$$= \sum_{k=0}^{N} (-1)^k (f_k(\xi), (g * \varphi)^{(k)}(\xi)) = \sum_{k=0}^{N} (-1)^k \int_c^d f_k(\xi) (g * \varphi)^{(k)}(\xi) d\xi$$

$$= \sum_{k=0}^{N} (-1)^k \int_c^d f_k(\xi) \left\{ \sum_{m=0}^{p} (-1)^m \int_a^b g_m(x - \xi) \varphi^{(k+m)}(x) dx \right\} d\xi$$

$$= \sum_{k=0}^{N} \sum_{m=0}^{p} (-1)^{k+m} \int_a^b \left\{ \int_c^d f_k(\xi) g_m(x - \xi) d\xi \right\} \varphi^{(k+m)}(x) dx$$

$$= \sum_{k=0}^{N} \sum_{m=0}^{p} (-1)^{k+m} ((f_k * g_m)(x), \varphi^{(k+m)}(x))$$

$$= \sum_{k=0}^{N} \sum_{m=0}^{p} ((f_k * g_m)^{(k+m)}(x), \varphi(x)).$$

Thus by the same considerations as before,

$$(f * g)(x) = \sum_{k=0}^{\infty} \sum_{m=0}^{p} (f_k * g_m)^{(k+m)}(x). \tag{15}$$

Formulas (14) and (15) could be placed at the foundation of convolution theory. Of course, if we were to proceed this way, we would have to prove that (14) and (15) are independent of the choice of f_k and g_m occurring in the representations of f and g.

2.5.6. It follows from formula (14) and (15) that in general

$$f * g = g * f, \tag{16}$$

i.e., the convolution of generalized functions is commutative. The relation (16) has the following interpretation. We have strictly speaking defined the convolution of generalized functions in different ways depending on whether the first or second functional has compact support. If say f has compact support, then $f * g$ and $g * f$ have different meanings.

Formula (16) asserts that the result is the same in both cases. Indeed, by virtue of (14) and (15), we know that if $f = \sum\limits_{m=0}^{p} f_m^{(m)}(x)$ and $g = \sum\limits_{k=0}^{\infty} g_k^{(k)}(x)$, then

$$
\left.
\begin{aligned}
f * g &= \sum_{m=0}^{p} \sum_{k=0}^{\infty} (f_m * g_k)^{(k+m)}, \\
g * f &= \sum_{k=0}^{\infty} \sum_{m=0}^{p} (g_k * f_m)^{(k+m)}.
\end{aligned}
\right\}
\tag{17}
$$

But as we have seen, the convolution is commutative for ordinary functions and so

$$f_m * g_k = g_k * f_m.$$

Therefore, if we reverse the order of summation (which is permissible since only finitely many non-zero terms remain when the expressions in (17) are applied to each test function $\varphi(x)$), we obtain (16), as asserted.

2.5.7. Let us describe the support of $f * g$. Suppose $(f * g, \varphi) = (f, g * \varphi) \neq 0$. It then follows that the supports of f and $g * \varphi$ have a point ξ in common:

$$\xi \in \operatorname{supp} f \quad \text{and} \quad \xi \in \operatorname{supp}(g * \varphi) \subset \operatorname{supp} \varphi - \operatorname{supp} g.$$

Set $\xi = \eta - \zeta$, with $\eta \in \operatorname{supp} \varphi$ and $\zeta \in \operatorname{supp} g$. Then, $\eta = \xi + \zeta \in \operatorname{supp} f + \operatorname{supp} g$. In other words, if the support of a test function φ lies outside the set $\operatorname{supp} f + \operatorname{supp} g$, we have $(f * g, \varphi) = 0$. Thus, it follows that

$$\operatorname{supp}(f * g) \subset \operatorname{supp} f + \operatorname{supp} g.$$

This last formula is useless if the support of g is the entire line. However, if we are interested in $f * g$ in just a bounded region G rather than everywhere, then complete information about g is unnecessary. It is sufficient for us to know g in a neighborhood of the set $G - \operatorname{supp} f$. For, consider a test function $e(x)$ equaling 1 on $G - \operatorname{supp} f$ and 0 outside some neighborhood V of this set. The determination of the functional eg requires the knowledge of g just on V. For any φ in K carried in G, we have

$$(f * (1 - e)g, \varphi) = ((1 - e)g, f * \varphi) = 0$$

since $\operatorname{supp}(f * \varphi) \subset G - \operatorname{supp} f$ and $(1 - e)g$ is carried in the complement of this set. Therefore, for the specified φ,

$$(f * g, \varphi) = (f * eg, \varphi),$$

as asserted.

2.5.8. In this section, we shall consider continuity and differentiation of the convolution.

THEOREM 1. (*Theorem on continuity of the convolution*). *If* $f_\nu \to f$, *in* K', *then*

$$f_\nu * g \to f * g$$

in the following cases:

(a) *g has compact support*;

(b) *the support of each* f_ν *lies in the same interval* $[a, b]$.

Proof. For any φ in K,

$$(f_\nu * g, \varphi) = (f_\nu, g * \varphi) \to (f, g * \varphi) = (f * g, \varphi),$$

since (a) $g * \varphi$ is a test function together with φ and (b) $g * \varphi$ may be replaced by a test function without changing it on $[a, b]$, where all f_ν have their supports. Hence, $f_\nu * g \to f * g$, as required.

Owing to (16), a similar theorem holds for the case where the second term of the convolution depends on ν.

THEOREM 2. (*Theorem on differentiating the convolution*). *If one of the functionals f or g has compact support, then*

$$(f * g)' = f' * g = f * g'.$$

Proof. According to the definitions of derivative and convolution,

$$((f * g)', \varphi) = -(f * g, \varphi') = -(f, g * \varphi');$$
$$(f' * g, \varphi) = (f', g * \varphi) = -(f, (g * \varphi)');$$
$$(f * g', \varphi) = (f, g' * \varphi).$$

The right-hand sides of the relations are meaningful if f or g has compact support and coincide on account of (10).

2.5.9. The definition and basic properties of convolution carry over to the case of several variables without difficulty.

Let $f(x_1, ..., x_n) = f(x)$ and $g(x_1, ..., x_n) = g(x)$ be ordinary functions. Their convolution is defined by

$$(f * g)(x) = \int_{R_n} f(\xi_1, ..., \xi_n) g(x_1 - \xi_1, ..., x_n - \xi_n) \, d\xi_1 ... d\xi_n.$$

$(f * g)(x)$ exists for almost every $x \in R_n$ and is again an ordinary function if at least one of the functions $f(x)$ or $g(x)$ has compact support. In that case,

$$f * g = g * f,$$

or the convolution is commutative.

Let f and g be two generalized functions. The expression

$$(g * \varphi)(\xi) = (g(x), \varphi(x + \xi))$$

is an infinitely differentiable function of ξ and is called the *convolution of g and φ*. The functional $f * g$ operating through

$$(f * g, \varphi) = (f(\xi), (g * \varphi)(\xi)),$$

is called the *convolution of f and g*. It is defined on K_n and continuous if either f or g has compact support.

In particular, for any f and g,

$$\delta * g = g, \quad f * \delta = f.$$

If $f = \sum_{|m| \leq p} D^m f_m(x)$ and $g = \sum_k D^k g_k(x)$ are representations of the functionals f and g in terms of ordinary functions (f has compact support, the $f_m(x)$ are carried in a fixed block, and the supports of the $g_k(x)$ recede further from the origin with increasing $|k|$), then

$$(f * g)(x) = \sum_{|m| \leq p} \sum_k D^{m+k}(f_m * g_k)(x) \tag{18}$$

is a representation for $f * g$ in terms of ordinary functions. Using relations such as (18), we find as in Sec. 2.5.6 that

$$f * g = g * f,$$

i.e., the convolution is commutative. The function $g * \varphi$ is carried in the set

$$\operatorname{supp} \varphi - \operatorname{supp} g.$$

The functional $f * g$ is carried in the set

$$\operatorname{supp} f + \operatorname{supp} g.$$

When f has compact support, the properties of $f * g$ in a region G are determined by those of g in an arbitrarily small neighborhood of $G - \operatorname{supp} f$.

The following theorem on the continuity of the convolution holds:

The limit relation $f_v * g \to f * g$ in K' is valid if

(a) $f_v \to f$ in K' and g has compact support, or

(b) $f_v \to f$ in K' and the supports of the f_v lie in the same bounded block.

Differentiation of the convolution is carried out according to the formula

$$\frac{\partial}{\partial x_j}(f * g) = \frac{\partial f}{\partial x_j} * g = f * \frac{\partial g}{\partial x_j}.$$

Problems

1. Let f_1 and f_2 be functionals with compact support and g an arbitrary functional. Show that $f_1 * (f_2 * g) = (f_1 * f_2) * g$ (associativity of the convolution).

Hint. Use the expression for convolutions in terms of ordinary functions (Sec. 2.5.5).

2. Show that the "measure type functional"

$$(\mu, \varphi) = \int \varphi(x) \, d\mu(x),$$

where $\mu(x)$ is a Stieltjes measure, has order of singularity not exceeding 1.

Hint. μ may be assumed to have compact support. Let $\mathscr{E}$ be a fundamental function for the Laplacian. Then $\mathscr{E} * \mu$ is an ordinary function having ordinary first order derivatives (Fubini's theorem). Use the relation $\varDelta(\mathscr{E} * \mu) = \varDelta\mathscr{E} * \mu = \mu$.

3. Consider the space D_m of functions $\varphi(x)$ having compact support and continuous derivatives up to order m (Sec. 2.3, Prob. 9). Show that every continuous linear functional on D_m represents a generalized function with order of singularity not exceeding $m + 1$ (but generally speaking at least m) when restricted to just the $\varphi(x)$ in K.

Hint. Use Prob. 2 and Prob. 11 of Sec. 2.3. $\delta(x)$ is a functional on D_0 and has order of singularity 1.

4. Suppose that the convolution of $g \in K'$ and any $\varphi(x) \in K$ is known to be a test function. Prove that g has compact support.

Hint. (E. A. Gorin). Suppose g does not have compact support. Then there is a $\varphi_1 \in K$ with support in the ball $|x| \leq 1$ for which for some x_1, $(g * \varphi_1)(x_1) \neq 0$. By assumption, $g * \varphi_1$ has support in a ball $|x| \leq r_1$. One can find a $\varphi_2 \in K$ with support in $|x| \leq 1$ such that $(g * \varphi_2)(x_2) \neq 0$

for some x_2 with $|x_2| > r_1 + 1$. Determine a $\psi_2 \in K$ of the form $\psi_2 = \varphi_2 - \lambda \varphi_1$ so that $(g * \psi_2)(x_1) = (g * \varphi_2)(x_1) - \lambda(g * \varphi_2)(x_1) = 0$. Continuing in this way, one can find a sequence of test functions $\psi_\nu(x)$ with supports in $|x| \leq 1$ and a sequence of points $x_\nu \to \infty$ for which $(g * \psi_\nu)(x_\mu) \neq 0$ just for $\nu = \mu$. Set $\psi(x) = \sum_{k=1}^{\infty} c_k \psi_k(x)$. For sufficiently small c_k, $\psi(x)$ is a test function but $g * \psi$ does not have compact support.

5. Show that the definition and basic properties of the convolution including commutativity and associativity carry over to functionals f and g with support in the interval $[0, \infty)$ and that the support of $f * g$ also lies in $[0, \infty)$.

6. (Continuation). The convolution of $x_+^{n-1}/(n-1)!$ and a function $f(x)$ with support in the interval $[0, \infty)$ is equivalent to the n-fold integral of $f(x)$ by virtue of Dirichlet's formula

$$\int_0^x \int_0^{t_n} \cdots \int_0^{t_3} \int_0^{t_2} f(t_1)\, dt_1\, dt_2 \ldots dt_n = \frac{1}{(n-1)!} \int_0^x f(\xi)(x - \xi)^{n-1}\, d\xi$$

$$= f * \frac{x_+^{n-1}}{(n-1)!}.$$

It is therefore natural to call the expression

$$I_\lambda f = f * \frac{x_+^{\lambda-1}}{\Gamma(\lambda)}$$

the λ-fold indefinite integral of f. The definition is applicable to any functional f with support in $[0, \infty)$ and for any complex λ since the generalized function $x_+^{\lambda-1}/\Gamma(\lambda)$ is defined for all λ (Sec. 2.4, Prob. 3). Show that

$$I_\lambda(I_\mu f) = I_{\lambda + \mu} f,$$
$$I_0 f = f,$$
$$I_{-k} f = f^{(k)}.$$

Hint. Use the relation

$$\int_0^x \xi^{\lambda-1}(x - \xi)^{\mu-1}\, d\xi = \frac{\Gamma(\lambda)\,\Gamma(\mu)}{\Gamma(\lambda + \mu)} x^{\lambda + \mu - 1}.$$

7. (Continuation). Show that the Abel integral equation

$$g(x) = \frac{1}{\Gamma(1-\alpha)} \int_0^x \frac{f(\xi)\,d\xi}{(x-\xi)^\alpha},\tag{1}$$

with $g(x)$ a given (differentiable) function, has the solution

$$f(x) = \frac{1}{\Gamma(\alpha)} \int_0^x \frac{g'(\xi)\,d\xi}{(x-\xi)^{1-\alpha}}.$$

Hint. Write (1) in the form $g = I_{1-\alpha}f$ and use Prob. 6.

2.6. Order of Singularity

2.6.1. The order of singularity of a generalized function has been considered repeatedly in examples. We now wish to revise the definition and to analyze it in more detail.

We shall say that *a generalized function f has order of singularity not exceeding p in a region G* (and we shall write $s(f) \leqq p$ or when necessary $s_G(f) \leqq p$), if f can be represented in G by

$$f = \sum_{|k|\leqq p} D^k f_k(x),\tag{1}$$

where the $f_k(x)$ are ordinary functions.

Moreover, if no representation of the form

$$f = \sum_{|k|\leqq p-1} D^k g_k(x)$$

in terms of ordinary functions $g_k(x)$ is possible, we say that *the order of singularity of f in G is exactly p* ($s(f) = p$ or $s_G(f) = p$). The results of Sec. 2.3 imply that every f has finite order of singularity in a bounded region G and a generalized function with compact support has finite order of singularity throughout R_n.

By its meaning, p is non-negative. A generalized function for which $s(f) = 0$ is an ordinary function. We shall now extend the concept of order of singularity to negative p. Namely, let $q = -p > 0$. We shall say that *a generalized function f has order of singularity in G not exceeding p if f and all its derivatives (in K') up to order q are ordinary functions in G. The order of singularity of f in G is exactly p* if among the $(q+1)$-st order derivatives of f there are functionals not reducing to ordinary functions in G.

The order of singularity of a generalized function is $-\infty$ *in G if its* derivatives (in K') of all orders throughout G are ordinary functions.

We wish to emphasize that the discussion here concerns derivatives in K'. In general, a function having negative order of singularity p does not have to be differentiable in the usual sense and need not even be continuous. Only for sufficiently large $|p|$ can we assert that such a function is continuous (see Sec. 2.6.6).

2.6.2. The order of singularity $s(f)$ defined above relates a functional f to ordinary (locally integrable) functions. At the same time, it is natural to introduce the concept in relation to continuous functions. We shall say that *a generalized function has order of c-singularity in G not exceeding* p and we shall write $c(f) \leqq p$ or $c_G(f) \leqq p$ if f has a representation

$$f = \sum_{|k| \leqq p} D^k f_k(x) \tag{2}$$

in G in terms of continuous functions $f_k(x)$. The order of c-singularity of f is exactly p $(c(f) = p$ or $c_G(f) = p)$ if no representation

$$f = \sum_{|k| \leqq p-1} D^k g_k(x)$$

in terms of continuous $g_k(x)$ is possible in G.

We can also extend this definition to negative p. Namely, if $q = -p > 0$, we say that $f = f(x)$ *has order of c-singularity in G not exceeding* p if $f(x)$ is continuous and has continuous derivatives up to q-th order in G. It is *exactly* p if among the $(q + 1)$-st order derivatives of $f(x)$ there are functionals which do not reduce to continuous functions in G.

Clearly, $c(f) \geqq s(f)$. The fact that $c(f) \leqq 0$ means that $f(x)$ is at least a continuous function and $c(f) = -\infty$ that it is continuous and has continuous derivatives of all orders in G.

We recall that having continuous derivatives in the sense of K' is already a guarantee that these derivatives exist in the customary sense (Sec. 1.7.4).

2.6.3. We now give several simple theorems concerning $s(f)$ and $c(f)$. All of the generalized functions occurring below are in a fixed region G.

1. A single differentiation may increase the order of singularity by at most 1. More generally, if $P\left(\dfrac{\partial}{\partial x}\right)$ is an m-th order linear differential operator, then

$$s\left(P\left(\frac{\partial}{\partial x}\right)f\right) \leq s(f) + m, \quad c\left(P\left(\frac{\partial}{\partial x}\right)f\right) \leq c(f) + m. \qquad (3)$$

It is enough to prove the first statement. If $f = \sum\limits_{|k|\leq p} D^k f_k(x)$ and the $f_k(x)$ are ordinary (continuous) functions, then $\dfrac{\partial f}{\partial x_j} = \sum\limits_{|k|\leq p} \dfrac{\partial}{\partial x_j} D^k f_k(x)$, i.e., $\dfrac{\partial f}{\partial x_j}$ is obtained from ordinary (continuous) functions by differentiations up to order $p + 1$, as asserted.

2. $s(f + g) \leq \max [s(f), s(g)]$ for every pair of generalized functions f and g.

Indeed, suppose for definiteness that $\max [s(f), s(g)] = s(f) = p \geq 0$. We can write

$$f = \sum_{|k|\leq p} D^k f_k(x), \quad g = \sum_{|k|\leq p} D^k g_k(x),$$

if we substitute zeros for the missing functions $g_k(x)$. But then

$$f + g = \sum_{|k|\leq p} D^k [f_k(x) + g_k(x)],$$

or $s(f + g) \leq p$, as asserted. But if $\max [s(f), s(g)] = s(f) = p = -q < 0$, then the derivatives up to q-th order of both f and g are ordinary functions. Thus the derivatives of $f + g$ up to q-th order are also ordinary functions, as required.

Similarly,

$$c(f + g) \leq \max [c(f), c(g)].$$

3. If a generalized function f is given by

$$f = f_0 + \sum_{j=1}^{n} \frac{\partial f_j}{\partial x_j}, \qquad (4)$$

where the order of singularity of $f_0, f_1, ..., f_n$ (in any sense) does not exceed m, then the order of singularity of f (in that sense) is obviously at most $m + 1$. Conversely, *if the order of singularity of f is at most $m + 1$, with $m \geq 0$, then it can be represented by (4), where $f_0, ..., f_n$ have order of* *singularity not exceeding m.* The representation is immediately obtained from (1) by separating the first derivatives from the operator D^k.

4. *The product of a functional f such that $s(f) \leq p$ (in any sense) and an infinitely differentiable function $\alpha(x)$ also has order of singularity at most p* (in the same sense). The assertion is trivial for $p \leq 0$. For $p > 0$, we argue inductively. Suppose the assertion is true for all functionals with order of singularity at most p. We represent a given functional of order of singularity at most $p + 1$ in the form (4), with $f_0, f_1, ..., f_n$ functionals with order of singularity at most p. We then have

$$\alpha f = \alpha f_0 + \sum_{j=1}^{n} \alpha \frac{\partial f_j}{\partial x_j} = \alpha f_0 + \sum_{j=1}^{n} \frac{\partial (\alpha f_j)}{\partial x_j} - \sum_{j=1}^{n} \frac{\partial \alpha}{\partial x_j} f_j.$$

Hence, using the induction assumption, we conclude that αf has order of singularity not exceeding $p + 1$, as asserted.

2.6.4. A number of important relations exist between the orders of singularity of generalized functions and their convolutions. Let us first consider f and g throughout R_n and let one of them have compact support. Then

$$s(f * g) \leq s(f) + s(g). \tag{5}$$

Indeed, if $s(f) = l \geq 0$, $s(g) = m \geq 0$, and for definiteness, g has compact support, then

$$f = \sum_{|j| \leq l} D^j f_j(x), \quad g = \sum_{|k| \leq m} D^k g_k(x)$$

with $f_j(x)$ ordinary functions and $g_k(x)$ ordinary functions with compact support. Hence,

$$f * g = \sum_{|j| \leq l} \sum_{|k| \leq m} D^{j+k} [f_j(x) * g_k(x)],$$

and thus, the order of singularity of $f * g$ is at most $l + m$. Suppose now that $l \geq 0$ and $m < 0$, so that g has ordinary derivatives up to order $|m|$. Then if $l + m \geq 0$,

$$f * g = \sum_{|j| \leq l} D^j f_j(x) * g(x)$$

$$= \sum_{|j| < l+m} D^j f_j(x) * g(x) + \sum_{l+m \leq |j| \leq l} D^{j+m} [f_j(x) * D^{-m} g(x)],$$

where the $f_j(x) * D^{-m} g(x)$ are again ordinary functions. Thus $s(f * g)$

$\leqq l + m$ in this case also. Now if $l + m < 0$,

$$f * g = \sum_{|j| \leqq l} D^j f_j(x) * g(x) = \sum_{|j| \leqq l} f_j(x) * D^j g(x)$$

has ordinary derivatives up to order $-m - l$ and so has order of singularity at most $m + l$.

Finally, if l and m are both negative, $f(x)$ has ordinary derivatives up to order $|l|$ and $g(x)$ has ordinary derivatives up to order $|m|$. The operator $D^{|l|+|m|}$ may be applied to $f * g$ by the rule

$$D^{|l| + |m|}[f * g] = D^{|l|} f * D^{|m|} g.$$

The result represents an ordinary function and so again $s(f * g) \leqq |l| + |m|$.

In particular, the convolution of a function $f(x)$ for which $s(f) = -\infty$ (a function with ordinary derivatives of all orders) and any functional $g(x)$ with finite order of singularity (when it makes sense, i.e., f or g has compact support) again has order of singularity $s(f) = -\infty$.

Let us see what can be said about the order of singularity of $f * g$ in a bounded region G. Since complete information about f and g is not needed, the result (5) can sometimes be improved. Namely, we know from Sec. 2.5.7. that if f has compact support, the behavior of $f * g$ in G is determined by the properties of g just in a neighborhood of the set $G - \operatorname{supp} f$. In other words, in G we have the representation

$$(f * g, \varphi) = (f * eg, \varphi),$$

where $e(x)$ is a test function equaling 1 on $G - \operatorname{supp} f$ and 0 outside some neighborhood V of this set. By what was proved before,

$$s_G(f * g) \leqq s(f) + s(eg) = s(f) + s_V(g), \qquad (6)$$

and this is the required formula.

2.6.5. We now clarify to what extent continuity and differentiability are preserved under convolution.

Consider first the convolution of ordinary functions,

$$f(x) * g(x) = \int_{R_n} f(\xi) g(x - \xi) \, d\xi,$$

assuming that one of them as always has compact support. Suppose that $g(x)$ is continuous. We wish to show that $f(x) * g(x)$ is also continuous. If $f(x)$ has compact support and is carried for instance in the ball $|x| \leqq r$, then we can write the estimate

$$|(f * g)(x_0) - (f * g)(x_1)| \leq \int_{|\xi| \leq r} |f(\xi)| \, |g(x_0 - \xi) - g(x_1 - \xi)| \, d\xi$$

$$\leq \max_{|\xi| \leq r} |g(x_0 - \xi) - g(x_1 - \xi)| \int_{|x| \leq r} |f(\xi)| \, d\xi,$$

and this implies the continuity of $(f * g)(x)$ at $x = x_0$. If $g(x)$ has compact support and is carried in say $|x| \leq r$, then similarly

$$|(f * g)(x_0) - (f * g)(x_1)| \leq \int_{|\xi| \leq r} |g(\xi)| \, |f(x_0 - \xi) - f(x_1 - \xi)| \, d\xi$$

$$\leq \max_{|\xi| \leq r} |g(\xi)| \int_{|\xi| \leq r} |f(x_0 - \xi) - f(x_1 - \xi)| \, d\xi \to 0 \quad \text{as} \quad x_1 \to x_0.$$

The integral here approaches zero as $x_1 \to x_0$ since the ordinary function $f(x)$ is "continuous in the mean".†

By the theorem on differentiating the convolution, we have

$$D^{k+m}(f * g) = D^k f * D^m g.$$

Therefore, if g is not only continuous but also has continuous derivatives up to order say q and f has ordinary derivatives up to order p, then their convolution has continuous derivatives up to order $p + q$.

This result can be formulated using the concept of order of singularity as follows:

If $s(f) \leq 0$ and $c(g) \leq 0$, then

$$c(f * g) \leq s(f) + c(g). \tag{7}$$

Inequality (7) actually holds for any $s(f)$ and $c(g)$ regardless of signs. Indeed, suppose for example that f has compact support, so that

$$f = \sum_{|j| \leq p} D^j f_j(x), \quad s(f) = p \geq 0,$$

where the $f_j(x)$ are ordinary functions with compact support. Suppose further that

$$g = \sum_{|k| \leq m} D^k g_k(x), \quad c(g) = m \geq 0,$$

where the $g_k(x)$ are continuous functions. Then

$$f * g = \sum_{|j| \leq p} \sum_{|k| \leq m} D^{j+k} [f_j(x) * g_k(x)], \tag{8}$$

† See L. Graves, *The Theory of Functions of Real Variables*, McGraw-Hill Book CO., New York, 2nd ed., 1956, Chapt. 9, p. 247. The latter case obviously requires merely the boundedness of $g(x)$.

with the $f_j * g_k$ continuous by the above. Hence, $c(f * g) \leq p + m$. Now let $c(g) < 0$. Instead of (8), we can write

$$f * g = \sum_{|j| \leq p} D^j [f_j * g].$$

Therefore, by what has been proved,

$$c(f * g) \leq p + c(f_j * g) \leq p + c(g).$$

If $s(f) < 0$ and $c(g) = m \geq 0$, then instead of (8) we can write

$$f * g = \sum_{|k| \leq m} f * D^k g_k(x) = \sum_{|k| \leq m} D^k [f * g_k]$$

and so
$$c(f * g) \leq m + s(f).$$

Thus, inequality (7) is valid in all cases.

COROLLARY 1. *The convolution* $f * g$ *is infinitely differentiable (in the customary sense) if at least one of the functionals* f *or* g *is infinitely differentiable.*
Proof. Suppose that f has compact support. If $g = g(x)$ is infinitely differentiable, then by (7)

$$c(f * g) \leq s(f) + c(g) = -\infty,$$

since $s(f) < \infty$ and $c(g) = -\infty$. If $f = f(x)$ is infinitely differentiable formula (7) does not work directly.† Representing $g(x)$ as the limit of functionals g_v with compact support whose supports recede from the origin, we have

$$f * g = \lim_{v \to \infty} (f * g_v), \quad c(f * g_v) \leq s(f) + c(g_v) = -\infty.$$

But since in any bounded region, the functions $f * g_v$ cease changing beginning with a certain subscript, it follows from this that $f * g$ is infinitely differentiable everywhere.

COROLLARY 2. *Every generalized function* f *can be represented as the limit (in* K') *of a sequence of ordinary functions* $f_1(x)$, $f_2(x)$, ... *which are infinitely differentiable and have compact support.*
Proof. Let $h_1, h_2, \ldots$ be a delta-defining sequence of infinitely differentiable functions with support in the ball $|x| \leq 1$ (Sec. 2.2.2). Form the functions $g_v = h_v * f$. Since $h_v \to \delta$ (in K'), by the theorem on the continuity of the convolution (Sec. 2.5.9), we have $g_v = h_v * f \to \delta * f = f$

---

† Since it is possible that $c(g) = +\infty$.

By Corollary 1, the $g_\nu(x)$ are infinitely differentiable. Now let $e_\nu(x)$ be a test function equaling 1 in $|x| \leq \nu$ ($\nu = 1, 2, \ldots$). Set $f_\nu = e_\nu g_\nu$. The functions $f_\nu(x)$ are also infinitely differentiable and have compact support. Finally, for any test function $\varphi(x)$, we have $\varphi = e_\nu \varphi$ beginning with sufficiently large ν. Therefore,

$$\lim_{\nu \to \infty} (f_\nu, \varphi) = \lim_{\nu \to \infty} (e_\nu g_\nu, \varphi) = \lim_{\nu \to \infty} (g_\nu, e_\nu \varphi) = \lim_{\nu \to \infty} (g_\nu, \varphi) = (f, \varphi),$$

or $f_\nu \to f$ in K', as asserted.

REMARK. If in Corollary 2, f is assumed to be a functional with support *inside* a bounded region G, then the supports of the functions $f_\nu(x)$ may be taken to be in the same region. For suppose any sequence of functions $g_1, g_2, \ldots$ converging to f has been found. By multiplying them by a test function $e(x)$ equal to 1 on supp f and to 0 outside G, we arrive at a sequence $eg_1, eg_2, \ldots$ also converging to f. At the same time, the supports of $eg_1, eg_2, \ldots$ are in G, as required.

2.6.6. Let us recall a result from Sec. 1.7 to be used in the considerations to follow. The equation

$$\Delta^m \mathscr{E}_n^{2m} = \delta(x) \quad \left(\Delta = \frac{\partial^2}{\partial x_1^2} + \cdots + \frac{\partial^2}{\partial x_n^2} \right)$$

has been shown to have the solution

$$\mathscr{E}_n^{2m}(x) = \begin{cases} a_{n,\,2m} r^{2m-n} & \text{for } n \text{ odd or } n \text{ even and } 2m < n; \\ b_{n,\,2m} r^{2m-n} \log r & \text{for } n \text{ even and } 2m \geq n. \end{cases}$$

$\mathscr{E}_n^{2m}$ is clearly an ordinary function having ordinary derivatives up to order $2m - 1$ and so

$$s(\mathscr{E}_n^{2m}) = 1 - 2m. \tag{9}$$

Let us show that

$$c(\mathscr{E}_n^{2m}) = 1 - 2m + n. \tag{10}$$

If $2m > n$, then $\mathscr{E}_n^{2m}(x)$ is continuous and has continuous derivatives up to order $2m - n - 1$, so that in this case $c(\mathscr{E}_n^{2m}) = 1 - 2m + n$. If $2m = n$, then $\mathscr{E}_n^n = a \log r$ and is no longer continuous at $x = 0$. However, it is equal to a sum of first derivatives of continuous functions. $\left(\text{Specifically, } \log r + b = c\Delta \, (r^2 \log r) = c \sum \frac{\partial}{\partial x_k} \left(\frac{\partial}{\partial x_k} r^2 \log r \right) \text{ and the functions } \frac{\partial}{\partial x_k} r^2 \log r \text{ are continuous.} \right)$ Therefore, $c(\mathscr{E}_n^n) = 1 = n - 2m$

$+ 1$. Finally, when $2m < n$, $\mathscr{E}_n^{2m} = cr^{2m-n}$ can be written as

$$a_1 \Delta^{\frac{n-2m+1}{2}} r \qquad \text{for } n \text{ odd,}$$

$$a_2 \Delta^{\frac{n-2m+2}{2}} r^2 \log r \qquad \text{for } n \text{ even.}$$

In every case, we also find that

$$c(r^{2m-n}) = 1 - 2m + n.$$

The next theorems are aimed at providing estimates for the order of singularity of a functional in terms of that of its derivatives.

THEOREM 1. *If f has compact support, then for any m*

$$s(f) \leq s(\Delta^m f) - 2m + 1, \tag{11}$$

$$c(f) \leq s(\Delta^m f) - 2m + 1 + n. \tag{12}$$

Proof. We apply estimates (5) and (7) to the relation

$$\mathscr{E}_n^{2m} * \Delta^m f = \Delta^m \mathscr{E}_n^{2m} * f = \delta * f = f.$$

This yields

$$s(f) \leq s(\mathscr{E}_n^{2m}) + s(\Delta^m f)$$

and

$$c(f) \leq c(\mathscr{E}_n^{2m}) + s(\Delta^m f).$$

On using (9) and (10), we arrive at the required inequalities (11) and (12).

COROLLARY 1. *For any functional f with compact support,*

$$c(f) \leq s(f) + n + 1. \tag{13}$$

This result is obtained from the estimate (12) upon the replacement of $s(\Delta^m f)$ by the quantity $s(f) + 2m$ which is at least as large.†

COROLLARY 2. *If $s(f) = -\infty$, then $c(f) = -\infty$. In other words, if all the derivatives of a functional f are ordinary functions, then all of them are also continuous ordinary derivatives.*

2.6.7. We now derive similar estimates for arbitrary functionals (without compact support) considered in a bounded region G.

---

† Inequality (13) may be strengthened to $c(f) \leq s(f) + n$ and no further refinement is possible. See G. E. Shilov, *On the theory of generalized functions.* Izv. Vysch. Uchebn. Zaved. Mathematika, Kazan. Univ., No. 5(54), 1966, pp. 124—128 (In Russian).

9*

As we know, every functional f has a finite order $s(f)$ in a bounded region G. Let G' be a region contained in G together with its closure, and let $e(x)$ be a test function with support in G equaling 1 on G'. Then ef has compact support and by Corollary 1,

$$c(ef) \leqq s(ef) + n + 1.$$

In G' we have $ef = f$. Therefore in G',

$$c(f) \leqq s(f) + n + 1.$$

Thus, Corollaries 1 and 2 continue to hold for arbitrary functionals in a region G' contained strictly in G.

The inequalities (11) and (12) of course remain valid for ef. But in themselves, they are of no further great interest since $\Delta^m(ef)$ can not only be expressed in terms of the $2m$-th derivatives of f but even in terms of intermediary derivatives and f itself. Nevertheless, the result is useful and we formulate it as a lemma:

LEMMA 1. *If* $s(D^j f) \leqq k$ *for all* $|j| \leqq 2m$, *then*

$$s(f) \leqq k - 2m + 1.$$

Indeed, by (11),

$$s(ef) \leqq s(\Delta^m ef) - 2m + 1 \leqq k - 2m + 1.$$

Since for any test function $\varphi(x)$ with support strictly in G, $(f, \varphi) = (ef, \varphi)$ for some $e(x)$, the functional f itself has order of singularity in G not exceeding $k - 2m + 1$, as asserted.

In order to express $s(f)$ solely in terms of $s(D^{2m}f)$, one can make use of the following lemma:

LEMMA 2. *If all the first-order derivatives of f in a bounded region G have order $s(f)$ not exceeding k, then f itself has order $s(f)$ in G not exceeding k.*

Proof. In any event, the order of f in G is finite and we can write

$$s(f) \leqq k + 2m - 1$$

for some m. By assumption, $s(Df) \leqq k$. Hence, $s(D^2 f) \leqq k + 1, ...,$ $s(D^{2m}f) \leqq k + 2m - 1$. We see that the order of f and its derivatives up to order $2m$ does not exceed $k + 2m - 1$. By Lemma 1, $s(f) \leqq k$, as asserted.

THEOREM **2.** *If all the 2m-th derivatives of a functional f in a bounded region G have order* $s(D^{2m}f) \leqq k$, *then*

$$s(f) \leqq k - 2m + 1.$$

Proof. Applying Lemma 2 sequentially, we find that $s(D^{2m-1}) \leqq k, \ldots,$ $s(D^j f) \leqq k$ for all $|j|$ from $2m$ down to 0. It now merely remains to refer to Lemma 1.

Problems

1. If $c(\partial f/\partial x_k) \leqq 0$ $(k = 1, \ldots, n)$, then $c(f) \leqq -1$.

Hint. See Theorem 2 of Sec. 1.7.4.

2. If $c(\partial f/\partial x_k) \leqq -m$ $(k = 1, \ldots, n)$, then $c(f) \leqq -m - 1$.

Hint. Analyze the structure of the solution to Prob. 1.

Note. There are functions $f(x_1, x_2)$ for which $\partial^2 f/\partial x_1^2$ and $\partial^2 f/\partial x_2^2$ are continuous while $\partial^2 f/\partial x_1 \, \partial x_2 = g(x_1, x_2)$ is not continuous (it becomes infinite at a point).† We have

$$c\left(\frac{\partial g}{\partial x_1}\right) = c\left(\frac{\partial}{\partial x_2} \frac{\partial^2 f}{\partial x_1^2}\right) = 1, \quad c\left(\frac{\partial g}{\partial x_2}\right) = c\left(\frac{\partial}{\partial x_1} \frac{\partial^2 f}{\partial x_2^2}\right) = 1,$$

whereas $c(g) > 0$. Thus, the assertion of Prob. 2 is no longer valid for negative values of m.

3. If $s(\partial f/\partial x_k) = 0$ $(k = 1, \ldots, n)$, then $s(f) \leqq -1$.

Hint. See Lemma 2 above and the definition of $s(f)$.

4. If $s(\partial f/\partial x_k) \leqq -m$ $(k = 1, \ldots, n)$, then $s(f) \leqq -m - 1$.

Hint. The same as for Prob. 3.

Note. It is possible to construct a function $f(x_1, x_2)$ for which $\partial^2 f/\partial x_1^2$ and $\partial^2 f/\partial x_2^2$ are locally integrable but $\partial^2 f/\partial x_1 \, \partial x_2 = g(x_1, x_2)$ is not.‡ Show in the same way as in the note of Prob. 2 that the assertion of Prob. 4 is not valid for negative values of m.

---

† B. S. Mityagin, *On the mixed second derivative*, Dokl. Akad. Nauk SSR, Vol. 123, No. 4 (1958), pp. 606–608 (In Russian).

‡ D. Ornstein, *A non-inequality for differential operators in the L_1 norm*. Arch. Rational Mech. and Analysis, Vol. 11, No. 1 (1962), pp. 40–49.

5. Is it possible to represent an ordinary function f in the form

$$f = f_0 + \sum_{k=1}^{n} \frac{\partial f_k(x)}{\partial x_k},$$

with $s(f_0) \leqq -1, ..., s(f_n) \leqq -1$? (In other words, does statement 3 of Sec. 2.6.3 hold when $m = -1$?)

Answer unknown.

6. It is possible to represent a continuous function f by

$$f = f_0 + \sum_{k=1}^{n} \frac{\partial f_k(x)}{\partial x_k},$$

where $c(f_0) \leqq -1, ..., c(f_n) \leqq -1$?

Answer unknown.

2.7. Fourier Transforms of Generalized Functions

2.7.1. Let $\varphi(x)$ $(-\infty < x < \infty)$ be a complex-valued test function Consider its Fourier transform

$$\psi(\sigma) \equiv F[\varphi(x)] \equiv \int_{-\infty}^{\infty} \varphi(x) e^{i\sigma x} dx. \tag{1}$$

Since $\varphi(x)$ has compact support, the integral (1) is over a finite interval say $-a \leq x \leq a$. Therefore the function $\psi(\sigma)$ can be extended to complex values of $s = \sigma + i\tau$ and

$$\psi(\sigma + i\tau) = \int_{-a}^{a} \varphi(x) e^{isx} dx = \int_{-a}^{a} \varphi(x) e^{i\sigma x} e^{-\tau x} dx. \tag{2}$$

The Fourier transform $\psi(s)$ of each $\varphi(x)$ in K vanishing for $|x| > a$ is an entire function of $s = \sigma + i\tau$ satisfying the inequality

$$\left| s^q \psi(s) \right| \leq C_q e^{a|\tau|} \tag{3}$$

for $q = 0, 1, 2, ...$

Indeed, the integral (2) may be differentiated with respect to the complex parameter s and so $\psi(s)$ is an entire function. On the other hand, taking the Fourier transform of $\varphi'(x)$ is equivalent to multiplying $\psi(s)$ by $-is$. We have

$$\int_{-\infty}^{\infty} \varphi'(x) e^{isx} dx = \varphi(x) e^{isx} \Big|_{-\infty}^{\infty} - \int_{-\infty}^{\infty} is\varphi(x) e^{isx} dx = -is\psi(s).$$

Further differentiation leads to

$$F[\varphi^{(q)}(x)] = (-is)^q F[\varphi(x)] \tag{4}$$

for $q = 1, 2, \ldots$ More generally,

$$F\left[P\left(\frac{d}{dx}\right)\varphi(x)\right] = P(-is)\,F[\varphi(x)], \tag{5}$$

where P is a polynomial with constant coefficients. At the same time, we have the estimate

$$|s|^q|\psi(s)| = \left| \int_{-a}^{a} \varphi^{(q)}(x)\,e^{isx}\,dx \right| \leqq C_q\,e^{a|\tau|},$$

thus completing the proof of the theorem.

The converse is also true:

Every entire function $\psi(s)$ satisfying (3) for each q is the Fourier transform of some infinitely differentiable function $\varphi(x)$ vanishing for $|x| > a$.

We naturally take $\varphi(x)$ to be the inverse Fourier transform

$$\varphi(x) = \frac{1}{2\pi}\int_{-\infty}^{\infty}\psi(\sigma)\,e^{-i\sigma x}\,d\sigma. \tag{6}$$

Since $\psi(\sigma)$, as is apparent from inequality (3) for $\tau = 0$, decreases faster than any negative power of $|\sigma|$ as $|\sigma| \to \infty$, the integral (6) is absolutely and uniformly convergent. It still remains absolutely and uniformly convergent when we differentiate under the integral sign.

Therefore $\varphi(x)$ has derivatives of all orders and

$$\varphi^{(q)}(x) = \frac{1}{2\pi}\int_{-\infty}^{\infty}(-i\sigma)^q\psi(\sigma)\,e^{-i\sigma x}\,d\sigma.$$

To establish that $\varphi(x)$ has compact support, we first observe that the integration over the real axis in the expression (6) can be replaced by integration over a line parallel to the real axis:

$$\frac{1}{2\pi}\int_{-\infty}^{\infty}\psi(\sigma)\,e^{-i\sigma x}\,d\sigma = \frac{1}{2\pi}\int_{-\infty}^{\infty}\psi(\sigma + i\tau)\,e^{-i(\sigma + i\tau)x}\,d\sigma. \tag{7}$$

For, by Cauchy's theorem, the integral of the analytic function $\psi(s)$, $s = \sigma + i\tau$, around the rectangle with vertices at $(\pm A, 0)$ and $(\pm A + i\tau)$ is equal to zero. By virtue of (3), the integrals along the vertical segments tend to zero as $A \to \infty$, and so (7) results. We can now write

$$\varphi(x) = \frac{1}{2\pi} e^{\tau x} \int_{-\infty}^{\infty} \psi(\sigma + i\tau) e^{-i\sigma x} \, d\sigma.$$

Let $|x| > a$. We prescribe a number $t > 0$ and determine τ by the condition $\tau x = -t|x|$. Applying inequality (3) with $q = 0$ and $q = 2$, we obtain

$$|\psi(s)| \leq e^{a|\tau|} \min\left\{C_0, \frac{C_2}{|s|^2}\right\} \leq C\frac{e^{a|\tau|}}{1 + |s|^2} \leq C\frac{e^{a|\tau|}}{1 + \sigma^2}.$$

Thus

$$|\varphi(x)| \leq \frac{1}{2\pi} e^{\tau x} \int_{-\infty}^{\infty} C\frac{e^{a|\tau|}}{1 + \sigma^2} \, d\sigma = C' e^{-t|x| + at} = C' e^{t(a - |x|)}.$$

Since C' is independent of t, on letting $t \to \infty$, we find that $\varphi(x) = 0$. Thus, $\varphi(x)$ vanishes for $|x| > a$. Hence $\varphi(x)$ satisfies all of the stated conditions. By the well-known Fourier integral theorem,[†] its Fourier transform coincides with $\psi(\sigma)$, as required.

2.7.2. The study of the Fourier transform of functions in K begun above leads us to introduce the following natural definition. The *space* Z is the set of all entire functions $\psi(s)$ such that

$$|s^q| |\psi(s)| \leq C_q e^{a|\tau|} \quad (q = 0, 1, 2, \ldots) \tag{8}$$

(for some constants a and C_q depending on ψ), with the linear operations of addition and scalar multiplication being defined in the obvious way.

As the preceding subsection shows, *the Fourier transformation establishes a one-to-one mapping between K and Z*. It is clear that the mapping preserves the specified linear operations and is thereby an isomorphism.

Multiplication by certain functions $G(s)$ is also admissible in Z and is distributive over addition. The admissible $G(s)$ by which one can multiply

† See S. Bochner, *Lectures on Fourier Integrals*, Princeton University Press, 1959, Chapt. III.

within the class Z are for instance entire functions satisfying the estimate

$$|G(s)| \leq C(1 + |s|)^m e^{b|\tau|} \tag{9}$$

(and no others; see Prob. 5 of Sec. 2.8).

A special role in Z is played by the *involutory* transformation mapping $\psi(s)$ into $\psi*(s) \equiv \overline{\psi(\bar{s})}$. In addition to this, all linear operations in K may be transferred to Z by using the existing isomorphism.

To each linear operation defined in K there corresponds a certain "dual" linear operation in Z. Thus, for example, to differentiation in K corresponds multiplication by $-is$ in Z by virtue of formula (4). On the other hand, the formula

$$\frac{d\psi(s)}{ds} = \int_{-\infty}^{\infty} ix\, e^{isx} \varphi(x)\, dx$$

shows that the dual of multiplication by ix in K is differentiation in Z. If we iterate the operation, we arrive at the following formula valid for $q = 0, 1, 2, \ldots$:

$$\frac{d^q}{ds^q} F[\varphi] = F[(ix)^q \varphi(x)]. \tag{10}$$

The existence of the right-hand side implies the existence of the left-hand side. Therefore, a function $\psi(s)$ in Z may be differentiated indefinitely without leaving the space. One can also write a slightly more general formula,

$$P\left(\frac{d}{ds}\right) F[\varphi] = F[P(ix)\, \varphi(x)], \tag{11}$$

in which P is an arbitrary polynomial with constant coefficients.

A reflection $\varphi(x) \to \varphi(-x)$ has the reflection $\psi(s) \to \psi(-s)$ as its dual in Z since

$$\int_{-\infty}^{\infty} \varphi(-x)\, e^{ixs}\, dx = \int_{-\infty}^{\infty} \varphi(x)\, e^{-ixs}\, dx.$$

A translation $\varphi(x) \to \varphi(x - h)$ in K has multiplication by e^{ish} as its dual in Z. For we have

$$F[\varphi(x - h)] = \int_{-\infty}^{\infty} e^{isx} \varphi(x - h)\, dx$$

$$= \int_{-\infty}^{\infty} e^{is(y + h)} \varphi(y)\, dy = e^{ish} F[\varphi(x)].$$

Conversely, multiplication by e^{ixh} in K (with h arbitrary and possibly complex) has a translation in Z as its dual. In fact

$$F[e^{ixh}\varphi(x)] = \int_{-\infty}^{\infty} e^{isx}\, e^{ihx}\, \varphi(x)\, dx = \int_{-\infty}^{\infty} e^{i(s + h)x} \varphi(x)\, dx = \psi(s + h).$$

We see particularly that every conceivable (real and complex) translate of $\psi(s)$ belongs to Z.

Convergence may be induced in Z by asserting that a sequence $\psi_\nu(s)$ has the limit zero if its image sequence $\varphi_\nu(x)$ has the limit zero in K. However, convergence in Z can be described intrinsically. Namely, a sequence $\psi_\nu(s)$ converges to zero in Z if first the inequalities

$$|s^a \varphi_\nu(s)| \leqq C_q\, e^{a|\tau|}$$

hold for some constants C_q and a not depending on ν, and if second $\psi_\nu(s)$ approaches zero uniformly in each interval of the line (see Prob. 1).

All of the operations specified above (addition, multiplication by a number or a function, involution, translation, differentiation) are continuous in Z under this convergence.

The Taylor series expansion

$$\sum_{q=0}^{\infty} \psi^{(q)}(s)\, \frac{h^q}{q!} = \psi(s + h)$$

with h any fixed (complex) number holds in the sense of the indicated convergence. This is a consequence of the fact that the dual formula

$$\sum_{q=0}^{\infty} (ix)^q\, \frac{h^q}{q!}\, \varphi(x) = e^{ixh}\varphi(x)$$

s valid under convergence in K.

2.7.3. Let us return for a moment to functionals on the (complex) space K.

With any ordinary complex-valued function $f(x)$, we associate a functional $f \in K'$ by the following rule:

$$(f, \varphi) = \cdot \int_{-\infty}^{\infty} \overline{f(x)}\, \varphi(x)\, dx.$$

This choice of rule is explained by the fact that an important part is played in Fourier transform theory by Parseval's equality. It has the form

$$\int_{-\infty}^{\infty} \overline{f(x)}\, \varphi(x)\, dx = \frac{1}{2\pi} \int_{-\infty}^{\infty} \overline{g(\sigma)}\, \psi(\sigma)\, d\sigma,$$

where $g(\sigma)$ and $\psi(\sigma)$ are the respective Fourier transforms of $f(x)$ and $\varphi(x)$ (if for instance $f(x)$ and $\varphi(x)$ are square integrable).†

Similarly to K, one can construct continuous linear functionals (generalized functions) on Z.

Every (complex-valued) function $g(\sigma)$ defined for $-\infty < \sigma < \infty$, locally integrable, and increasing no faster than a power of $|\sigma|$ at infinity, defines a functional on Z through the formula‡

$$(g, \psi) = \int_{-\infty}^{\infty} \overline{g(\sigma)}\, \psi(\sigma)\, d\sigma. \tag{12}$$

Functionals such as (12) will be called *regular* and all others *singular*. Among the singular functionals are the delta-function

$$(\delta(s), \psi(s)) = \psi(0),$$

and its translates (possibly complex)

$$(\delta(s - s_0), \psi(s)) = \psi(s_0).$$

The collection of all generalized functions on Z will be denoted by Z'. One can also perform operations on the generalized functions on Z analogous to those introduced for generalized functions on K. It is clear that nothing new is involved in the definitions of the linear operations of addition and multiplication by a scalar and taking limits.

† See S. Bochner, *ibid.*, Chapt. VIII, Sec. 40.
‡ Of course, any one of the four functionals analogous to (6)–(9) of Sec. 3 could be assigned here to $g(\sigma)$.

In particular, convergence in Z' is defined as follows: $g_\nu \to g$ in Z' if

$$(g_\nu, \psi) \to (g, \psi)$$

for each test function $\psi \in Z$. Multiplication by a function $G(s)$ is defined formally by

$$(G(s) g, \psi) = (g, G^*(s) \psi), \tag{13}$$

where $G^*(s)$ stands for $\overline{G(\bar{s})}$. It is admissible for any entire function which satisfies inequality (9).

Multiplication by each such function $G(s)$ is distributive over addition and continuous in the sense that if $g_\nu(s) \to g(s)$ in Z', then $G(s)g_\nu(s) \to G(s)g(s)$ in Z'. Indeed,

$$(Gg_\nu, \psi) = (g_\nu, G^*\psi) \to (g, G^*\psi) = (Gg, \psi). \tag{14}$$

The derivative of a functional $g \in Z'$ is defined by

$$(g', \psi(s)) = -(g, \psi'(s)).$$

As in K', differentiation is continuous under convergence in Z'. If $g_\nu \to g$ in Z', then for any $\psi \in Z$, we can write

$$(g'_v, \psi) = -(g_\nu, \psi') \to -(g, \psi') = (g', \psi)$$

and so $g'_\nu \to g'$ in Z'.

Similarly to K', each generalized function in Z' can be differentiated indefinitely.

But in contrast to the functionals in K', the generalized functions $g \in Z'$ are not only infinitely differentiable but also *analytic*. This means that for each $g \in Z'$, we can write

$$\sum_{q=0}^{\infty} g^{(q)}(s) \frac{h^q}{q!} = g(s + h), \tag{15}$$

in which the series on the left-hand side is convergent in Z' and $g(s + h)$ is a translate of $g(s)$ by the amount h defined by

$$(g(s + h), \psi(s)) = (g(s), \psi(s - h)).$$

To this end, we note that for any $\psi(s) \in Z$,

$$\left(g^{(q)}(s) \frac{h^q}{q!}, \psi(s) \right) = \left(g(s), \frac{(-h)^q}{q!} \psi^{(q)}(s) \right).$$

The series $\sum \dfrac{(-h)^q}{q!}\, \psi^{(q)}(s)$ converges to $\psi(s-h)$ in the sense of convergence in Z, as we have already pointed out. Thus

$$\left(\sum g^{(q)}(s)\,\frac{h^q}{q!}, \psi(s)\right) = (g(s),\, \psi(s-h)) = (g(s+h),\, \psi(s)),$$

as asserted.

In particular, the expansion

$$\delta(s+h) = \sum_{q=0}^{\infty} \delta^{(q)}(s)\,\frac{h^q}{q!} \tag{16}$$

is valid for any (complex) h.

2.7.4. Since a one-to-one mapping exists between K and Z which preserves linear operations and convergence, an analogous mapping can be established between the continuous linear functionals defined on these spaces. We shall do this so that for a functional corresponding to an absolutely integrable function it turns into a correspondence between the function and its classical Fourier transform.

Let $f(x)$ be an absolutely integrable function and $g(\sigma)$ its Fourier transform. Then for any test function $\varphi(x)$ with Fourier transform $\psi(\sigma)$, we have

$$(f, \varphi) = \int_{-\infty}^{\infty} \overline{f(x)}\, \varphi(x)\, dx = \frac{1}{2\pi}\int_{-\infty}^{\infty} \overline{f(x)}\left\{\int_{-\infty}^{\infty}\psi(\sigma)\, e^{-ix\sigma}\, d\sigma\right\} dx$$

$$= \frac{1}{2\pi}\int_{-\infty}^{\infty}\psi(\sigma)\left\{\int_{-\infty}^{\infty}\overline{f(x)}\, e^{ix\sigma}\, dx\right\} d\sigma = \frac{1}{2\pi}\int_{-\infty}^{\infty}\overline{g(\sigma)}\,\psi(\sigma)\, d\sigma.$$

This relation is generally called *Parseval's equality*. Parseval's equality shows that as a generalized function, $g(\sigma)$ is operating on a test function $\psi(s)$ according to

$$(g, \psi) = 2\pi(f, \varphi). \tag{17}$$

In this form, it serves to define a generalized function g on Z for any generalized function f prescribed on K. We shall call the functional g *defined* by (17) the *Fourier transform of f* and we shall denote it by $F[f]$.

We wish to emphasize that the functional $F[f]$ operates in the space Z dual to K.

The Fourier transform $F[f]$ mapping K' into Z' is a continuous linear operator. Let us prove this. For any complex α_1 and α_2 and any f_1 and $f_2 \in K'$, we have

$$
\begin{aligned}
(F[\alpha_1 f_1 + \alpha_2 f_2], F[\varphi]) &= 2\pi(\alpha_1 f_1 + \alpha_2 f_2, \varphi) \\
&= 2\pi\alpha_1(f_1, \varphi) + 2\pi\alpha_2(f_2, \varphi) \\
&= \alpha_1(F[f_1], F[\varphi]) + \alpha_2(F[f_2], F[\varphi]) \\
&= (\alpha_1 F[f_1] + \alpha_2 F[f_2], F[\varphi]).
\end{aligned}
$$

Moreover, if $f_\nu \to f$ in K', then

$$
(F[f_\nu], F[\varphi]) = 2\pi(f_\nu, \varphi) \to 2\pi(f, \varphi) = (F[f], F[\varphi]).
$$

The formulas for differentiating the usual Fourier transform continue to hold for the Fourier transform of generalized functions namely

$$
P\left(\frac{d}{ds}\right) F[f] = F[P(ix)f], \tag{18}
$$

$$
F\left[P\left(\frac{d}{dx}\right)f\right] = P(-is)F[f]. \tag{19}
$$

Thus, multiplication by ix in K' has differentiation as its dual in Z', and differentiation in K' has multiplication by $-is$ as its dual in Z'.

To this end, it is sufficient to consider the case where $P(d/dx) \equiv d/dx$. We have

$$
(F[ixf], F[\varphi]) = 2\pi(ixf, \varphi) = 2\pi(f, -ix\varphi)
$$

$$
= (F[f], F[-ix\varphi]) = \left(F[f], -\frac{d}{ds}F[\varphi]\right) = \left(\frac{d}{ds}F[f], F[\varphi]\right),
$$

which yields (18). Equation (19) is established similarly.

The inverse operator F^{-1} is defined on Z' and transforms a functional g into a functional f by the same formula (17) (read from right to left), so that

$$
\left.
\begin{aligned}
F^{-1}[F[f]] &= f, \quad F[F^{-1}[g]] = g, \\
(F^{-1}[g], \varphi) &= \frac{1}{2\pi}(g, F[\varphi]).
\end{aligned}
\right\} \tag{20}
$$

We point out two simple and frequently used rules.

(a) *Fourier transform of* $f(-x)$. Let $f(x) \in K'$. The functional $f(-x)$ is defined by

$$(f(-x), \varphi(x)) = (f(x), \varphi(-x)).$$

Let $F[f(x)] = g(\sigma)$ and $F[f(-x)] = g_1(\sigma)$. According to Sec. 2.7.2, we have

$$(g_1(\sigma), \psi(\sigma)) = 2\pi(f(-x), \varphi(x)) = 2\pi(f(x), \varphi(-x))$$
$$= (g(\sigma), \psi(-\sigma)) = (g(-\sigma), \psi(\sigma)).$$

Hence, $g_1(\sigma) = g(-\sigma)$. In other words, to a reflection about the origin on the x-axis there corresponds a reflection about the origin on the σ-axis.

As a corollary, we conclude that an *even* generalized function has an *even* Fourier transform and an *odd* generalized function an *odd* Fourier transform.

(b) *Iterated Fourier transform.* If $\varphi(x)$ is a test function, then as we know

$$F^{-1}F[\varphi] = \varphi(x).$$

However, the direct Fourier transform may be applied to $\psi = F[\varphi]$ once again. We obtain

$$F[\psi] = FF[\varphi] = 2\pi\varphi(-x),$$

since the operation F^{-1} differs from F by the factor $1/2\pi$ and the sign of the independent variable.

We now define the iterated Fourier transform of a functional $f \in K'$ (and the direct Fourier transform of the functional $Ff = g \in Z'$) by the analogue of formula (17),

$$(FFf, FF\varphi) \equiv (F(g), F(\psi)) = 2\pi(g, \psi).$$

To establish the connection between the functionals f and FFf, we substitute $g = F[f]$ and $\psi = F[\varphi]$ in the last expression. This yields

$$(FF[f], FF[\varphi]) = 2\pi(F[f], F[\varphi]) = (2\pi)^2 (f, \varphi(x)).$$

Now $2\pi\varphi(-x)$ may be substituted for $FF[\varphi]$ on the left-hand side. Canceling 2π and replacing x by $-x$, we arrive at

$$(FF[f], \varphi(x)) = (2\pi f, \varphi(-x)),$$

or

$$FF[f] = 2\pi f(-x). \tag{21}$$

Similarly,

$$F^{-1}F^{-1}[g(\sigma)] = \frac{1}{2\pi} g(-\sigma).$$

Suppose for example that the direct Fourier transform of a function $f(x)$ is known to be $g(\sigma)$, i.e., $g(\sigma) = F[f(x)]$. We wish to find the inverse Fourier transform of $f(x)$. We have

$$F^{-1}[f(x)] = F^{-1}[F^{-1}[g]] = \frac{1}{2\pi} g(-\sigma). \tag{22}$$

Similarly if it is known that $g(\sigma) = F^{-1}[f(x)]$ then

$$F[f(x)] = FF[g(\sigma)] = 2\pi g(-\sigma). \tag{23}$$

2.7.5. EXAMPLES. 1°. We determine $F[\delta]$. By definition,

$$(F[\delta], F[\varphi]) = 2\pi(\delta, \varphi) = 2\pi\varphi(0) = \int_{-\infty}^{\infty} \psi(\sigma)\, d\sigma = (1, \psi),$$

and so

$$F[\delta] = 1, \quad F^{-1}[1] = \delta. \tag{24}$$

2°. We determine $F[1]$.

$$(F[1], F[\varphi]) = 2\pi(1, \varphi) = 2\pi \int_{-\infty}^{\infty} \varphi(x)\, dx$$

$$= 2\pi \int_{-\infty}^{\infty} \varphi(x)\, e^{ix0}\, dx = 2\pi\psi(0) = 2\pi(\delta, \psi).$$

Hence,

$$F[1] = 2\pi\delta, \quad F^{-1}[\delta] = \frac{1}{2\pi}. \tag{25}$$

This result could have been obtained directly from (24) by applying formula (21) for the iterated Fourier transform.

3°. Fourier transform of a polynomial. Using (18) and (19), we find that

$$F[P(x)] = F[P(x) \cdot 1] = 2\pi P\left(-i\frac{d}{ds}\right)\delta(s), \tag{26}$$

$$F\left[P\left(\frac{d}{dx}\right)\delta(x)\right] = P(-is)F[\delta] = P(-is) \cdot 1 = P(-is). \tag{27}$$

In particular,

$$\left.\begin{array}{l} F[\delta^{(2m)}(x)] = (-1)^m s^{2m}, \\ F[\delta^{(2m+1)}(x)] = (-1)^m i s^{2m+1}. \end{array}\right\} \tag{28}$$

4°. Fourier transform of x^{-m}. Let $F[1/x] = g(\sigma)$. Since $1/x$ is an odd generalized function, so is $g(\sigma)$. From the identity $x\dfrac{1}{x} \equiv 1$ it follows by (18) and (25) that

$$F\left[x\frac{1}{x}\right] = -i\frac{d}{d\sigma}F\left[\frac{1}{x}\right] = F[1] = 2\pi\delta(\sigma).$$

Hence

$$g(\sigma) = F\left[\frac{1}{x}\right] = 2\pi i[\theta(\sigma) + C].$$

Since $g(\sigma)$ is an odd function, $C = -\dfrac{1}{2}$ and therefore

$$g(\sigma) = \pi i \operatorname{sgn} \sigma.$$

Now

$$\frac{1}{x^m} = \frac{(-1)^{m-1}}{(m-1)!}\frac{d^{m-1}}{dx^{m-1}}\frac{1}{x}.$$

Hence

$$F\left[\frac{1}{x^m}\right] = \frac{(-1)^{m-1}}{(m-1)!}(-i\sigma)^{m-1}F\left[\frac{1}{x}\right] = \frac{i^m \pi}{(m-1)!}\sigma^{m-1}\operatorname{sgn}\sigma.$$

5°. Fourier transform of e^{bx}. We can use the fact that the series

$$e^{bx} = \sum_{k=0}^{\infty}\frac{b^k x^k}{k!}$$

10   Shilov

is convergent in K'. This permits us to calculate $F[e^{bx}]$ by applying F to the series termwise. We obtain (see (16))

$$F[e^{bx}] = \sum_{k=0}^{\infty} F\left[\frac{b^k x^k}{k!}\right] = 2\pi \sum_{k=0}^{\infty} \frac{b^k}{k!}\left(-i\frac{d}{ds}\right)^k \delta(s) = 2\pi\delta(s - ib). \quad (29)$$

This result easily leads to Fourier transforms for the commonly occurring functions $\sin bx$, $\cos bx$, $\sinh bx$, and $\cosh bx$:

$$F[\sin bx] = F\left[\frac{e^{ibx} - e^{-ibx}}{2i}\right] = -i\pi[\delta(s + b) - \delta(s - b)],$$

$$F[\cos bx] = F\left[\frac{e^{ibx} + e^{-ibx}}{2}\right] = \pi[\delta(s + b) + \delta(s - b)],$$

$$F[\sinh bx] = F\left[\frac{e^{bx} - e^{-bx}}{2}\right] = \pi[\delta(s - ib) - \delta(s + ib)],$$

$$F[\cosh bx] = F\left[\frac{e^{bx} + e^{-bx}}{2}\right] = \pi[\delta(s - ib) + \delta(s + ib)].$$

6°. Fourier transform of the shifted delta-function. We wish to determine $F[\delta(x - h)]$. By definition,

$$(F[\delta(x - h)], F[\varphi]) = 2\pi(\delta(x - h), \varphi(x)) = 2\pi\varphi(h)$$

$$= \int_{-\infty}^{\infty} \psi(\sigma) e^{-i\sigma h} \, d\sigma = (e^{i\sigma h}, \psi),$$

and so

$$F[\delta(x - h)] = e^{i\sigma h}.$$

It is now easy to obtain the formulas

$$F\left[\frac{\delta(x - h) + \delta(x + h)}{2}\right] = \frac{e^{i\sigma h} + e^{-i\sigma h}}{2} = \cos h\sigma,$$

$$F\left[\frac{\delta(x - h) - \delta(x + h)}{2i}\right] = \frac{e^{i\sigma h} - e^{-i\sigma h}}{2i} = \sin h\sigma.$$

Problems

1. To a sequence $\varphi_\nu \in K$ converging to 0 in K there corresponds a sequence $\psi_\nu(\sigma) = F[\varphi_\nu(x)]$ with the following property: $|s^q \psi_\nu(s)| \leq C_q e^{a|\tau|}$

(*a* and C_q independent of *v*). Show that $\psi_v(\sigma) \to 0$ uniformly in each interval of the σ-axis and conversely.

Hint. The direct statement is a consequence of the definition of Fourier transform. To prove the converse, verify that all $\varphi_v(x)$ vanish outside the interval $|x| < a$ and that their derivatives are bounded by fixed constants. Then apply the compactness principle.

2. Find the Fourier transform of the generalized function x_+^λ (Sec. 2.4).

Hint. For $-1 < \operatorname{Re}\lambda < 0$,

$$\int_0^\infty x^\lambda e^{i\sigma x}\,dx = \lim_{\tau\to 0+}\int_0^\infty x^\lambda e^{i(\sigma+i\tau)x}\,dx.$$

Make the substitution $isx = \xi$ in the right-hand integral and apply Cauchy's formula and the definition of the Gamma function. Then use analytic continuation with respect to λ.

Ans.

$$F[x_+^\lambda] = i e^{i\frac{\pi}{2}\lambda}\,\Gamma(\lambda+1)\lim_{\tau\to 0+}(\sigma+i\tau)^{-\lambda-1}$$

$$= i\Gamma(\lambda+1)\left[e^{i\frac{\pi}{2}\lambda}\sigma_+^{-\lambda-1} + e^{-i\frac{\pi}{2}\lambda}\sigma_-^{-\lambda-1}\right].$$

Note. The functional $\lim_{\tau\to 0+}(\sigma+i\tau)^\lambda$ is also denoted by $(\sigma+i0)^\lambda$.

3. Let $f_\lambda(x) = 2^{-\lambda/2}\dfrac{|x|^\lambda}{\Gamma\left(\dfrac{\lambda+1}{2}\right)}$. Show that

$$F[f_\lambda(x)] = \sqrt{2\pi}\,f_{-\lambda-1}(\sigma).$$

4. Find the Fourier transform of $\log x_+$.

Hint. Differentiate the answer to Prob. 2 with respect to λ and set $\lambda = 0$.

Ans.

$$F[\log x_+] = i\left\{\left(F'(1) + i\frac{\pi}{2}\right)(\sigma+i0)^{-1} - (\sigma+i0)^{-1}\log(\sigma+i0)\right\},$$

where $\log(\sigma+i0) = \lim_{\tau\to 0+}(\log(\sigma+i\tau))$.

10*

2.8. Fourier Transforms of Generalized Functions (Continuation)

2.8.1. Our constructions can be carried over to the case of n independent variables almost without change. The Fourier transform of a test function $\varphi(x) = \varphi(x_1, ..., x_n)$ is given by

$$\psi(\sigma) = \psi(\sigma_1, ..., \sigma_n)$$

$$= \int_{-\infty}^{\infty} \cdots \int_{-\infty}^{\infty} \varphi(x_1, ..., x_n)\, e^{i(x_1\sigma_1 + \cdots + x_n\sigma_n)}\, dx_1 \cdots dx_n,$$

or more concisely,

$$\psi(\sigma) = \int_{R_n} \varphi(x)\, e^{i(x, \sigma)}\, dx, \tag{1}$$

where (x, σ) stands for the quantity $x_1\sigma_1 + \cdots + x_n\sigma_n$. Owing to the fact that $\varphi(x)$ has compact support, ψ can be extended to complex values of $s = (s_1, ..., s_n) = (\sigma_1 + i\tau_1, ..., \sigma_n + i\tau_n)$ and we can write

$$\psi(s) = \int_{R_n} \varphi(x)\, e^{i(x, s)}\, dx. \tag{2}$$

$\psi(s)$ is now defined in n-dimensional complex space C_n and is continuous and analytic with respect to each of its arguments $s_1, ..., s_n$. If $\varphi(x)$ vanishes for $|x_k| > a_k$ $(k = 1, ..., n)$, then its Fourier transform $\psi(s)$ can be estimated by

$$\left| s_1^{q_1} \cdots s_n^{q_n} \psi(\sigma_1 + i\tau_1, ..., \sigma_n + i\tau_n) \right| \leq C_q\, e^{a_1|\tau_1| + \cdots + a_n|\tau_n|}. \tag{3}$$

Conversely, every entire function $\psi(s_1, ..., s_n)$ which satisfies (3) is the Fourier transform of a test function $\varphi(x_1, ..., x_n)$ vanishing for $|x_k| > a_k$ $(k = 1, ..., n)$. Namely,

$$\varphi(x_1,...,x_n) = \frac{1}{(2\pi)^n} \int_{-\infty}^{\infty} \cdots \int_{-\infty}^{\infty} \psi(\sigma_1, ..., \sigma_n)\, e^{-i(x_1\sigma_1 + \cdots + x_n\sigma_n)}\, d\sigma_1 \cdots d\sigma_n.$$

The function is infinitely differentiable with respect to each of its arguments. To prove that it has compact support in say x_1 one has to go from σ_1 into the complex plane and apply Cauchy's theorem as in the single variable case.

The following analogues of formulas (5) and (11) of Sec. 2.7 are valid:

$$F\left[P\left(\frac{\partial}{\partial x_1}, \dots, \frac{\partial}{\partial x_n}\right)\varphi(x)\right] = P(-is_1, \dots, -is_n)\, F[\varphi], \qquad (4)$$

$$P\left(\frac{\partial}{\partial s_1}, \dots, \frac{\partial}{\partial s_n}\right)F[\varphi] = F[P(ix_1, \dots, ix_n)\, \varphi(x)]. \qquad (5)$$

Here $P(\cdot)$ is any polynomial in n variables with constant coefficients.

As before, Z (or Z_n) will denote the linear space of all entire functions $\psi(s)$ satisfying inequality (3) with addition and scalar multiplication defined in it in the natural way. The Fourier transform establishes a one-to-one mapping between K_n and Z_n which preserves linear operations. Convergence in Z_n is definable as follows. A sequence $\psi_\nu(s)$ ($\nu = 1, 2, \dots$) *converges to zero* if the sequence of corresponding test functions $\varphi_\nu(x)$ converges to zero in K_n. The convergence can also be described intrinsically. The conditions are that the inequality

$$|s^q \psi_\nu(s)| \leqq C_q\, e^{a_1|\tau_1| + \dots + a_n|\tau_n|} \quad (|q| = 0, 1, 2, \dots)$$

should hold for some constants C_q and a_j not depending on ν and the sequence $\psi_\nu(\sigma)$ should converge uniformly to zero on each bounded set in R_n.

The Fourier transform of a functional f on K_n is defined to be the functional g on Z_n operating according to

$$(g, \psi) = (2\pi)^n (f, \varphi), \qquad (6)$$

where $\psi = F[\varphi]$ is the Fourier transform of $\varphi(x)$. The functional g is linear and continuous and is denoted by $F[f]$. The results of Sec. 2.7 carry over to the present case without any special changes. Formulas (18) and (19) of Sec. 2.7 become

$$P\left(\frac{\partial}{\partial s_1}, \dots, \frac{\partial}{\partial s_n}\right)F[f] = F[P(ix_1, \dots, ix_n)\, f], \qquad (7)$$

$$F\left[P\left(\frac{\partial}{\partial x_1}, \dots, \frac{\partial}{\partial x_n}\right)f\right] = P(-is_1, \dots, -is_n)\, F[f]. \qquad (8)$$

Here, P is a polynomial in n variables with constant coefficients.

The inverse Fourier transform F^{-1} maps Z'_n into K'_n by the formula

$$(F^{-1}[g], \varphi) = \frac{1}{(2\pi)^n} (g, F[\varphi]). \tag{9}$$

If $f(x)$ is a function having a classical Fourier transform, then $F[f]$ is a regular functional corresponding to the Fourier transform of $f(x)$.

The simplest of singular functionals, namely, the delta-function satisfies the relations

$$F[\delta(x)] = 1, \quad F[1] = (2\pi)^n \, \delta(s), \tag{10}$$

which are the analogues of (24) and (25) of Sec. 2.7.

An expression for the *Fourier transform of a polynomial* is

$$F[P(x_1, \ldots, x_n)] = F[P(x_1, \ldots, x_n) \cdot 1] =$$

$$= (2\pi)^n \, P\left(-i \frac{\partial}{\partial s_1}, \ldots, -i \frac{\partial}{\partial s_n} \right) \delta(s) \tag{11}$$

obtained by combining (8) and (10).

2.8.2. We next point out a formula for the Fourier transform of the "rotation" of a generalized function. Let $Ux = y$ be a rotation in R_n (an orthogonal transformation with determinant 1). Each test function $\varphi(x)$ is carried into a test function $\varphi(Ux)$ by the rotation. The Fourier transform $\psi_U (\sigma)$ of $\varphi(Ux)$ is given by

$$\psi_U(\sigma) = \int_{R_n} \varphi(Ux) \, e^{i(x,\sigma)} \, dx = \int_{R_n} \varphi(y) \, e^{i(U^{-1}y, \sigma)} \, dy$$

$$= \int_{R_n} \varphi(y) \, e^{i(y, U\sigma)} \, dy = \psi(U\sigma) \tag{12}$$

and is thus equal to the rotation of the Fourier transform of the initial function.

Let $f(x)$ be an ordinary function and $f(Ux)$ its rotation. The functional corresponding to $f(Ux)$ can be written as

$$(f(Ux), \varphi(x)) = \int_{R_n} \overline{f(Ux)} \, \varphi(x) \, dx$$

$$= \int_{R_n} \overline{f(y)} \, \varphi(U^{-1}y) \, dy = (f(x), \varphi(U^{-1}x)).$$

This motivates us to *define* the rotation of any generalized function f by

$$(f(Ux), \varphi(x)) = (f(x), \varphi(U^{-1}x)). \tag{13}$$

Let us find the Fourier transform of $f(Ux)$. Set $F[\varphi(x)] = \psi(\sigma)$ and $F[f(x)] = g(\sigma)$. We have

$$(F[f(Ux)], \psi(\sigma)) = (2\pi)^n (f(Ux), \varphi(x)) = (2\pi)^n (f(x), \varphi(U^{-1}x)) =$$
$$= (F[f(x)], F[\varphi(U^{-1}x)]) = (g(\sigma), \psi(U^{-1}\sigma)) = (g(U\sigma), \psi(\sigma)).$$

This yields the formula

$$F[f(Ux)] = g(U\sigma), \tag{14}$$

thus showing that the Fourier transform of the rotation of a generalized function is the rotation of its Fourier transform. As a corollary, we obtain the following:

The Fourier transform of a spherically symmetric generalized function is a spherically symmetric generalized function.

2.8.3. *Fourier transform of a functional with compact support.* First let the functional f correspond to an ordinary function $f(x)$ with support in the block $B = \{x : |x_j| \leq a_j\}$. Then the customary Fourier transform of $f(x)$,

$$g(\sigma) = \int_B f(x) e^{i(\sigma, x)} \, dx,$$

can be continued analytically into complex s-space, $s = \sigma + i\tau = (\sigma_1 + i\tau_1, \ldots, \sigma_n + i\tau_n)$, by the formula

$$g(s) = \int_B f(x) e^{i(s, x)} \, dx. \tag{15}$$

The function $g(s)$ has the estimate

$$|g(s)| \leq \int_B |f(x)| e^{-(\tau, x)} \, dx \leq e^{\Sigma a_j |\tau_j|} \int_B |f(x)| \, dx = C \, e^{\Sigma a_j |\tau_j|}. \tag{16}$$

Thus, $g(s)$ is an entire function of at most first order. Moreover, for real $s = \sigma$, $g(\sigma)$ is bounded (and even tends to 0 as $|\sigma| \to \infty$) by virtue of the general properties of the Fourier transform of a summable function.

If f is an arbitrary functional with compact support, we can represent it by

$$f = \sum_{|k| \leq m} D^k f^k(x),$$

where the $f_k(x)$ are ordinary functions whose supports lie in a certain block $B_\varepsilon = \{x : |x_1| \leq a_1 + \varepsilon, \ldots, |x_n| \leq a_n + \varepsilon\}$.

By the foregoing considerations,

$$g = F[f] = \sum_{|k| \leq m} (- is)^k g_k(s),$$

$g_k(x)$ being the Fourier transform of $f_k(x)$ and $(- is)^k$ symbolizing the product $(- is_1)^{k_1} (- is_2)^{k_2} ... (- is_n)^{k_n}$. Thus $g(s) = F[f]$ is also an entire function of first order. But now it generally speaking has power growth in the real domain since

$$\left. \begin{array}{l} |g(s)| \leq C(1 + |s|)^m e^{\Sigma(a_j + \varepsilon)|\tau_j|}, \\ |g(\sigma)| \leq C(1 + |\sigma|)^m. \end{array} \right\} \tag{17}$$

The power growth of $g(\sigma)$ is evidently fixed by the order of singularity of f.

In certain instances, the estimates (17) can be improved by making special assumptions about the disposition of the support of f. For example, suppose that the supports of f and $f_k(x)$ lie in the octant $\{x: x_1 \geq \alpha_1 > 0, ..., x_n \geq \alpha_n > 0\}$. Then for $\tau_1 > 0, ..., \tau_n > 0$, we have

$$e^{-(\tau,x)} \leq e^{-(\tau_1\alpha_1 + ... + \tau_n\alpha_n)}.$$

Hence for such τ,

$$|g(s)| \leq C(1 + |s|)^m e^{-\Sigma\tau_j\alpha_j}, \tag{18}$$

or $g(s)$ is actually exponentially decreasing.

The expression (15) may be written as

$$g(s) = \overline{(f(x), e^{i(s,x)})} \tag{19}$$

(of course since $e^{i(s,x)}$ does not have compact support, the right-hand side of (19) has to be interpreted here in the sense indicated in Sec. 2.1.1.

The formula (19) turns out to be valid not only for a function $f(x)$ with compact support but also for any functional f with compact support.

Indeed, since such a functional is a sum of derivatives of functions with compact support, it is sufficient to show that whenever (19) is valid for $f(x)$ it remains valid for any of its derivatives $\partial f/\partial x_j$. But the latter is a consequence of the equations

$$F\left[\frac{\partial f}{\partial x_j}\right] = -is_j F[f] = -is_j(\bar{f}, e^{i(s,x)}) = (\bar{f}, -is_j e^{i(s,x)})$$

$$= \left(\bar{f}, -\frac{\partial}{\partial x_j} e^{i(s,x)}\right) = \left(\frac{\overline{\partial f}}{\partial x_j}, e^{i(s,x)}\right).$$

The assertion is proved.

2.8.4. In multi-dimensional problems especially of a spherically symmetric nature, one often encounters Bessel functions. Let us point out some definitions and formulas we shall need.

One of the representations of the Bessel function $J_p(z)$ is†

$$J_p(z) = \frac{1}{\sqrt{\pi}\,\Gamma\left(p + \frac{1}{2}\right)} \left(\frac{z}{2}\right)^p \int_0^\pi e^{iz\cos\theta} \sin^{2p}\theta \, d\theta. \tag{20}$$

When p is half an odd integer, the integral is expressible in terms of elementary functions. In particular,

$$J_{\frac{1}{2}}(z) = \sqrt{\frac{2}{\pi z}} \sin z, \tag{21}$$

. . . .

$$J_{m+\frac{1}{2}}(z) = (-1)^m z^{m+\frac{1}{2}} \left(\frac{1}{z}\frac{d}{dz}\right)^m \sqrt{\frac{2}{\pi}} \frac{\sin z}{z}. \tag{22}$$

An inversion formula for the latter is

$$\left(\frac{1}{z}\frac{d}{dz}\right)^m \left[z^{m+\frac{1}{2}} J_{m+\frac{1}{2}}(z)\right] = \sqrt{\frac{2}{\pi}} \sin z. \tag{23}$$

For illustrative purposes, we consider the Fourier transform

$$F[f(r)] = \int_{R_n} f(r) e^{i(x,\sigma)} \, dx$$

of a spherically symmetric integrable function $f(r)$. We convert the integral to spherical coordinates directing the polar axis along the vector σ. If $|\sigma| = \varrho$, $|x| = r$, and θ is the angle between the vectors x and σ, we obtain

$$F[f(r)] = \Omega_{n-1} \int_0^\infty \left[e^{ir\varrho\cos\theta} \sin^{n-2}\theta \, d\theta\right] f(r) r^{n-1} \, dr,$$

---

† See E. T. Whittaker and G. N. Watson, *Modern Analysis*, Cambridge Univ. Press, 4th ed., Chapt. 17.

where Ω_{n-1} is the surface area of the $(n\text{-}1)$-dimensional unit sphere. Therefore, by (20),

$$F[f(r)] = \Omega_{n-1} \sqrt{\pi}\, \Gamma\left(\frac{n-1}{2}\right) \int_0^\infty \left(\frac{r\varrho}{2}\right)^{1-\frac{n}{2}} J_{\frac{n}{2}-1}(r\varrho)\, r^{n-1} f(r)\, dr$$

$$= 2\pi^{\frac{n}{2}} \left(\frac{\varrho}{2}\right)^{1-\frac{n}{2}} \int_0^\infty r^{\frac{n}{2}} f(r) J_{\frac{n}{2}-1}(r\varrho)\, dr. \tag{24}$$

The integral (24) has been evaluated for many specific functions $f(r)$. For example, when $f(r) = e^{-tr}$ it is known that†

$$\int_0^\infty e^{-tr} r^{\frac{n}{2}} J_{\frac{n}{2}-1}(r\varrho)\, dr = \frac{2^{\frac{n}{2}} t\varrho^{\frac{n}{2}-1} \Gamma\left(\frac{n+1}{2}\right)}{\sqrt{\pi}\,(t^2 + \varrho^2)^{\frac{n+1}{2}}}.$$

Thus, the Fourier transform of e^{-tr} is given by

$$F[e^{-tr}] = 2\pi^{\frac{n}{2}} \left(\frac{\varrho}{2}\right)^{1-\frac{n}{2}} \frac{2^{\frac{n}{2}} t\varrho^{\frac{n}{2}-1} \Gamma\left(\frac{n+1}{2}\right)}{\sqrt{\pi}\,(t^2 + \varrho^2)^{\frac{n+1}{2}}}$$

$$= 2^n \pi^{\frac{n-1}{2}} \Gamma\left(\frac{n+1}{2}\right) \frac{t}{(t^2 + \varrho^2)^{\frac{n+1}{2}}}. \tag{25}$$

2.8.5. Let us find the Fourier transform of $\delta(r - a)$. This is a functional with compact support defined by

$$(\delta(r - a), \varphi(x)) = \int_{|x|=a} \varphi(x)\, dx$$

and therefore assigns to each $\varphi(x)$ in K its integral over a sphere of radius a centered at the origin.

By Sec. 2.8.3, we have

$$F[\delta(r - a)] = (\delta(r - a), e^{i(x,\,\sigma)}) = \int_{|x|=a} e^{i(x,\,\sigma)}\, dx.$$

† See S. Bochner, *ibid.*, Chapt. IX, p. 235.

Introducing spherical coordinates ($r = |x| = a$, $\varrho = |\sigma|$, and θ the angle between x and σ), we find that

$$F[\delta(r-a)] = \int_{\Omega_n} e^{ia\varrho\cos\theta} a^{n-1} \sin^{n-2}\theta \, d\theta \, d\omega_{n-1}$$

$$= a^{n-1}\Omega_{n-1} \int_0^\pi e^{ia\varrho\cos\theta}\sin^{n-2}\theta \, d\theta,$$

where $d\omega_{n-1}$ is the surface element of a sphere in the $(n-1)$-dimensional subspace orthogonal to the vector σ.

The resulting integral can be expressed in terms of the Bessel function yielding

$$F[\delta(r-a)] = a^{n-1}\Omega_{n-1}\sqrt{\pi}\,\Gamma\left(\frac{n-1}{2}\right)\left(\frac{a\varrho}{2}\right)^{1-\frac{n}{2}} J_{\frac{n}{2}-1}(a\varrho)$$

$$= \left(\frac{2\pi a}{\varrho}\right)^{\frac{n}{2}} \varrho J_{\frac{n}{2}-1}(a\varrho). \tag{26}$$

For odd $n = 2m + 3$, $m \geq 0$, we can use (22) to express the Bessel function in terms of elementary functions obtaining

$$F[\delta(r-a)] = \left(\frac{2\pi a}{\varrho}\right)^{\frac{n}{2}} \varrho \,(-1)^m (a\varrho)^{m+\frac{1}{2}} \left(\frac{1}{z}\frac{d}{dz}\right)^m \sqrt{\frac{2}{\pi}}\frac{\sin z}{z}\bigg|_{z=a\varrho}$$

$$= (-1)^m (2\pi)^{\frac{n}{2}} a^{n-1} \left(\frac{1}{z}\frac{d}{dz}\right)^m \sqrt{\frac{2}{\pi}}\frac{\sin z}{z}\bigg|_{z=a\varrho}.$$

When $n = 3$,

$$F[\delta(r-a)] = 4\pi a\frac{\sin a\varrho}{\varrho}. \tag{27}$$

2.8.6. Let us next find the Fourier transform of the convolution $f = f_0 * f_1$ where for definiteness f_0 is the functional with compact support. Let g, g_0, and g_1 be the corresponding Fourier transforms. In particular, $g_0 = g_0(s)$ is an entire function of first order having power growth for real $s = \sigma$. Consider first the case where both f_0 and f_1 are functions with compact support. Their convolution

$$f_0 * f_1 = \int_{R_n} f_0(\xi) f_1(x-\xi)\, d\xi$$

is also a function with compact support. Hence the Fourier transform of $f_0 * f_1$ can be determined by the classical formula to obtain

$$g(s) = \int_{R_n} e^{i(x,s)} \left[\int_{R_n} f_0(\xi) f_1(x - \xi) \, d\xi \right] dx$$

$$= \int_{R_n} \int_{R_n} e^{i(x,s)} f_0(\xi) f_1(x - \xi) \, d\xi \, dx$$

$$= \int_{R_n} e^{i(\xi,s)} f_0(\xi) \, d\xi \int_{R_n} e^{i(x-\xi,s)} f_1(x - \xi) \, dx = g_0(s) \, g_1(s).$$

Thus, the Fourier transform of the convolution of $f_0(x)$ and $f_1(x)$ is equal to the product of their Fourier transforms.

Now let f_0 and f_1 be arbitrary functionals with compact support. Then

$$f_0 = \sum_{|j| \leq m} D^j f_{0j}(x), \quad f_1 = \sum_{|k| \leq p} D^k f_{1k}(x), \quad g_0(s) = \sum_{|j| \leq m} (- is)^j g_{0j}(s),$$

$$g_1(s) = \sum_{|k| \leq p} (- is)^k g_{1k}(s),$$

where $g_{0j}(s)$ and $g_{1k}(s)$ are the respective Fourier transforms of the ordinary functions $f_{0j}(x)$ and $f_{1k}(x)$ having compact support. Therefore,

$$F[f_0 * f_1] = F\left[\sum_{|j| \leq m} \sum_{|k| \leq p} D^{j+k}(f_{0j}(x) * f_{1k}(x)) \right]$$

$$= \sum_{|j| \leq m} \sum_{|k| \leq p} (- is)^{j+k} g_{0j}(s) g_{1k}(s) = g_0(s) g_1(s).$$

Finally, let f_0 have compact support and f_1 be any functional. We know that

$$f_1 = \lim_{\nu \to \infty} f_\nu$$

(in K') of functionals having compact support. By the theorem on the continuity of the convolution,

$$f_0 * f_1 = \lim_{\nu \to \infty} (f_0 * f_\nu).$$

Let g_ν and g_1 be the respective Fourier transforms of f_ν and f_1. Then owing to the continuity of the Fourier operator,

$$g_1 = \lim_{\nu \to \infty} g_\nu(s) \text{ (in } Z').$$

Now

$$F[f_0 * f_1] = \lim_{\nu \to \infty} F[f_0 * f_\nu] = \lim_{\nu \to \infty} g_0(s) \, g_\nu(s)$$

$$= g_0(s) \lim_{\nu \to \infty} g_\nu(s) = g_0(s) g_1(s).$$

This completes the proof of the *convolution theorem*.

2.8.7. Our definition of the Fourier transform of generalized functions corresponded to the case where each ordinary function $f(x)$ is assigned a functional on K by means of

$$(f, \varphi) = \int \overline{f(x)}\, \varphi(x)\, dx.$$

And similarly, each ordinary function $g(\sigma)$ is assigned a functional on Z by means of

$$(g, \psi) = \int \overline{g(\sigma)}\, \psi(\sigma)\, d\sigma.$$

But it is just as easy to compose definitions of the Fourier transform for the other assignments of functionals to $f(x)$ (Sec. 1.3.4). With the existing definition and Parseval's equality in mind, we can write the following definitions:

$$(\bar{g}, \psi)_1 = (2\pi)^n\, (\bar{f}, \varphi)_1,$$

$$(g, \psi)_2 = (2\pi)^n\, (f, \varphi)_2, \quad \text{(the prevailing one),}$$

$$(\bar{g}, \bar{\psi})_3 = (2\pi)^n\, (\bar{f}, \bar{\varphi})_3;$$

$$(g, \bar{\psi})_4 = (2\pi)^n\, (f, \bar{\varphi})_4.$$

Problems

1. A functional

$$f = \sum_{|k| \le m} D^k f_k(x),$$

with $f_k(x)$ an ordinary function, $|k| \le m$, having no more than power growth at infinity (the space S'_n discussed in Prob. 4 of Sec. 2.3) is defined both in K_n and in Z_n. Show that the Fourier transformation maps S'_n into itself.

Hint. $f_k(x) = (1 + r^2)^p f_{0k}(x)$, where $f_{0k}(x) \in L_2(R_n)$.

2. Show that the Fourier transform of a homogeneous generalized function of degree p (Sec. 1.6, Prob. 3) is a homogeneous generalized function of degree $-n-p$.

3. Show that the Fourier transform of r^λ ($\lambda \ne -n, -n-2, \ldots$) is $C_\lambda \varrho^{-\lambda-n}$ and find C_λ.

Hint. r^λ is a spherically symmetric homogeneous function of degree λ. Use Probs. 1 and 2 and Prob. 1 of Sec. 2.4. To evaluate C_λ, apply the functional r^λ to e^{-r^2}, which belongs to S.

Ans.

$$\frac{F[r^\lambda]}{\Gamma\left(\dfrac{\lambda + n}{2}\right)} = 2^{\lambda+n}\pi^{\frac{n}{2}}\,\frac{\varrho^{-\lambda-n}}{\Gamma\left(-\dfrac{\lambda}{2}\right)}\,.$$

4. Show that a homogeneous spherically symmetric $f \in K'_n$ of degree $-n - 2k$ has the form $c\Delta^k\,\delta(x)$.

Hint. The Fourier transform of f is a spherically symmetric homogeneous generalized function of degree $2k$. Apply Prob. 11 of Sec. 2.4.

5. Suppose multiplication by $g(s)$ is a continuous operation in Z_n. Show that $g(s)$ is the Fourier transform of a functional $f \in K'_n$ for which the convolution $(f * \varphi)(x) = (f(\xi), \varphi(x + \xi))$ is a well-defined and continuous operation.

Hint. Set $f = F^{-1}[g]$ and use the formula $F[\varphi(x + \xi)] = e^{-ix\sigma}F[\varphi(\xi)]$.

Note. By Prob. 4 of Sec. 2.5, f has compact support. Therefore $g(s)$ is an entire function of first order and has at most power growth for real σ.

6. Let A_m denote the collection of entire functions $\psi(s)$ for which the inequalities

$$|s^k\psi(\sigma + i\tau)| \leqq C e^{b|\tau|}, \quad |k| \leqq m,$$

hold.

Further, let C_m be the space of functions $\varphi(x)$ with compact support having continuous derivatives up to order m. Show that

$$F[C_m] \subset A_m, \quad F^{-1}[A_m] \subset C_{m-n-1} \quad (m \geqq n + 1).$$

Part 2

PROBLEMS IN THE GENERAL THEORY
OF PARTIAL DIFFERENTIAL EQUATIONS

Chapter 3

Fundamental Functions of Differential Operators and Local Properties of Solutions

3.1. A Poisson Type Formula

3.1.1. Consider the differential equation in K_n'

$$P\left(\frac{\partial}{\partial x}\right)u \equiv \sum_{|k| \leq p} a_k D^k u \equiv \sum_{k_1 + \cdots + k_n \leq p} a_{k_1 \ldots k_n} \frac{\partial^{k_1 + \cdots + k_n} u(x)}{\partial x_1^{k_1} \ldots \partial x_n^{k_n}} = 0 \qquad (1)$$

with constant coefficients a_k. If $n = 1$, we have seen in Sec. 1.5 that all of its solutions are ordinary functions, the classical solutions of the equation. The matter is quite different for $n > 1$. The solutions may now include generalized functions that do not reduce to ordinary functions. For example, for $n = 2$ the equation $\dfrac{\partial u}{\partial x_1} = 0$ has among its solutions the singular function $\delta(x_2)$ defined by

$$(\delta(x_2), \ \varphi(x_1, x_2)) = \int_{-\infty}^{\infty} \varphi(x_1, 0) \, dx_1.$$

Let $L(P)$ denote the collection of all solutions of (1) in K_n'. Since P is a linear operator, $L(P)$ is a subspace of K_n'. Let us point out several more or less obvious properties of the subspace $L(P)$.

1. $L(P)$ is closed in K_n'.

For if $Pu_v = 0$ and $u_v \to u$, then $Pu = \lim_{v \to \infty} Pu_v = 0$ on account of the continuity of the operator P in K_n'.

2. If $u \in L(P)$ and f is a functional with compact support, then $f * u \in L(P)$.

For if $Pu = 0$, then

$$P(f * u) = f * Pu = 0.$$

In particular, both $\dfrac{\partial u}{\partial x_j} = \dfrac{\partial \delta(x)}{\partial x_j} * u$ and $u(x + h) = \delta(x + h) * u$ belong to $L(P)$, the latter for any vector $h \in R_n$.

3. Each solution $u \in L(P)$ is the limit in K' of ordinary and even infinitely differentiable solutions, namely, $f_v(x) * u$, where $f_v(x)$ is a delta-defining sequence (Sec. 2.2.2.) of infinitely differentiable functions with compact support.

4. Each solution $u \in L(P)$ having finite order of singularity is a finite sum of derivatives of functions having continuous derivatives up to an assigned order. Indeed, we can always represent the delta-function in the form

$$\delta(x) = \sum_{|k| \leq p} D^k \omega_k(x),$$

where $\omega_k(x)$ has continuous derivatives up to a given order q and vanishes for $|x| \geq \varepsilon.$† Therefore each solution $u = L(P)$ can be respresented by

$$u = u * \delta = \sum_{|k| \leq p} D^k(u * \omega_k).$$

Let the order of singularity of the generalized function u be m. Then by Sec. 2.6.4 the order of $u * \omega_k$ does not exceed $m - q$ and thus is less than any preassigned number for q sufficiently large, as required.

Generally, speaking the solutions of (1) may include ones having infinite order of singularity. An example is $u(x_1, x_2) = \displaystyle\sum_{q=0}^{\infty} \dfrac{\partial^q \delta(x_2 - q)}{\partial x_2^q}$ for the equation $\dfrac{\partial u(x_1, x_2)}{\partial x_1} = 0.$

3.1.2. Of great importance in the study of the properties of the solutions of (1) are the fundamental functions of the operator P. It will be recalled that a fundamental function is a (generalized) function $\mathscr{E}(x)$ satisfying the equation

$$P\left(\dfrac{\partial}{\partial x}\right)\mathscr{E}(x) = \delta(x).$$

† For instance, we have $\delta = \Delta^m \mathscr{E}_n^{2m}(x)$ with $\mathscr{E}_n^{2m}$ continuously differentiable up to order $2m - n - 1$ by Sec. 2.6.6 ($2m > n$). Now if $e(x)$ is a test function equaling 1 for $|x| \leq \varepsilon/2$ and 0 for $|x| \geq \varepsilon$, then $\Delta^m(e\mathscr{E}_n^{2m}) = \delta + e_0(x)$, where $e_0(x)$ is infinitely differentiable and different from zero just for $\varepsilon/2 < |x| < \varepsilon$. Hence, $\delta = \Delta^m(e\mathscr{E}_n^{2m}) - e_0(x)$ and so is obtained from functions with continuous derivatives up to order $2m - n - 1$ by differentiating $2m$ times.

Thus,

$$\mathscr{E}_n^2(x) = \frac{-1}{(n-2)\,\Omega_n}\,r^{2-n} \tag{2}$$

is a fundamental function of the Laplacian Δ in R_n for $n > 2$ (see Sec. 1.7.2, 8°). Later on in Sec. 3.2 we shall prove the existence of a fundamental function for every operator $P\left(\dfrac{\partial}{\partial x}\right)$.

It is well-known that the fundamental function (2) can be used to construct an integral representation for any solution of $\Delta u = 0$ in a region G in terms of its values on the boundary of G (Poisson's formula).†

It turns out that a similar integral representation can be obtained for the solutions of $P\left(\dfrac{\partial}{\partial x}\right) u = 0$ by using a corresponding fundamental function. Let us carry out the construction.

By definition, a generalized function $u(x)$ defined in $G \subset R_n$ (see Sec. 1.6.3) is a solution of $P\left(\dfrac{\partial}{\partial x}\right) u = 0$ in G if‡

$$\left(P\left(\frac{\partial}{\partial x}\right) u, \varphi\right) \equiv \left(u, \overline{P}\left(-\frac{\partial}{\partial x}\right)\varphi\right) = 0$$

for each test function $\varphi(x)$ with support in G. We pointed out in Sec. 2.1.2 that a generalized function u in G can be extended to all of R_n so that it remains unchanged in a given region V interior to G. Thus a solution u in G may be thought of as being a generalized function defined on all of K_n and solving (1) in $V \subset G$. We observe that if G is a bounded region, then the functional u may be considered to have compact support and to be carried in G. If necessary, we could multiply it by an infinitely differentiable function $\beta(x)$ equaling 0 outside G and 1 in V. Let $\mathscr{E}(x)$ be a fundamental solution of equation (1). Further, let $\alpha(x)$ be an infinitely differentiable function equaling 0 outside a neighborhood U of the origin in R_n and 1 inside a smaller neighborhood of the origin.

---

† See Courant-Hilbert, *Methods of Mathematical Physics*, Vol. II, Interscience, New York, 1962, Chapt. 4, Sec. 2.1.

‡ Recall that if $P(\xi) = \sum_k a_k \xi^k$, then $\overline{P}(D) = \sum_k \bar{a}_k D^k$, where $\bar{a}_k$ is the complex conjugate of a_k.

Let W denote a region with the property that the arithmetic difference $W - U$ is a subset of V. W will necessarily be non-vacuous if U is of sufficiently small size.

LEMMA. *The following representation holds in W:*

$$P\left(\frac{\partial}{\partial x}\right)[(1 - \alpha)\,\mathscr{E}] * u(x) = u(x). \tag{3}$$

Proof. Let $\varphi(x)$ be a test function with support in W. Then we have

$$\left(P\left(\frac{\partial}{\partial x}\right)(\alpha\mathscr{E}) * u, \varphi\right) = \left(\alpha\mathscr{E} * P\left(\frac{\partial}{\partial x}\right)u, \varphi\right)$$

$$= \left(P\left(\frac{\partial}{\partial x}\right)u, \alpha\mathscr{E} * \varphi\right) = \left(P\left(\frac{\partial}{\partial x}\right)u, \psi\right),$$

where by Sec. 2.5.3 $\psi = \alpha\mathscr{E} * \varphi$ is an infinitely differentiable function with support in $W - U \subset V$. Since $u(x)$ is a solution of (1) in V, the result is equal to zero. Therefore for the stipulated $\varphi(x)$ we have

$$P\left(\frac{\partial}{\partial x}\right)[(1 - \alpha)\,\mathscr{E}] * u = P\left(\frac{\partial}{\partial x}\right)\mathscr{E} * u - P\left(\frac{\partial}{\partial x}\right)(\alpha\mathscr{E}) * u = \delta * u = u,$$

as asserted.

Formula (3) is our desired analogue of the Poisson integral formula. Indeed, let us see what values of $u(x)$ go into the make-up of the left-hand side in determining u in a region $W_0 \subset V$ (Fig. 6). The functional

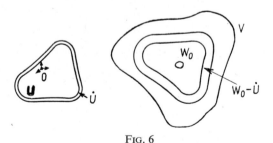

FIG. 6

$f = P\left(\dfrac{\partial x}{\partial x}\right)[(1 - \alpha)\,\mathscr{E}]$ is carried in a region where $\alpha(x)$ differs from 0 and

1, i.e., in a region $\dot{U} \subset U$ adjacent to the boundary of U. The considerations of Sec. 2.5 show that to determine $u(x)$ in W_0 one needs to know $u(x)$

for the left-hand side of (3) in $W_0 - \dot{U}$. This region may be pictured as follows. Translate the origin to any point of W_0 observing what the effect is on the region $-\dot{U}$. Then consider the union of all such effects as the image of the origin varies over all of W_0. If W_0 is sufficiently small and $\dot{U}$ is sufficiently far from the origin, $W_0 - \dot{U}$ will be disjoint from W_0 and completely encompass it. At the same time W_0 lies inside V where $u(x)$ is a solution of (1). Thus the value of $u(x)$ in W_0 is expressed in terms of its values near the boundary of V (since the shape of U is up to us, the boundary of V may be approached arbitrarily closely).

In particular, we have deduced the following uniqueness theorem:

THEOREM 1. *If two solutions of the equation* $P\left(\dfrac{\partial}{\partial x}\right) u = 0$ *coincide near the boundary of* V, *they coincide throughout* V. *If a solution of* $P\left(\dfrac{\partial}{\partial x}\right) = 0$ *vanishes near the boundary of* V, *it vanishes identically in* V.

3.1.3. Applying inequality (7) of Sec. 2.6.5 to (3), we arrive at the following relation between orders of singularity:

$$c_{W_0}(u) \leqq s\left[P\left(\frac{\partial}{\partial x}\right)(1-\alpha)\,\mathscr{E} \right] + c_{W_0-\dot{U}}(u). \tag{4}$$

If $P(\xi)$ is an m-th degree polynomial, we further have

$$P\left(\frac{\partial}{\partial x}\right)(1-\alpha)\,\mathscr{E} = [1-\alpha(x)]\,P\left(\frac{\partial}{\partial x}\right)\mathscr{E}(x)$$

$$+ \sum_{|q|\leqq m-1} \alpha_q(x)\,D^q\mathscr{E}(x) = \sum_{|q|\leqq m-1} \alpha_q(x)\,D^q\mathscr{E}(x),$$

where the $\alpha_q(x)$ are infinitely differentiable functions. Hence by Property 1 of Sec. 2.6.3.,

$$c_{W_0}(u) \leqq m - 1 + s_{\dot{U}}(\mathscr{E}) + c_{W_0-\dot{U}}(u). \tag{5}$$

COROLLARY 1. *If a solution* $u(x)$ *is infinitely differentiable near the boundary of* V, *then it is infinitely differentiable at every interior point of* V.

COROLLARY 2. *If the operator* $P\left(\dfrac{\partial}{\partial x}\right)$ *possesses a fundamental function* $\mathscr{E}(x)$ *which is infinitely differentiable everywhere except at the origin, then every solution of* $P\left(\dfrac{\partial}{\partial x}\right)u = 0$ *in* V, *assumed initially to be only a generalized function, is an ordinary infinitely differentiable function.*

Both of these corollaries follow from (5): the hypothesis of the first corollary means that $c_{W-\dot{U}}(u) = -\infty$ and that of the second that $s_{\dot{U}}(\mathscr{E}) = -\infty$.

COROLLARY 3. *If* $P\left(\dfrac{\partial}{\partial x}\right)$ *has a fundamental function* $\mathscr{E}(x)$ *which is infinitely differentiable everywhere except at the origin, then every fundamental function* $\mathscr{E}_1(x)$ *of the operator is also infinitely differentiable except at the origin.*

In fact, $\mathscr{E}_1(x)$ is a solution of the equation $P\left(\dfrac{\partial}{\partial x}\right)\mathscr{E}_1(x) = 0$ everywhere except at the origin and so it merely remains to apply Corollary 2.

An operator $P\left(\dfrac{\partial}{\partial x}\right)$ for which all solutions of the equation $P\left(\dfrac{\partial}{\partial x}\right)u = 0$ are infinitely differentiable is called *hypoelliptic*. A fundamental function for a hypoelliptic operator is infinitely differentiable everywhere except at the origin. The Laplacian and its iterates are hypoelliptic operators.

REMARK. An equation $P\left(\dfrac{\partial}{\partial x}\right)u = 0$ whose solutions in V all have order of singularity not exceeding a fixed number q is also hypoelliptic.

For as we know from Sec. 3.1.1, each derivative $\dfrac{\partial u}{\partial x_j}$, $\dfrac{\partial^2 u}{\partial x_j\,\partial x_k}$, etc., of a solution of $P\left(\dfrac{\partial}{\partial x}\right)u = 0$ in K'_n is likewise a solution of the equation. If a solution of $P\left(\dfrac{\partial}{\partial x}\right)u = 0$ in V has order of singularity not exceeding q, then by Theorem 1 of Sec. 2.6.6 each solution has order of singularity $q - 2m + 1$ at the most for arbitrary m. In other words, it is an ordinary infinitely differentiable function.

3.1.4. When an equation is hypoelliptic (i.e., the operator in it is hypoelliptic), formula (3) can be converted into the usual Poisson type formula expressing a solution inside a region in terms of the boundary of values the solution and its derivatives.

To see this, we note that the relation (3) is now one involving ordinary functions and can therefore be written in integral form. We have

$$u(x) = P\left(\frac{\partial}{\partial x}\right)[(1-\alpha)\,\mathscr{E}] * u(x) = (1-\alpha)\,\mathscr{E} * P\left(\frac{\partial}{\partial x}\right)u(x)$$

$$= \int_{R_n} (1-\alpha)\,\mathscr{E}(\xi)\, P\left(\frac{\partial}{\partial x}\right)u(x-\xi)\,d\xi. \tag{6}$$

This relation recall is valid in any W for which the arithmetic difference of W and any U containing the support of $\alpha(\xi)$ lies inside V, the region in which $u(x)$ is a solution of (1).

Let us pass to the limit in (6) letting $\alpha(\xi)$ approach the characteristic function for the region U. We obtain in the limit

$$u(x) = \int_{R_n - U} \mathscr{E}(\xi)\, P\left(\frac{\partial}{\partial x}\right)u(x-\xi)\,d\xi = \int_{R_n - U} \mathscr{E}(\xi)\, P\left(-\frac{\partial}{\partial \xi}\right)u(x-\xi)\,d\xi. \tag{7}$$

As we have already ascertained, $u(x)$ may be considered infinitely differentiable outside V and vanishing outside $G \supset V$. But, generally speaking, $u(x)$ will no longer be a solution of (1) outside V. We now integrate by parts in (7) gradually transferring all derivatives from the factor $u(x-\xi)$ to $\mathscr{E}(\xi)$. Since $P\left(\dfrac{\partial}{\partial \xi}\right)\mathscr{E}(\xi) = 0$ in $R_n - U$, only the boundary terms remain and we obtain

$$u(x) = \int_\Gamma \sum_k P_k\left(\frac{\partial}{\partial \xi}\right)\mathscr{E}(\xi)\, Q_k\left(-\frac{\partial}{\partial \xi}\right)u(x-\xi)\,d\xi$$

$$= \sum_k \int_\Gamma c_k(\xi)\, Q_k\left(-\frac{\partial}{\partial \xi}\right)u(x-\xi)\,d\xi. \tag{8}$$

Here, P_k and Q_k are polynomials of degree less than m, $c_k(\xi) = P_k\left(\dfrac{\partial}{\partial \xi}\right)\mathscr{E}(\xi)$, and Γ is the boundary of U.

This in turn can be written as

$$u(x) = \int_{x-\Gamma} \sum_k c_k(x-\eta)\, Q_k\left(\frac{\partial}{\partial \eta}\right)u(\eta)\,d\eta.$$

The region U may be changed to a great extent in an arbitrary way as x changes. If x_0 is a fixed arbitrary point, it is possible therefore to replace the boundary $x - \Gamma$ by $x_0 - \Gamma$ at least for small values of $x - x_0$. This leads to the formula

$$u(x) = \int_{x_0 - \Gamma} \sum_k c_k(x - \eta) \, Q_k\left(\frac{\partial}{\partial\eta}\right) u(\eta) \, d\eta. \tag{9}$$

The expression (9) shows particularly that a solution of an m-th order hypoelliptic equation is uniquely determined inside a region by the assignment of the values of the solution and its derivatives up to $(m-1)$-st order on the boundary of the region.

The derivatives of $u(x)$ satisfy expressions of the form

$$R\left(\frac{\partial}{\partial x}\right) u(x) = \int_{x_0 - \Gamma} \sum_k R\left(\frac{\partial}{\partial x}\right) c_k(x - \eta) \, Q_k\left(\frac{\partial}{\partial\eta}\right) u(\eta) \, d\eta. \tag{10}$$

This formula makes it possible to estimate the growth of the derivatives of $u(x)$ in terms of the growth of the derivatives of the fundamental solution $\mathscr{E}(x)$, as we shall see.

3.1.5. A convenient description of the growth of the derivatives of a function may be given by means of the notion of Gevrey class.

We shall first define Gevrey's class for functions of a single variable.

Let β be a fixed non-negative number. A function $f(x)$ $(a \leq x \leq b)$ is said to belong to Gevrey's class G_β (or more precisely $G_\beta[a, b]$) if it has derivatives of all orders in $[a, b]$ and satisfies the inequalities

$$|f^{(q)}(x)| \leq CB^q q^{q\beta} \quad (q = 0, 1, 2, \ldots)$$

for some constants B and C (depending possibly on f).

If $\beta \leq 1$, $f(x)$ is an analytic function. To show this, we expand the difference $f(x + h) - f(x)$ by Taylor's theorem obtaining

$$f(x + h) - f(x) - hf'(x) - \frac{1}{2}h^2 f''(x) - \cdots$$

$$- \frac{h^{m-1}}{(m-1)!} f^{(m-1)}(x) = R_m,$$

where

$$|R_m| = \frac{|h|^m}{m!} |f^{(m)}(x + \theta h)| \leq \frac{C|h|^m}{m!} B^m m^{m\beta} \leq C \frac{(|h| \, Bm)^m}{m!}.$$

The quantity $d_m = C \dfrac{(|h| \, Bm)^m}{m!}$ approaches zero as $m \to \infty$ if $|h| \, Be < 1$

since

$$\frac{d_{m+1}}{d_m} = |h| \, B \left(1 + \frac{1}{m}\right)^m < |h| \, Be.$$

Thus for $|h| < \dfrac{1}{Be}$, R_m approaches zero uniformly in x and $f(x + h)$ is

therefore the sum of its Taylor series. In other words, $f(x)$ is analytic at each point of $[a, b]$.

Conversely, if $f(x)$ is an analytic function in $[a, b]$, it may be continued analytically to a domain V of the complex plane containing the interval inside it. We can then write down Cauchy's formula

$$f^{(q)}(x) = \frac{q!}{2\pi i} \int_L \frac{f(\zeta) \, d\zeta}{(\zeta - x)^{q+1}},$$

where the closed contour L lies in V at a minimum distance ϱ from the points of $[a, b]$. Estimating the modulus of the derivative, we find that

$$|f^{(q)}(x)| \leq q! \frac{C}{\varrho^q} \max |f(\zeta)| \leq C_1 \left(\frac{1}{\varrho}\right)^q q^q,$$

or $f \in G_1[a, b]$, as required.

When $\beta > 1$, the class G_β contains functions which are not analytic. Such a function is $\exp(-|x|^{-p})$ which belongs to class $G_{1+1/p}$ (see Prob. 4).

In the n variable case, Gevrey's class G_β is defined as follows:

Gevrey's class $G_\beta = G_\beta(V)$ $(\beta \geq 0)$ is the collection of all functions $u(x_1, ..., x_n) = u(x)$ defined and continuous in a closed region V and having derivatives of all orders satisfying the inequalities

$$\left| \frac{\partial^{q_1 + \cdots + q_n} u(x)}{\partial x_1^{q_1} \cdots \partial x_n^{q_n}} \right| \leq C B_1^{q_1} \cdots B_n^{q_n} q_1^{q_1 \beta} \cdots q_n^{q_n \beta} \quad (q_1, ..., q_n = 0, 1, ...). \quad (11)$$

The class G_1 consists of functions analytic in each of the variables $x_1, ..., x_n$.

It is clear that the class G_β is closed under addition and under multiplication by scalars. Let us show that it is also closed under differentiation. Let $u \in G_\beta$ and $u_1(x) = \dfrac{\partial u}{\partial x_1}$. Then

$$\left| \frac{\partial^{q_1 + \cdots + q_n} u_1(x)}{\partial x_1^{q_1} \ldots \partial x_n^{q_n}} \right| = \left| \frac{\partial^{q_1 + 1 + \cdots + q_n} u(x)}{\partial x_1^{q_1 + 1} \ldots \partial x_n^{q_n}} \right|$$

$$\leqq C B_1^{q_1 + 1} \ldots B_n^{q_n} (q_1 + 1)^{(q_1 + 1)\beta} \ldots q_n^{q_n \beta}$$

$$= C B_1^{q_1} \ldots B_n^{q_n} q_1^{q_1 \beta} \ldots q_n^{q_n \beta} \cdot B_1 \frac{(q_1 + 1)^{(q_1 + 1)\beta}}{q_1^{q_1 \beta}}. \qquad (12)$$

The last factor can be estimated as follows:

$$\frac{(q_1 + 1)^{(q_1 + 1)\beta}}{q_1^{q_1 \beta}} = \left(1 + \frac{1}{q_1} \right)^{q_1 \beta} (q_1 + 1)^\beta \leqq e^\beta C_1 2^{q_1}.$$

Thus, the estimate (11) holds for $\dfrac{\partial u}{\partial x_1}$ with different C and B_1.

Suppose that the fundamental function $\mathscr{E}(x)$ for $P\left(\dfrac{\partial}{\partial x} \right)$ belongs to class G_β in an arbitrary bounded region not containing the origin within or on its boundary. We can then assert that any solution of

$$P\left(\frac{\partial}{\partial x} \right) u(x) = 0 \qquad (13)$$

will also belong to class G_β in a bounded region. To show this, we apply formula (9). Here the functions $c_k(x)$ are certain derivatives of $\mathscr{E}(x)$ in a region not containing the origin. They therefore belong to class G_β together with $\mathscr{E}(x)$. This leads to the following estimate for $D^q u = \dfrac{\partial^{q_1 + \cdots + q_n} u}{\partial x^{q_1} \ldots \partial x^{q_n}}$:

$$|D^q u(x)| \leqq \int_{x_0 - \Gamma} \sum_k |D^q c_k(x - \eta)| \left| Q_k\left(\frac{\partial}{\partial \eta} \right) u(\eta) \right| d\eta$$

$$\leqq \sum_k \max_{\eta \in x_0 - \Gamma} |D^q c_k(x - \eta)| \int_{x_0 - \Gamma} \left| Q_k\left(\frac{\partial}{\partial \eta} \right) u(\eta) \right| d\eta$$

$$\leqq C \sum_k \max_{\eta \in x_0 - \Gamma} |D^q c_k(x - \eta)|.$$

It follows from this that $u(x)$ is of class G_β whenever $\mathscr{E}(x)$ is.

The converse is of course also valid: If in any bounded region all of the solutions of equation (13) are of class G_β, then the fundamental solution $\mathscr{E}(x)$ is also of class G_β in any bounded region not containing the origin within or on its boundary. This is so because $\mathscr{E}(x)$ is obviously a solution of (13) in any such region.

Thus for all m, every solution of the iterated Laplace equation $\varDelta^m u = 0$ is an analytic function (and so is of class G_1) since the fundamental function of $\varDelta^m$ is analytic except at the origin (Sec. 1.7.2, 9°).

Henceforth, the equation $P\left(\dfrac{\partial}{\partial x}\right)u = 0$ (as well as the operator $P\left(\dfrac{\partial}{\partial x}\right)$

and polynomial P) will be called β-*hypoelliptic* if all solutions of $P\left(\dfrac{\partial}{\partial x}\right)u = 0$

are locally of class G_β. A 1-hypoelliptic equation (operator, polynomial) will be called simply *elliptic*.

Problems

1. If an m-th order operator $P\left(\dfrac{\partial}{\partial x}\right)$ has a fundamental function $\mathscr{E}(x)$ having ordinary derivatives up to m-th order everywhere except at the origin (i.e., $s(\mathscr{E}) \leqq -m$), then P is hypoelliptic.

Hint. The inequality (5) becomes in the present case

$$c_W(u) \leqq c_{W-\dot{v}}(u) - 1. \tag{1}$$

Thus, the order of singularity of a solution u in W is less than at the boundary. Iterate (1) to show that the order of singularity of u in an interior region is less than any arbitrary integer and so is equal to $-\infty$.

Note. A stronger result is presented in Prob. 5 of Sec. 3.5.

2. Given that every solution of $P\left(\dfrac{\partial}{\partial x}\right)u = 0$ in V having ordinary derivatives up to a fixed order N is infinitely differentiable. Show that P is hypoelliptic.

Hint. Use the fourth property of Sec. 3.1.1.

3. Given that every solution of $P\left(\dfrac{\partial}{\partial x}\right)u = 0$ in V having order $s(u) \leqq p$ has order $s(u) \leqq p - 1$ (p is fixed). Prove that $P\left(\dfrac{\partial}{\partial x}\right)$ is hypoelliptic.

Hint. All of the derivatives of $u(x)$ also have order $s(u) \leqq p$. Apply Theorem 1 of Sec. 2.6.6 and Prob. 2.

4. Show that $f(x) = e^{-|x|^{-p}}$ $(p > 0)$ belongs to Gevrey's class $G_{1+\frac{1}{p}}$

Hint. $f(z) = e^{-z^{-p}}$ is analytic for $z = x + iy \neq 0$. Use Cauchy's formula to write an expression for $f^{(q)}(x)$ in terms of the values of $f(z)$ on a circle centered at x and tangent to the lines $y = \pm \dfrac{x}{p}$. Estimate the modulus and then maximize over all x.

5. Given that a polynomical $P(\sigma) \equiv P(\sigma_1, ..., \sigma_n)$ has

(a) no real zeros,
(b) a single real zero at $\sigma = 0$,
(c) a bounded set of real zeros.

Show that every solution of $P\left(i\dfrac{\partial}{\partial x}\right) u(x) = 0$ with at most power growth at infinity (for a more general kind of growth see Prob. 4 of Sec. 2.3 dealing with S') is

(a) identically zero,
(b) a polynomial in x,
(c) an entire function of $z = x + iy$ of at most first order.

Hint. By Prob. 1 of Sec. 2.8, the Fourier transform $v(\sigma)$ of $u(x)$ belongs to S' (and therefore to K'). Find its support (see Prob. 2 of Sec. 2.1). Apply the theorem of Sec. 2.3.6 for case (b) and the theorem of Sec. 2.8.3 for case (c).

6. Show that every differentiable solution of $(\Delta + k^2) u = 0$ (for $n = 3$) satisfying Sommerfeld's conditions,

$$u(x) = O\left(\frac{1}{r}\right), \quad \frac{\partial u}{\partial x} - iku = o\left(\frac{1}{r}\right), \quad |x| \to \infty,$$

is identically zero.

Hint. Sommerfeld's conditions are satisfied in particular by the fundamental function $\mathscr{E}(x) = -\dfrac{e^{ikr}}{4\pi r}$ of the operator $\Delta + k^2$ (Sec. 1.7, Prob. 7). Now let $\mathscr{E}_R(x) = \mathscr{E}(x)$ for $|x| \leq R$ and 0 for $|x| > R$. Then by Green's formula (Sec. 1.7.2, 7°),

$$0 = (\mathscr{E}_R, (\Delta + k^2) u) = ((\Delta + k^2) \mathscr{E}_R, u)$$

$$= (\delta, u) + \int_{|x| = R} \left[\mathscr{E}_R \frac{\partial u}{\partial r} - \frac{\partial \mathscr{E}_R}{\partial r} u\right] dx.$$

Hence

$$u(0) = \int\limits_{|x|=R} \left[\frac{\partial \mathscr{E}_R}{\partial r} u - \mathscr{E}_R \frac{\partial u}{\partial r} \right] dx$$

$$= \int\limits_{|x|=R} \left[\left(\frac{\partial}{\partial r} - ik \right) \mathscr{E}_R u - \mathscr{E}_R \left(\frac{\partial}{\partial r} - ik \right) u \right] dx$$

$$= \int\limits_{|x|=R} o\left(\frac{1}{R^2} \right) dx = o(1) \to 0$$

and therefore $u(0) = 0$. Since any point may be chosen as the origin, $u(x) \equiv 0$.

3.2. Existence of a Fundamental Function

3.2.1. In this section, we shall prove that the equation

$$P\left(\frac{\partial}{\partial x} \right) \mathscr{E}(x) = \delta(x) \tag{1}$$

has a solution in K' for every given polynomial $P \neq 0$. With a view to simplifying the subsequent computations, we shall write (1) in the form†

$$\bar{P}\left(i\frac{\partial}{\partial x} \right) \mathscr{E}(x) = \delta(x). \tag{2}$$

If we formally take Fourier transforms in (2), we obtain

$$\bar{P}(s) E(s) = 1, \tag{3}$$

an equation for functionals on Z. By proving there exists a continuous linear functional $E(s)$ in Z' satisfying (3), we shall also be able to conclude the existence of a solution to (2) in K' by applying the inverse Fourier transformation.

Equation (3) is equivalent to

$$(\bar{P}(s) E(s), \psi(s)) = (1, \psi(s)) = \int\limits_{R_n} \psi(\sigma) \, d\sigma$$

for any $\psi(s) \in Z$.

---

† By definition, $\overline{\sum\limits_k a_k D^k} = \sum\limits_k \bar{a}_k D^k$. By the same token, $\bar{P}(s) = \sum\limits_k \bar{a}_k s^k = \overline{\sum\limits_k a_k \bar{s}^k}$
$= \overline{P(\bar{s})} = P^*(s)$ (Sec. 2.7.3).

If the polynomial $P(s)$ had no roots for real values of $s = \sigma$ and its modulus had a positive lower bound, then we could always take

$$(E(s), \psi(s)) = \int_{R_n} \frac{\psi(\sigma)\, d\sigma}{P(\sigma)}. \tag{4}$$

For then formula (4) would define a continuous linear functional on Z and

$$(\bar{P}(s)\, E(s), \psi(s)) = (E(s), P(s)\, \psi(s)) = \int_{R_n} \frac{P(\sigma)\, \psi(\sigma)\, d\sigma}{P(\sigma)} = \int_{R_n} \psi(\sigma)\, d\sigma,$$

as asserted. However, the expression (4) is unsuitable for the general case since it is not even defined for every $\psi(s)$ in Z.

For $n = 1$, we can write

$$(E(s), \psi(s)) = \int_{-\infty}^{\infty} \frac{\psi(\sigma + i\tau)}{P(\sigma + i\tau)}\, d\sigma.$$

The integration is over any line parallel to the real axis that does not pass through any root of the polynomial $P(s)$. We have here also

$$(\bar{P}(s)\, E(s), \psi(s)) = (E(s), P(s)\, \psi(s)) = \int_{-\infty}^{\infty} \frac{P(\sigma + i\tau)\, \psi(\sigma + i\tau)}{P(\sigma + i\tau)}\, d\sigma$$

$$= \int_{-\infty}^{\infty} \psi(\sigma + i\tau)\, d\sigma = \int_{-\infty}^{\infty} \psi(\sigma)\, d\sigma.$$

The last simplification is the result of applying Cauchy's theorem in conjunction with the rapid decrease of $\psi(\sigma + i\tau)$ as $|\sigma| \to \infty$.

Thus, the problem is solved for $n = 1$ by going off into the complex plane. It would be natural to expect to be able to solve it in this way also for the case of any n. It turns out that for general n it is sufficient to go into the complex plane of just one coordinate.

3.2.2. We first show that a given m-th order operator $P\left(i \dfrac{\partial}{\partial x}\right)$ can always be written in the *normal form*

$$a_0\left(i \frac{\partial}{\partial x_1}\right)^m + \sum_{k=0}^{m-1} P_k\left(i \frac{\partial}{\partial x_2}, \dots, i \frac{\partial}{\partial x_n}\right)\left(i \frac{\partial}{\partial x_1}\right)^k, \tag{5}$$

where a_0 is a non-vanishing constant, by performing a rotation of axes if necessary.

According to Sec. 2.8.2, each rotation $x' = Ux$ of R_n where the functions $\varphi(x)$ are defined induces an analogous rotation $\sigma' = U\sigma$ in the space where their Fourier transforms are defined.

It is therefore sufficient to find a rotation of R_n under which our prescribed m-th degree polynomial $P(\sigma)$ assumes the form

$$P(\sigma) = a_0 \sigma_1^m + \sum_{k=0}^{m-1} P_k(\sigma_2, \ldots, \sigma_n)\, \sigma_1^k, \quad a_0 \neq 0. \tag{6}$$

Suppose that $P(\sigma)$ is not of this form at the outset. We perform a real linear transformation

$$\sigma_j = \sum_{k=1}^{n} c_{jk}\xi_k$$

with a temporarily unspecified matrix $C = \|c_{jk}\|$. As a result, $P(\sigma)$ will go over into a new m-th degree polynomial in $\xi_1, \ldots, \xi_n$. The coefficient of ξ_1^m comes from the highest degree terms of P. Specifically, if

$$P_0(\sigma) = \sum_r a_r \sigma_1^{q_{1r}} \ldots \sigma_n^{q_{nr}}, \quad q_{1r} + \cdots + q_{nr} = m,$$

is the set of highest degree terms of $P(\sigma)$, then the coefficient of ξ_1^m will be

$$P_0(c_1) = \sum_r a_r c_{11}^{q_{1r}} \ldots c_{n1}^{q_{nr}}.$$

We see that it is expressed just in terms of the elements of the first column of the matrix C. We now choose $c_{11}, \ldots, c_{n1}$ to be any normalized set of numbers $(\Sigma\, c_{j1}^2 = 1)$ assuring the non-vanishing of $P_0(c_1)$, or in other words, any point of the (real) unit sphere not lying on the cone $P_0(\sigma) = 0$. The remaining columns of C may be chosen arbitrarily except that they must be orthogonal and normalized as the matrix of a rotation requires.

Thus a sufficient (and necessary) condition for writing P as (6) is that the σ_1-axis of the $\sigma_1 \ldots \sigma_n$-coordinate system should not be directed along the cone $P_0(\sigma) = 0$. This cone is called the *characteristic cone* of the polynomial P.

After $P(\sigma)$ has been written as (6), the polynomial $P(s)$ assumes the same form

$$P(s) = a_0 s_1^m + \sum_{k=0}^{m-1} P_k(s_2, \ldots, s_n)\, s_1^k$$

in the entire complex space C_n.

The result obtained plus the fact that the δ-function is invariant under rotation makes it possible to henceforth confine our considerations to polynomials in normal form.

3.2.3. Consider the real $(n + 1)$-dimensional space in which the coordinates are the real parts $\sigma_1, \ldots, \sigma_n$ and the imaginary part τ_1. We shall construct a certain discontinuous manifold H in this space, which we shall call "Hörmander's staircase". To this end, we decompose the $(n - 1)$-dimensional $\sigma_2 \ldots \sigma_n$-space into a locally finite family of subsets $\Delta_1, \ldots, \Delta_r, \ldots$ (i.e., locally finite in each ball) by $(n - 2)$-dimensional hyperplanes parallel to the coordinate hyperplanes. With Δ_j, we then associate a value of $\tau_1 = \tau_1^{(j)}$. Hörmander's staircase is defined to be the set of all points $(\sigma_1, \ldots, \sigma_n, \tau_1)$ such that $-\infty < \sigma_1 < \infty$ and $\tau_1 = \tau_1^{(j)}$ if $(\sigma_2, \ldots, \sigma_n) \in \Delta_j$, $(j = 1, 2, \ldots)$.

Fig. 7 depicts the staircase for the case of three coordinates σ_1, σ_2, and τ_1.

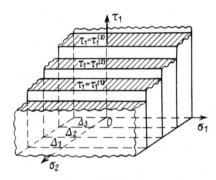

FIG. 7

We shall prove shortly that for any polynomial $P(s)$, one can construct a staircase H on which $|P(s)| \geq C > 0$ and all $|\tau_1^{(j)}|$ are uniformly bounded say by the constant C_0. Assuming we have such a staircase, we define

$$(E(s), \psi(s)) = \int_H \frac{\psi(\sigma_1 + i\tau_1, \sigma_2, \ldots, \sigma_n)}{P(\sigma_1 + i\tau_1, \sigma_2, \ldots, \sigma_n)} \, d\sigma_1 \ldots d\sigma_n. \qquad (7)$$

The integral exists since the denominator exceeds C in modulus and $\psi(\sigma_1 + i\tau_1, \sigma_2, \ldots, \sigma_n)$ is integrable; recall that ψ is a test function in Z

and so approaches zero uniformly in τ_1 (for $|\tau_1| \leq C_0$) faster than any power of $1/|\sigma|$ as $|\sigma| \to \infty$.

We wish to show that this functional multiplied by $\bar{P}(s)$ is equal to the functional 1. We have

$$(\bar{P}(s)\, E(s),\, \psi(s)) = (E(s),\, P(s)\, \psi(s)) = \int_H \psi(\sigma_1 + i\tau_1,\, \sigma_2,\, ...,\, \sigma_n)\, d\sigma_1\, ...\, d\sigma_n$$

$$= \sum_j \int_{(\sigma_2,\, ...,\, \sigma_n)\in\Delta_j} \cdots \int \left[\int_{-\infty}^{\infty} \psi(\sigma_1 + i\tau_1^{(j)},\, \sigma_2,\, ...,\, \sigma_n)\, d\sigma_1 \right] d\sigma_2\, ...\, d\sigma_n.$$

The inner integral extends over a line parallel to the real axis in the $(\sigma_1 + i\tau_1)$-plane at a distance $|\tau_1^{(j)}|$ from it. By Cauchy's theorem, it may be replaced without changing its value by an integral along the real axis itself. This yields

$$(\bar{P}(s)\, E(s),\, \psi(s)) = \sum_j \int_{\Delta_j} \cdots \int \int_{-\infty}^{\infty} \psi(\sigma_1,\, ...,\, \sigma_n)\, d\sigma_1\, ...\, d\sigma_n = \int_{R_n} \psi(\sigma)\, d\sigma$$

from which (3) follows, as required.

3.2.4. Finally, let us show that a staircase exists for every polynomial $P(s)$ in normal form on which $|P(s)| \geq C > 0$.

Consider the polynomial

$$P(s) = as_1^m + \sum_{k=0}^{m-1} P_k(s_2,\, ...,\, s_n)\, s_1^k, \qquad a \neq 0,$$

for fixed arbitrary values of $s_2 = \sigma_2,\, ...,\, s_n = \sigma_n$. It has at most m roots $s_1^{(1)},\, ...,\, s_1^{(m)}$ in the s_1-plane and can be factored in the form

$$P(s) = a(s - s_1^{(1)})\, (s - s_1^{(2)}) \cdots (s - s_1^{(m)}). \tag{8}$$

It is always possible to draw a line $\tau_1 = \text{const.}$ in the region $|\tau_1| \leq m + 1$, a strip of width $2m + 2$, whose distance from each of the roots of the polynomial is greater than 1. From (8) it is apparent that $|P(s)|$ is greater than $|a|$ on any such line.

The roots of a polynomial with a constant leading coefficient depend continuously on the remaining coefficients. Thus the roots $s_1^{(1)},\, ...,\, s_1^{(m)}$ corresponding to a sufficiently small neighborhood of the point $(\sigma_2, ..., \sigma_n)$ will be located in the s_1-plane in arbitrarily small circles about their original positions. Therefore the inequality $|P(s)| > |a|$ will be preserved on $\tau_1 = \text{const.}$ in some neighborhood of the point $(\sigma_2, ..., \sigma_n)$ we select.

Each point of $\sigma_2 \ldots \sigma_n$-space can be assigned a neighborhood by the indicated rule. The neighborhoods may be considered bounded by hyperplanes parallel to the coordinate hyperplanes. Using the Heine-Borel Theorem, we can extract from the whole neighborhood cover of $\sigma_2 \ldots \sigma_n$-space a locally finite cover $\overline{\varDelta}_1, \overline{\varDelta}_2, \ldots, \overline{\varDelta}_j, \ldots$ On replacing each $\overline{\varDelta}_j$ by $\varDelta_j = \overline{\varDelta}_j - (\varDelta_1 \cup \cdots \cup \varDelta_{j-1})$, we obtain disjoint regions which together with the values $\tau_1^{(j)}$ determine our desired staircase.

Thus, the existence of a functional $E(s)$ in Z' satisfying

$$\overline{P(s)}\, E(s) = 1$$

has been proved in general and with it the existence of a solution of

$$\overline{P}\left(i\frac{\partial}{\partial x} \right)\mathscr{E}(x) = \delta(x),$$

as required.

3.2.5. Our construction allows considerable freedom in the choice of Hörmander's staircase and this can be used to improve various results concerning the properties of solutions.

For fixed $\sigma_2, \ldots, \sigma_n$, the corresponding cross-section of the staircase was a line in the s_1-plane parallel to the real axis. It is possible to construct a staircase using curved cross-sections not receding from the real axis too quickly. For example, the cross-section L may be chosen in the form†

$$\tau_1 = (a_1 \log^+ |\sigma_1| - a_2)^+ \quad (a_1 > 0). \tag{9}$$

In the strip $|a_2| = \left| \tau_1 - a_1 \log^+ |\sigma_1| \right| \leq m + 1$ there is automatically a curve of the form (9) on which $|P(s)| \geq C > 0$ everywhere. Let us show that

$$\int_{-\infty}^{\infty} \frac{\psi(\sigma_1 + i\tau_1, \sigma_2, \ldots, \sigma_n)}{P(\sigma_1 + i\tau_1, \sigma_2, \ldots, \sigma_n)}\, d\sigma_1, \quad (\sigma_1, \tau_1) \in L,$$

exists. For any k and large $|\sigma_1|$, we have on L and below L

$$|s_1^k \psi(\sigma_1 + i\tau_1, \sigma_2, \ldots, \sigma_n)| \leq C_k\, e^{b|\tau_1|} \leq C_k\, e^{b[a_1 \log^+ |\sigma_1| + |a_2|]} \leq C_k'|\sigma_1|^{ba_1}.$$

Thus the integral will be convergent if we take $k > ba_1 + 1$. $\tag{10}$

---

† As usual, $f^+(x) = \max \{f(x), 0\}$.

On the other hand, an integral along a vertical segment:

$$\int_0^{a_1 \log^+ |\sigma_1| - a_2} \psi(\sigma_1 + i\tau_1, \sigma_2, ..., \sigma_n) \, d\tau_1$$

approaches 0 with increasing $|\sigma_1|$ since by the inequality (10),

$$\left| \int_0^{a_1 \log^+ |\sigma_1| - a_2} \psi(\sigma_1 + i\tau_1, \sigma_2, ..., \sigma_n) \, d\tau_1 \right| \leq (a_1 \log^+ |\sigma_1| + |a_2|) \, C_k' |\sigma_1|^{ba_1 - k}.$$

Therefore by Cauchy's theorem,

$$\int_{-\infty}^{\infty} \psi(\sigma_1 + i\tau_1, \sigma_2, ..., \sigma_n) \, ds_1 = \int_{-\infty}^{\infty} \psi(\sigma_1, \sigma_2, ..., \sigma_n) \, d\sigma_1.$$

Hence it follows that the product of $\overline{P}(s)$ and the functional $E(s)$ defined by

$$(E(s), \psi(s)) = \int_H \frac{\psi(\sigma_1 + i\tau_1, \sigma_2, ..., \sigma_n)}{P(\sigma_1 + i\tau_1, \sigma_2, ..., \sigma_n)} \, ds_1 \, ... \, d\sigma_n, \quad (\sigma_1, \tau_1) \in L,$$

is equal to the functional 1. Thus we obtain a fundamental function also when integrating over a curved staircase. In addition to this, we could likewise go off into the complex space of any of the other variables including several at once.

Problems

1. Show that a fundamental function for any m-th order differential operator has order of singularity at least $1 - m$.

Hint. Use Sec. 2.6.3.

2. Show that the fundamental function $\mathscr{E}(x)$ constructed in Sec. 3.2.3 has order of singularity at most $n + 2$.

Hint. The functional $E(s)$ can be extended to the space A_{n+1} (Sec. 2.8, Prob. 6) with preservation of continuity and hence $\mathscr{E}(x)$ to the space C_{n+1} (since $C_{n+1} \subset F^{-1}(A_{n+1})$). Use Prob. 3 of Sec. 2.5.

3. Show that the fundamental function $\mathscr{E}(x)$ constructed in Sec. 3.2.5 using Hörmander's staircase in the region $|\tau| \leq a_1 \log^+ |\sigma| + a_2$ has order of singularity at most $n + ba_1 + 4$ in the region $|x| \leq b$.

Hint. To prove this, one has to use the integrability not only of $\psi(s)$ but also that of its product with a certain power of s.

12*

3.3. An Equation with Right-Hand Side

3.3.1. In this section, we shall show that the equation

$$\bar{P}\left(i\,\frac{\partial}{\partial x}\right)u(x) = f \tag{1}$$

has a solution in K' for any right-hand side $f \in K'$. If f has compact support, one such solution is

$$u = \mathscr{E}(x) * f, \tag{2}$$

where $\mathscr{E}(x)$ is a fundamental function for the operator $P\left(i\,\dfrac{\partial}{\partial x}\right)$. Formula (2) does not work in the general case since the convolution is undefined when neither of the functionals has compact support.

We note that because of the result of Sec. 3.2.2, we may restrict our attention to operators in normal form.

If we take Fourier transforms in (1), our problem goes over into the problem of solving the equation

$$\bar{P}(s)\,v(s) = g \tag{3}$$

in Z'. The Fourier transform g of the functional f is a certain functional on Z. When f has compact support, $g(s)$ is an entire function having first order exponential growth in the complex domain and power growth in the real domain. To (2), there corresponds a solution of (3) of the form

$$v(s) = E(s)\,g(s), \tag{4}$$

with the functional $E(s)$ given by

$$(E,\,\psi) = \int\limits_{H} \frac{\psi(s)\,ds}{P(s)} \tag{5}$$

for some suitable Hörmander staircase H. Moreover

$$(v,\,\psi) = (gE,\,\psi) = (E,\,g^*\psi) = \int\limits_{H} \frac{\overline{g(\bar{s})}\,\psi(s)\,ds}{P(s)}. \tag{6}$$

To solve (2) in the general case, we observe the following. Suppose that f has been respresented by a convergent series (in K')

$$f = \sum_{k=0}^{\infty} f_k \tag{7}$$

of functionals f_k with compact support. Correspondingly, we have

$$g = \sum_{k=0}^{\infty} g_k(s),$$

where the $g_k(s)$ are entire functions of the above-mentioned form. Suppose further that

$$v_k(s) = g_k(s) E_k,$$

where E_k is a functional given by (5) for some Hörmander staircase H_k. Specifically

$$(E_k, \psi) = \int_{H_k} \frac{\psi(s)\, ds}{P(s)} \tag{8}$$

so that

$$(v_k, \psi) = \int_{H_k} \frac{g_k^*(s)\, \psi(s)}{P(s)}\, ds, \qquad g_k^*(s) = \overline{g_k(\bar{s})}.$$

If we could now select the functionals E_k (i.e., the staircases H_k) so as to make the series

$$v = \sum_{k=0}^{\infty} v_k(s) \tag{9}$$

converge in Z', then we could assert that v is a solution of (3). For, multiplying (9) by $\overline{P}(s)$, we find that

$$\overline{P(s)}\, v = \sum_{k=0}^{\infty} \overline{P}(s)\, v_k = \sum_{k=0}^{\infty} g_k(s) = g,$$

as required. Thus the problem is reduced to selecting the expansion (7) and the functionals (8) so as to assure the convergence of the series (9). In turn, in order for (9) to converge in Z', it is sufficient that the series

$$\sum_{k=0}^{\infty} (v_k, \psi)$$

converge absolutely for each $\psi \in Z$.

3.3.2. We proceed to fulfill this plan. Let a_k be a sequence of positive numbers approaching infinity. We first break up f so that the term

f_0 in (7) is carried in a ball with center at the origin and for $k \geq 1$ the support of each functional f_k falls inside the octant $O_k = \{x: x_j \leq -a_k\}$ $(j = 1, ..., n)$ if a suitable rotation is performed. Assuming that the support of f_k does lie in the specified octant and knowing that $k > 0$, we shall construct a staircase in the strip

$$|\tau_j - \alpha_k \log^+ |\sigma_j|| \leq m + 1.$$

The α_k are positive constants to be chosen below.

We now estimate the term

$$(v_k, \psi) = \int\limits_{H_k} \frac{g_k^*(s)\, \psi(s)}{P(s)}\, ds.$$

As the Fourier transform of a functional with compact support in the octant O_k, $g_k(s)$ satisfies the inequality

$$|g_k(s)| \leq M_k(1 + |s|)^{p_k}\, e^{+a_k \Sigma \tau_j}$$

for $\tau_j < 0$ (Sec. 2.8.3). Correspondingly, an estimate for the function $g_k^*(s) = \overline{g_k(\bar{s})}$ for $\tau_j > 0$ is

$$|g_k^*(s)| \leq M_k(1 + |s|)^{p_k}\, e^{-a_k \Sigma \tau_j}.$$

We also know that $\psi(s)$ satisfies the inequality

$$|\psi(s)| \leq C\, e^{b|\tau|} \leq C_1\, e^{b \Sigma \tau_j} \quad (\tau_j \geq 0).$$

The polynomial $P(s)$ is bounded from below on Hörmander's staircase by constant $|a|$. Combining the estimates, we obtain

$$|(v_k, \psi)| \leq \frac{C_1}{|a|} \int\limits_{H_k} M_k(1 + |s|)^{p_k}\, e^{(b - a_k)\Sigma \tau_j}\, |ds|$$

$$\leq C_1' M_k \int\limits_{R_n} \prod_{j=1}^{n} (1 + |\sigma_j|)^{p_k}\, (1 + |\sigma_j|)^{\alpha_k(b - a_k)}\, d\sigma_1 \ldots d\sigma_n.$$

We now take $\alpha_k = p_k + k^{2/n} M_k^{1/n} + 1$. For any fixed $b > 0$, we have

$$1 + p_k + k^{\frac{2}{n}} M_k^{\frac{1}{n}} + \alpha_k(b - a_k) \leq \alpha_k(1 + b - a_k) \leq 0$$

for sufficiently large k. Hence

$$p_k + \alpha_k(b - a_k) \leq -k^{\frac{2}{n}} M_k^{\frac{1}{n}} - 1$$

for large k and therefore

$$|(v_k, \psi)| \leq C_2 M_k \prod_{j=1}^{n} \int_0^{\infty} (1 + \sigma_j)^{-k^{\frac{2}{n}} M_k^{\frac{1}{n}} - 1} d\sigma_j$$

$$\leq C_2 M_k \left(\frac{1}{k^{\frac{2}{n}} M_k^{\frac{1}{n}}} \right)^n = \frac{C_2}{k^2}.$$

Suppose now that the support of f_k falls in the octant O_k only after a certain rotation U_k is performed, i.e.,

$$\text{supp}\,(f_k(U_k x)) \subset O_k.$$

Set $f_k(U_k x) = \hat{f}_k(x)$. By the foregoing discussion, there exists a Hörmander staircase H_k for the polynomial $P(U_k s)$ for which

$$\int_{H_k} \frac{\hat{g}_k^*(s)\,\psi(U_k s)\,ds}{P(U_k s)} \leq \frac{C_2}{k^2}$$

for any $\psi \in Z$. Here $\hat{g}_k(s)$ is the Fourier transform of $f_k(x)$ and so by formula (14) of Sec. 2.8 is given by

$$\hat{g}_k(s) = F[f_k(U_k x)] = g_k(U_k s).$$

Substituting s' for $U_k s$ in the integral over H_k, we obtain an integral over the staircase $U_k^{-1} H_k$ and for sufficiently large k,

$$\int_{U_k^{-1} H_k} \frac{g_k^*(s')\,\psi(s')}{P(s')}\,ds' \leq \frac{C_2}{k^2}.$$

Thus for $k > 0$, we have determined a staircase (H_k or $U_k^{-1} H_k$) on which

$$|(v_k, \psi)| \leq \frac{C_2}{k^2}$$

for any $\psi \in Z$ for sufficiently large k, C_2 being a constant depending on ψ but not on k. The absolute convergence of the series of (v_k, ψ) is thereby guaranteed.

This completes the proof of the existence of a solution to equation (1) for any right-hand side $f \in K'$.

Problems

1. Prove that two solutions of $P\left(i\dfrac{\partial}{\partial x}\right)u = f$ that coincide for $|x| \geq R$ (for some R) coincide everywhere.

Hint. Apply the uniqueness theorem of Sec. 3.1.2.

2. If polynomial $P(\sigma)$ has no real zeros, then the equation $P\left(i\dfrac{\partial}{\partial x}\right)u = f$ can have only one solution with power growth at infinity (for a more general manner of growth pertaining to S', see Prob. 5 of Sec. 3.1).

3. If polynomial $P(\sigma)$ has a single real zero at $\sigma = 0$, then $P\left(i\dfrac{\partial}{\partial x}\right)u = f$ can have only one solution approaching zero at infinity (see Prob. 8 of Sec. 1.7 and Prob. 5 of Sec. 3.1).

4. Prove that the equation $\Delta u + k^2 u = f$ for $n = 3$ has just one differentiable solution satisfying Sommerfeld's conditions

$$u(x) = O(1). \quad \frac{\partial u}{\partial r} - iku = o\left(\frac{1}{r}\right) \quad (|x| \to \infty).$$

Hint. Use Prob. 6 of Sec. 3.1.

5. Show that $P\left(i\dfrac{\partial}{\partial x}\right)u = f$ always has at least one infinitely differentiable solution if $f(x)$ is an infinitely differentiable function with compact support.

Hint. See Sec. 3.3.1.

6. (Continuation). If *every* solution of $P\left(i\dfrac{\partial}{\partial x}\right)u = f$, where f is an infinitely differentiable function with compact support, is itself infinitely differentiable, then $P\left(i\dfrac{\partial}{\partial x}\right)$ is a hypoelliptic operator.

Hint. Use the relationship between the general and particular solutions.

7. If P is a hypoelliptic operator and $f(x)$ is an infinitely differentiable function (not necessarily having compact support), then every solution of the equation $P\left(i\dfrac{\partial}{\partial x}\right)u = f$ is infinitely differentiable.

Hint. Employ an expansion of unity (Sec. 2.1.4).

3.4. A Condition for Hypoellipticity Based on the Zeros of $P(s)$ (Necessity)

3.4.1. The equation

$$P\left(i\frac{\partial}{\partial x}\right)u(x) = 0 \tag{1}$$

has solutions of the form $e^{-i(x,\,s)} = e^{-i(x_1 s_1 + \cdots + x_n s_n)}$ for certain complex $s_1, \ldots, s_n$. It is easy to derive a condition on the complex vector $s = (s_1, \ldots, s_n) \in C_n$ so that $e^{-i(x,\,s)}$ will be a solution of (1). We have

$$i\frac{\partial}{\partial x_j} e^{-i(x,\,s)} = s_j e^{-i(x,\,s)}, \quad P\left(i\frac{\partial}{\partial x}\right)e^{-i(x,\,s)} = P(s)\,e^{-i(x,\,s)}, \tag{2}$$

and so a necessary and sufficient condition for the last expression to vanish is that s be a root of the equation $P(s) = 0$.

When $n > 1$ the locus of all points a satisfying the equation $P(s) = 0$ is a hypersurface in n-dimensional complex space ($2n$-dimensional real space) having real dimensionality $2n - 2$ (one complex equation is equivalent to two real ones).

It is of course difficult to picture a surface in n-dimensional space. To get an idea of the surface $P(s) = 0$, we shall construct a modulus graph – a certain plane geometric configuration – that will help us judge its structure on the whole from our standpoint of interest.

DEFINITION. The locus of all points in the $\xi\varrho$-plane, where $\xi = |\sigma| = \sqrt{\sum_{j=1}^{n} \sigma_j^2}$, $\varrho = |\tau| = \sqrt{\sum_{j=1}^{n} \tau_j^2}$, and $P(\sigma + i\tau) = 0$, is called the *modulus graph* of the surface $P(s) = 0$.

For $n = 1$, the modulus graph reduces to a finite number of points. To illustrate the various possibilities for $n > 1$, we shall discuss two examples. Let $n = 2$ and $P(s) = s_1^2 - s_2^2$. The condition $P(s) = 0$ leads to the relations $\sigma_1 = \pm\sigma_2$ and $\tau_1 = \pm\tau_2$. Hence $\xi = |\sigma| = \sqrt{2}|\sigma_1|$ and $\varrho = |\tau| = \sqrt{2}|\tau_1|$. Since σ_1 and τ_1 are independent (s_1 may be chosen arbitrarily and s_2 determined from it), *the set of points* $(\xi, \varrho) \in D$ *covers the entire first quadrant of the $\xi\varrho$-plane.*

Again let $n = 2$ and $P(s) = s_1^2 + s_2^2$. In this case, the condition $P(s) = 0$ yields the relations $\sigma_1 = \pm\tau_2$ and $\tau_1 = \mp\sigma_2$. Hence $\xi = |\sigma| = \sqrt{\sigma_1^2 + \sigma_2^2} = \sqrt{\sigma_1^2 + \tau_2^2} = |s_1|$ and $\varrho = |\tau| = \sqrt{\tau_1^2 + \tau_2^2} = \sqrt{\sigma_1^2 + \tau_1^2} = |s_1|$. Thus the modulus graph is the bisector $\varrho = \xi$ of the first quadrant of the $\xi\varrho$-plane.

In general, we shall be interested in the modulus graph only from the standpoint of whether the point (ξ, ϱ) of D can approach the ξ-axis when $\xi \to \infty$ or else must recede from the ξ-axis and if so in what way. In other words, what function of the form $\varrho = \varrho(\xi)$ serves as a lower boundary in the set D?

From this standpoint, consider an arbitrary polynomial $P(s_1, s_2)$ of two variables s_1 and s_2. Performing a real rotation if necessary, we may assume the equation $P(s) = 0$ to be in normal form (Sec. 3.2.2) with its principal part explicitly isolated as follows:

$$P(s_1, s_2) \equiv a_0 s_1^m + \sum_{k=1}^{m} a_k s_1^{m-k} s_2^k + \sum_{j<m} a_{jk} s_1^{j-k} s_2^k = 0. \qquad (3)$$

Consider first the case where s_1 is actually present in each term of (3) so that $P(0, s_2) \not\equiv 0$. Then every point $\sigma = (0, \sigma_2)$ and $\tau = (0, \tau_2)$ lies on the surface of zeros and the modulus graph D covers the entire first quadrant of the $\xi\varrho$-plane. Now suppose that $P(0, s_2) \neq 0$. Let s_2 be an infinitely distant point. Then in a neighborhood of s_2, each of the roots of (3) can be represented by

$$s_1 = A s_2^\gamma E, \quad A \neq 0, \qquad (4)$$

where γ is a real number and E has the limit 1 as $|s_2| \to \infty$.†

Take $s_2 = \sigma_2$ to be real and positive. Then there is a root $s_1 = (\sigma_1, \tau_1)$ for which

$$|\tau| = |\tau_1| \leq |s_1| \leq 2|A|\,|\sigma_2|^\gamma \leq 2|A|\,|\sigma|^\gamma$$

for sufficiently large $|\sigma_2|$. Thus the modulus graph will contain a curve on which $\varrho \leq C\xi^\gamma$.

To determine γ, we substitute the expression (4) in (3) obtaining

$$a_0 A^m s_2^{m\gamma} + \sum_{k=1}^{m} a_k A^{m-k} s_2^{k+\gamma(m-k)} + \sum_{j<m} a_{jk} A^{j-k} s_2^{k+\gamma(j-k)} \equiv 0.$$

† See E. Hille, *Analytic Function Theory*, Vol. 2, Ginn and CO., New York, 1962, Chapt. 12.

The highest power of s_2 in this identity must appear at least twice. Otherwise dividing by the highest power of s_2 and making $|s_2|$ increase indefinitely, we would conclude that $A = 0$.

Let $n(\gamma)$ denote the exponents on s_2 as a function of γ. The graph of $n(\gamma)$ has the form depicted in Fig. 8. The values of γ for which there are two equal highest exponents are the abscissas of the corners of the polygonal path $\mu(\gamma) = \max n(\gamma)$.

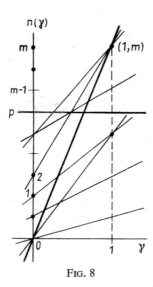

FIG. 8

Since $a_0 \neq 0$, the line $n(\gamma) = m\gamma$ is necessarily part of the graph of exponents. Moreover, since $P(0, s_2) \not\equiv 0$, there is a term in (3) of the form $b_p s_2^p$ with $p \leq m$. Thus the graph of exponents also includes the horizontal line $n(\gamma) = p$. The intersection of the two indicated lines is a point with abscissa $\gamma = \dfrac{p}{m} \leq 1$. Therefore *the smallest exponent γ in* (4) *does not exceed* 1 and (3) has a root representable for $|s_2| \to \infty$ by

$$s_1 = A s_2^\gamma E, \quad A \neq 0, \quad \gamma \leq 1.$$

Let $s_2 = \sigma_2$ be real and positive. Then there is a root $s_1 = (\sigma_1, \tau_1)$ for which

$$|\tau| = |\tau_1| \leq |s_1| \leq 2|A| \, |\sigma_2|^\gamma \leq 2|A| \, |\sigma|^\gamma$$

for sufficiently large $|\sigma_2|$. Thus the modulus graph always contains a curve of the form $\varrho = C\xi^\gamma E$, where $\gamma \leq 1$ and $E \to 1$.

We point out one case where it is possible to say at once that the modulus graph contains such a curve with $\gamma < 1$. This will happen when the term $a_m s_2^m$ is absent from (3) (i.e., $a_m = 0$). Then p is certainly less than m and $\gamma = \dfrac{p}{m} < 1$.

3.4.2. It will be recalled that a differential operator $P\left(\dfrac{\partial}{\partial x}\right)$ is *hypoelliptic* if all solutions of $P\left(\dfrac{\partial}{\partial x}\right) u = 0$ are infinitely differentiable functions.

It turns out that a hypoelliptic operator may be completely characterized in terms of its modulus graph. Namely, the modulus graph of a hypoelliptic operator recedes from the ξ-axis with increasing ξ. More precise statements will be given below.

We first establish a lemma concerning the derivatives of the solution of a hypoelliptic equation. The lemma is of great importance in its own right.

LEMMA **1.** *If the solutions $u(x)$ of a hypoelliptic equation are uniformly bounded in a region V, then their partial derivatives $D^k u(x)$ are uniformly bounded in any region W lying in V together with its closure.*

Proof. From formula (3) of Sec. 3.1 it follows that a solution $u(x)$ satisfies the relation

$$u(x) = P\left(\frac{\partial}{\partial x}\right)[(1 - \alpha)\,\mathscr{E}] * u(x) = A(x) * u(x) = \int_G A(x - \xi)\,u(\xi)\,d\xi$$

(5)

in the region W. Here $A(x) = P\left(\dfrac{\partial}{\partial x}\right)[(1 - \alpha)\,\mathscr{E}]$ is an infinitely differentiable function, and the integration extends over a bounded region $G \supset V$ outside of which $u(x)$ vanishes. The equation (5) may be differentiated with respect to x to obtain

$$D^k u(x) = \int_G D^k A(x - \xi)\,u(\xi)\,d\xi.$$

Hence

$$\max_{x \in W} |D^k u(x)| \leq C \max_{x \in G} |u(x)|,$$

(6)

where

$$C = \int_G |D^k A(x - \xi)|\,d\xi.$$

There is no loss of generality if in (6) $\max\limits_{x\in G} |u(x)|$ is replaced by $\max\limits_{x\in V} |u(x)|$. Indeed, $u(x)$ was determined in G by multiplying a solution of (1) in G by an infinitely differentiable function $\beta(x)$ equaling 1 in V and 0 outside G. By choosing $\beta(x)$ so that it falls off from 1 to 0 sufficiently fast, we can obtain for the product the inequality

$$\max_{x\in G} |u(x)| \leqq \max_{x\in V} |u(x)| + \varepsilon$$

for any $\varepsilon > 0$. Since G is independent of ε, on letting $\varepsilon \to 0$ we find using (6) that

$$\max_{x\in W} |D^k u(x)| \leqq C \max_{x\in V} |u(x)|, \tag{7}$$

thus proving the lemma.

3.4.3. We are now in a position to deduce a characteristic property of the surface of zeros of a hypoelliptic operator $P\left(i\dfrac{\partial}{\partial x}\right)$ in terms of its modulus graph.

Suppose that $P\left(i\dfrac{\partial}{\partial x}\right)$ is hypoelliptic. We shall also say that the polynomial $P(s)$ is hypoelliptic. Let W and V be two balls centered at the origin with $W \subset V$. The idea is to apply the inequality (7) with $|k| = 1$ to the solution $u(x) = e^{-i(x,s)}$ $[P(s) = 0]$. Let r and R denote the respective radii of W and V. Further let $s_j = \sigma_j + i\tau_j$. Then we can write

$$\max_{x\in V} |u(x)| = \max_{|x|\leqq R} |e^{-i(x,s)}| = \max_{|x|\leqq R} e^{x_1\tau_1 + \cdots + x_n\tau_n} = e^{R|\tau|},$$

where $|\tau| = \sqrt{\Sigma\tau_j^2}$. Further

$$\max_{x\in W} \left|\frac{\partial u}{\partial x_j}\right| = |s_j| \max_{|x|\leqq r} |e^{-i(x,s)}| = |s_j| e^{r|\tau|}.$$

Thus for any j, we have

$$|s_j| e^{r|\tau|} \leqq C e^{R|\tau|}$$

or

$$|s_j| \leqq C e^{(R-r)|\tau|}.$$

Squaring and summing over j, we obtain $|s| \leqq C_1 e^{(R-r)|\tau|}$. Hence $|\sigma| \leqq C_1 e^{(R-r)|\tau|}$ and so

$$|\tau| \geqq \frac{1}{R-r} (\log |\sigma| - \log C_1) = A \log |\sigma| - B, \tag{8}$$

where $A = 1/(R - r)$ and $B = \log C_1/(R - r)$. We have deduced the very important result that *the inequality* (8) *holds on the surface* $P(s) = 0$. Stated otherwise, if we go off to infinity along the surface in such a way that the absolute value of the real part of $s = \sigma + i\tau$ becomes infinite, then the absolute value of the imaginary part will also become infinite and moreover faster than $A \log |\sigma|$. We note that the coefficient A in (8) may be made as large as desired by choosing R and r to be arbitrarily close.

Since $\overline{P}(s)$ and $P(\bar{s})$ are equal in absolute value, the surface of zeros of the hypoelliptic polynomial $\overline{P}(s)$ possesses the same property.

The modulus graph of the surface $P(s) = 0$ thus lies above the curve $\varrho = A \log \xi - B$ for some positive A (Fig. 9).

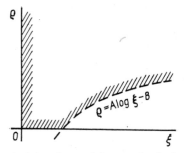

FIG. 9. The region containing the modulus graph of a hypoelliptic operator.

3.4.4. Suppose now that $P\left(i\dfrac{\partial}{\partial x}\right)$ is a β-hypoelliptic operator. Then all the solutions of $P\left(i\dfrac{\partial}{\partial x}\right)u = 0$ are locally of Gevrey class G_β so that $|D^q u(x)| \leqq CB^q q^{q\beta}$ (Sec. 3.1.5). In this instance, it is possible to make a stronger statement about how fast the modulus graph recedes from the real axis.

Theorem. *If* $P(s)$ *is a* β-*hypoelliptic polynomial, then its modulus graph lies above the curve* $\varrho = b\xi^{1/\beta} - C$ *for some positive constants* b *and* C.

The proof rests on the following refinement of Lemma 1:

LEMMA 2. *If* $u(x)$ *is any solution of a* β-*hypoelliptic equation in a region* V, *then its derivatives satisfy the estimate*

$$\max_{x \in W} |D^q u(x)| \leq CB^q q^{q\beta} \max_{x \in V} |u(x)|$$

in any region W lying in V together with its closure.

Proof. By virtue of formula (10) of Sec. 3.1,

$$|D^q u(x)| \leq \sum_k \max_\Gamma \left| Q_k \left(\frac{\partial}{\partial \eta} \right) u(\eta) \right| \int_\Gamma D^q c_k(x - \eta) \, d\eta$$

for any $x \in W$, where Γ is the boundary of W. By Sec. 3.1.5, $c_k(x)$ is o class G_β together with the fundamental solution $\mathscr{E}(x)$ and so

$$\int_\Gamma |D^q c_k(x - \eta)| \, d\eta = \int_{\Gamma_1} |D^q c_k(\xi)| \, d\xi \leq CB^q q^{q\beta}.$$

On the other hand, Lemma 1 implies that

$$\max_\Gamma \left| Q_k \left(\frac{\partial}{\partial \eta} \right) u(\eta) \right| \leq C_1 \max_V |u(\eta)|.$$

Combining these estimates, we obtain

$$\max_W |D^q u(x)| \leq C_2 B^q q^{q\beta} \max_V |u(x)|, \tag{9}$$

as required.

3.4.5. Again let W and V be two balls of radius r and R centered at the origin with $W \subset V$. Take $u(x)$ to be $e^{-i(x,s)}$, where $P(s) = 0$. As usual, let $s = \sigma + i\tau$. Then

$$\max_{|x| \leq R} |u(x)| = e^{R|\tau|}$$

and

$$\max_{|x| \leq r} |D^q u(x)| = |s_1^{q_1} \ldots s_n^{q_n}| e^{r|\tau|}.$$

It therefore follows from (9) that

$$|s_1^{q_1} \ldots s_n^{q_n}| e^{r|\tau|} \leq CB_1^{q_1} \ldots B_n^{q_n} q_1^{q_1\beta} \ldots q_n^{q_n\beta} e^{R|\tau|}.$$

Hence it further follows that

$$e^{(r-R)|\tau|} \leq C \frac{B_1^{q_1} \ldots B_n^{q_n}}{|s_1|^{q_1} \ldots |s_n|^{q_n}} q_1^{q_1\beta} \ldots q_n^{q_n\beta}.$$

Taking the infimum over all $q_1, \ldots, q_n$, we obtain (see the next subsection)

$$e^{(r-R)|\tau|} \leqq C_1 e^{-\Sigma b_j|s_j|^{1/\beta}}. \tag{10}$$

In other words,

$$|\tau| \geqq \sum b'_j|s_j|^{1/\beta} - C_2 \geqq b|s|^{1/\beta} - C_2 \geqq b|\sigma|^{1/\beta} - C_2,$$

as required.

3.4.6. We still have to show that for any $B > 0$ and $d > 0$

$$\inf_q \frac{B^q q^{q\beta}}{d^q} \leqq C e^{-bd^{1/\beta}} \quad (q = 0, 1, 2, \ldots).$$

Temporarily considering q to be a continuously varying quantity, let us determine the minimum of the function $f(q) = a^q q^{q\beta}$ by the usual procedures of the differential calculus. Differentiating $\log f$ and equating the result to zero, we obtain

$$[\log f(q_0)]' = \frac{f'(q_0)}{f(q_0)} = \log a + \beta \log q_0 + \beta = 0, \tag{11}$$

where q_0 is the value of q where the minimum is attained. From (11), we find that $q_0 = 1/ea^{1/\beta}$ and hence

$$\min_q \log f(q) = -\frac{\beta}{ea^{1/\beta}}, \quad \min_q f(q) = e^{-\frac{\beta}{ea^{1/\beta}}}.$$

Since q actually takes on only positive integral values, $\min f(q)$ is slightly higher than the one found. Now, we have

$$[\log f(q)]'' = \frac{\beta}{q}.$$

Therefore at the integer q_1 closest to q_0 on the right,

$$\log f(q_1) = \log f(q_0) + \frac{(q_1 - q_0)^2}{2} [\log f(\bar{q})]'' \leqq \log f(q_0) + \frac{\beta}{2q_0},$$

where $q_0 \leqq \bar{q} \leqq q_1$.

Thus

$$\min_q \log f(q) \leqq \log f(q_1) \leqq \log f(q_0) + \frac{\beta}{2q_0} = -\frac{\beta}{ea^{1/\beta}} + \frac{\beta ea^{1/\beta}}{2}$$

and so

$$\min_q f(q) \leq e^{\frac{\beta e a^{1/\beta}}{2}} e^{-\frac{\beta}{ea^{1/\beta}}}.$$

The first factor is bounded by $e^{\beta e/2}$ for $a \leq 1$. But if $a > 1$, then

$$\min_q a^q q^{a\beta} \leq 1 \leq e^{\frac{\beta}{e}} e^{-\frac{\beta}{ea^{1/\beta}}}.$$

Thus for any positive a,

$$\min_q a^q q^{a\beta} \leq C e^{-\frac{\beta}{ea^{1/\beta}}},$$

as required.

Problems

1. Show that the lower boundary of the modulus graph of the polynomial

$$P(s_1, s_2) = a_0 s_1^m + \sum_{k=1}^{m} a_k s_1^{m-k} s_2^k + \sum_{j<k} a_{jk} s_1^{j-k} s_2^k$$

contains the curve $\varrho = C\xi^\gamma$ with $\gamma < 0$ if and only if $P(s_1, s_2)$ has a term of the form $s_1^r s_2^k$ where k exceeds the degree of the polynomial $P(0, s_2)$ and $r \neq 0$.

Hint. Consider the graph of exponents (see Sec. 3.4).

2. Show that the lower boundary of the modulus graph (in the class of monotonic curves $\varrho = f(\xi)$) is $\varrho = 0$ if and only if the equation $P(\sigma_1, \sigma_2) = 0$ defines a curve going off to infinity.

3. If the modulus graph of a polynomial $P(s) = P(s_1, ..., s_n)$ lies entirely in the region $\varrho \geq \xi t(\xi)$ where $t(\xi) \to \infty$ when $\xi \to \infty$, then $n = 1$. In other words, P is a polynomial in one variable and so has only a finite number of zeros.

3.5. A Condition for Hypoellipticity Based on the Zeros of $P(s)$ (Sufficiency)

3.5.1. In Sec. 3.4, it was shown that the points of the modulus graph of a hypoelliptic polynomial $P(s)$ satisfy an inequality of the form

$$|\tau| \geq A \log |\sigma| - A_1 \tag{1}$$

for $A > 0$. And if $P(s)$ is also β-hypoelliptic, i.e., the solutions of $P\left(i\dfrac{\partial}{\partial x}\right)u$ $= 0$ have derivatives with growth governed by the condition

$$|D^a u(x)| \leqq CB^a q^{a\beta},$$

then the points of the modulus graph satisfy the more stringent inequality

$$|\tau| \geqq A_2 |\sigma|^{1/\beta} - A_3. \tag{2}$$

We shall now show that the inequalities (1) and (2) not only are necessary but are also respectively sufficient conditions for the hypoellipticity and β-hypoellipticity of the polynomial P.

We first establish the following lemma:

LEMMA 1. *If the modulus graph of the surface $P(s) = 0$ lies entirely in the region $\varrho \geqq f(\xi)$, with $f(\xi) \to \infty$ as $\xi \to \infty$ and $f'(\xi)$ bounded, then for sufficiently large ξ, $|P(s)| \geqq R$ in the region $\varrho \leqq \dfrac{1}{2} f(\xi)$ for some arbitrarily prescribed constant R.*

Proof. We already know that $P(s)$ can always be assumed to be in normal form

$$P(s) = a_0 s_1^m + \sum_{k=0}^{m-1} P_k(s_2, \ldots, s_n)\, s_1^k, \qquad a_0 \neq 0.$$

Let $\lambda_j = \lambda_j(s_2, \ldots, s_n)$ $(j = 1, 2, \ldots, m)$ be the roots of $P(s)$ (relative to s_1). Further let $\lambda_j = \xi_j + i\eta_j$. Then by hypothesis,

$$\sqrt{\eta_j^2 + \tau_2^2 + \cdots + \tau_n^2} \geqq f\left(\sqrt{\xi_j^2 + \sigma_2^2 + \cdots + \sigma_n^2}\right),$$

for $j = 1, \ldots, m$. Consider the value of $|P(s)|$ in the region $|\tau| \leqq \dfrac{1}{2} f(\xi)$, in other words, where

$$\sqrt{\tau_1^2 + \tau_2^2 + \cdots + \tau_n^2} \leqq \frac{1}{2} f\left(\sqrt{\sigma_1^2 + \sigma_2^2 + \cdots + \sigma_n^2}\right).$$

We have

$$|P(s)| = \left| a_0 \prod_{j=1}^m (s_1 - \lambda_j) \right| = |a_0| \prod_{j=1}^m \sqrt{(\sigma_1 - \xi_j)^2 + (\tau_1 - \eta_j)^2}.$$

Let us show that each of the m factors on the right-hand side exceeds a positive constant for sufficiently large $|\sigma|$. For conciseness, set $\tau_1 = t$,

$\tau_2^2 + \cdots + \tau_n^2 = q^2$, $\sigma_1 = \theta$, $\sigma_2^2 + \cdots + \sigma_n^2 = p^2$, $\eta_j = \eta$, and $\xi_j = \xi$.
From the relations

$$\sqrt{t^2 + q^2} \leqq \frac{1}{2} f\left(\sqrt{\theta^2 + p^2}\right), \tag{3}$$

$$\sqrt{\eta^2 + q^2} \geqq f\left(\sqrt{\xi^2 + p^2}\right)$$

we must deduce that for any p and q,

$$\sqrt{(\theta - \xi)^2 + (t - \eta)^2} \geqq C \tag{4}$$

for some prescribed constant $C \geqq 0$ when $|\sigma|$ is sufficiently large. If $|\theta - \xi| \geqq C$, the inequality (4) is clearly satisfied. Therefore it is enough to restrict our attention to the case $|\theta - \xi| \leqq C$. We have

$$\left|\sqrt{\theta^2 + p^2} - \sqrt{\xi^2 + p^2}\right| \leqq C,$$

$$\left|f\left(\sqrt{\xi^2 + p^2}\right) - f\left(\sqrt{\theta^2 + p^2}\right)\right| \leqq C \max f'(u) \leqq BC.$$

Now from the inequalities of (3) it is apparent that $\eta > t$ when $|\sigma|$ is sufficiently large. Since $\sigma^2 = \theta^2 + p^2 \to \infty$, $f\left(\sqrt{\theta^2 + p^2}\right) \to \infty$ and $f\left(\sqrt{\xi^2 + p^2}\right) \to \infty$. Hence

$$\sqrt{(\theta - \xi)^2 + (t - \eta)^2} \geqq \eta - t \geqq \sqrt{\eta^2 + q^2} - \sqrt{t^2 + q^2}$$

$$\geqq f\left(\sqrt{\xi^2 + p^2}\right) - \frac{1}{2} f\left(\sqrt{\theta^2 + p^2}\right)$$

$$= \frac{1}{2} f\left(\sqrt{\xi^2 + p^2}\right) + \frac{1}{2}\left[f\left(\sqrt{\xi^2 + p^2}\right) - f\left(\sqrt{\theta^2 + p^2}\right)\right]$$

$$\geqq \frac{1}{2} f\left(\sqrt{\xi^2 + p^2}\right) - \frac{1}{2} BC \to \infty,$$

which is even more than was asserted.

3.5.2. We now proceed to prove the main theorem:

THEOREM 1. *If the manifold* $N(P) = \{s: P(s) = 0\}$ *lies in the region*

$$|\tau| \geqq C \log |\sigma| - C_1 \tag{5}$$

for any C, then a fundamental function $\mathscr{E}(x)$ *for the operator* $\overline{P}\left(i\dfrac{\partial}{\partial x}\right)$ *is an ordinary infinitely differentiable function everywhere outside the origin.*

13*

Proof. Let $\varphi(x_1, ..., x_n)$ be a test function with support lying in a small region G not containing the origin. We wish to show that the expression $(\mathscr{E}(x), \varphi(x))$ can then be written as an integral over G of a product of $\varphi(x)$ with a certain infinitely differentiable function.

As we already know, the operator $P\left(i\dfrac{\partial}{\partial x}\right)$ may be assumed to be in normal form,† so that

$$P(s) = a_0 s_1^m + \sum_{k=0}^{m-1} P_k(s_2, ..., s_m)\, s_1^k.$$

As we saw in Sec. 3.2, the Fourier transform of a fundamental function is a functional E on Z defined by

$$(E, \psi) = \int_H \frac{\psi(s)\, ds}{P(s)}, \tag{6}$$

where H is a suitable Hörmander staircase. For simplicity, we shall consider the case $n = 2$. Hörmander's staircase is then the manifold defined by the conditions

(a) $s_1 = \sigma_1 + i\tau_1$, $-\infty < \sigma_1 < \infty$, $\tau_1 = l(\sigma_2)$ a piecewise constant function of σ_2;

(b) $s_2 = \sigma_2$, $-\infty < \sigma_2 < \infty$.

We now resort to the inequality (5) to help us refine the surface of integration in (6).

By Lemma 1, the inequality

$$|P(\sigma)| \geq 1$$

holds in the $\sigma_1\sigma_2$-plane for sufficiently large $|\sigma| > a$ (depending on just the constants C and C_1 in (5)).

Therefore all of the steps of Hörmander's staircase whose projections on the $\sigma_1\sigma_2$-plane lie entirely outside the disc $|\sigma| \leq a$ may be lowered to the plane. In other words, the piecewise constant function $l(\sigma_2)$ may be assumed to vanish for $|\sigma_2| \geq a$. Thus there remain just a finite number of steps on the staircase with projections intersecting the disc $|\sigma| \leq a$.

Consider any one of these steps. Within the step, the integration with respect to the variable σ_1 is carried out along the line $s_1 = \sigma_1 + i\tau_1$, $\tau_1 = \text{const}$. However the portions of the line outside the segment $|\sigma_1| \leq a$

---

† By using the theorem of Sec. 3.6.2, one can show that a hypoelliptic polynomial has a normal form in any other argument.

may be replaced by vertical segments and portions of the σ_1-axis. Thus we can reduce the entire Hörmander staircase to a figure standing above the $\sigma_1\sigma_2$-plane within the square $|\sigma_1| \leq a$, $|\sigma_2| \leq a$ and coinciding with the $\sigma_1\sigma_2$-plane outside it.

We now split (6) into a sum of two integrals, one over the part corresponding to the square $Q_1 = \{\sigma: |\sigma_1| \leq a, |\sigma_2| \leq a\}$ with its associated collection of vertical segments specified above and one over the exterior Q of the square. Consider first the integral corresponding to the square Q_1 given by

$$(E_1, \psi) = \int_{Q_1} \frac{\psi(s)\,ds}{P(s)}.$$

We have

$$(E_1, \psi) = \int_{Q_1} \frac{1}{P(s)} \left\{ \int_V \varphi(x)\, e^{i(x,\,s)}\,dx \right\} ds$$

$$= \int_V \left\{ \int_{Q_1} \frac{e^{i(x,\,s)}}{P(s)}\,ds \right\} \varphi(x)\,dx = \int_V \mathscr{E}_1(x)\,\varphi(x)\,dx, \qquad (7)$$

where V contains the support of $\varphi(x)$. The function

$$\mathscr{E}_1(x) = \int_{Q_1} \frac{e^{i(x,\,s)}}{P(s)}\,ds \qquad (8)$$

is evidently infinitely differentiable (and even analytic) for all x.

The second term

$$(\hat{E}, \psi) = \int_{\hat{Q}} \frac{\psi(\sigma)}{P(\sigma)}\,d\sigma$$

cannot be so transformed by reversal of the order of integration. The region of integration is no longer bounded and the integral analogous to (8) is divergent. We can bring the second term into the form (7) by replacing the plane portions of the staircase H by curved ones.

Let us first examine the fundamental function $\mathscr{E}(x)$ in a neighborhood V of a point (x_1, x_2) in the first quadrant lying at a positive distance d from both coordinate axes; the function $\varphi(x)$ to which $\mathscr{E}(x)$ is applied is carried in V. We divide the region of integration $\hat{Q}$ into eight parts Q_2,

$Q_2, \ldots, Q_9$ as shown in Fig. 10. Correspondingly, we write

$$(\hat{E}, \psi) = (E_2, \psi) + \cdots + (E_9, \psi).$$

We proceed now to suitably deform the regions of integration of the respective integrals.

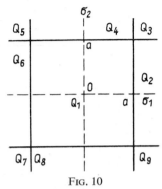

FIG. 10

Consider the functional

$$(E_2, \psi) = \int_{Q_2} \frac{\psi(\sigma)\, d\sigma}{P(\sigma)}. \tag{9}$$

In Q_2, the coordinate σ_2 is bounded while σ_1 varies from the positive value a to $+\infty$.

For fixed σ_2, we move into the s_1-plane and replace the integral along the half-line $a \leq \sigma_1 < \infty$, $\tau_1 = 0$, by an integral along the curve $2\tau_1 = C \log^+ \sigma_1 - C_1$ plus an appropriate segment of the line $\sigma_1 = a$ (Fig. 11). Let us prove that the integral along the original half-line equals the integral along the new contour. This amounts to showing that the integral along a vertical segment AA_1 (A on the σ_1-axis and A_1 on the specified curve) approaches zero when the segment moves off to the

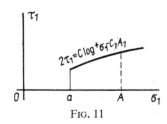

FIG. 11

right indefinitely. By Lemma 1, the quantity $\dfrac{1}{|P(s)|}$ remains bounded on the segment. Therefore using inequality (3) of Sec. 2.8.1, we find that

$$\left| \int_A^{A_1} \frac{\psi(s_1, \sigma_2)\, ds_1}{P(s_1, \sigma_2)} \right| \leqq C_0 \max_{s_1 \in AA_1} |\psi(s_1, \sigma_2)|\, \overline{AA_1}$$

$$\leqq C_k' \sigma_1^{\frac{bC}{2} - k} \log \sigma_1 .$$

Thus for sufficiently large k, the latter approaches zero as $\sigma_1 \to \infty$.

Since the estimate obtained is uniform with respect to all σ_2, the double integral

$$\int_{H_2} \frac{\psi(s)\, ds}{P(s)}$$

over the region H_2 determined by the conditions

$$s_1 = \sigma_1 + i\tau_1, \quad 2\tau_1 = C \log^+ \sigma_1 - C_1, \quad a < \sigma_1 < \infty,$$

$$s_2 = \sigma_2, \qquad |\sigma_2| \leqq a,$$

is equal to the original double integral (9). We now have

$$(E_2, \psi) = \int_{H_2} \frac{1}{P(s)} \left\{ \int_V e^{i(x,s)}\, \varphi(x)\, dx \right\} ds .$$

If we formally reverse the integrations with respect to s and x, we obtain

$$(E_2, \psi) = \int_V \varphi(x) \left\{ \int_{H_2} \frac{e^{i(x,s)}}{P(s)}\, ds \right\} dx . \tag{10}$$

To justify this, we must show that

$$\int_{H_2} \frac{e^{i(x,s)}}{P(s)}\, ds \tag{11}$$

is absolutely and uniformly convergent. But owing to the positiveness of

τ_1 on H_2, we have

$$|e^{i(x,\,s)}| = e^{-(x,\,\tau)} = e^{-x_1\tau_1} \leqq e^{-d\tau_1} = B\,e^{-dA\,\ln+\sigma_1} = B\sigma_1^{-dA},$$

$$|ds| = |ds_1|\,|ds_2| \leqq B_1\,d\sigma_1\,d\sigma_2 \quad \left(A = \frac{C}{2}\right).$$

This in conjunction with the boundedness of $\dfrac{1}{|P(s)|}$ leads to the absolute and uniform convergence of the integral (11). From (10) it follows that

$$\mathscr{E}_2(x) = \int_{H_2} \frac{e^{i(x,\,s)}}{P(s)}\,ds.$$

If we formally differentiate $\mathscr{E}_2(x)$ with respect to x_1 and x_2, we obtain factors of s_1 and s_2 under the integral sign. Clearly, the convergence of the integral will not be disturbed by the factors s_2, no matter how many, nor by the factors s_1 provided their number is such that the expression

$$|s_1^k\sigma_1^{-dA}| \leqq C'\sigma_1^k\sigma_1^{-dA}$$

remains integrable. But by hypothesis, A may be chosen arbitrarily large. Therefore independently of the specific choice of A, $\mathscr{E}_2(x)$ can be differentiated any number of times with respect to x_1. Thus $\mathscr{E}_2(x)$ is also an infinitely differentiable function.

A similar discussion can be carried out for the regions Q_4, Q_6, and Q_8. Consider now a region of type Q_3, where $a \leqq \sigma_1 < \infty$ and $a \leqq \sigma_2 < \infty$. Here we proceed successively from the original region of integration Q_3 to a region H_3 defined by

$$a \leqq \sigma_1 < \infty, \quad 2\tau_1 = C\log^+ \sigma_1 - C_1, \quad a \leqq \sigma_2 < \infty, \quad \tau_2 = 0,$$

and then to a region $\hat{H}_3$ defined by

$$a \leqq \sigma_1 < \infty, \quad 2\tau_1 = C\log^+ \sigma_1 - C_1,$$
$$a \leqq \sigma_2 < \infty, \quad 2\tau_2 = C\log^+ \sigma_2 - C_1.$$

The double integral

$$\int_{\hat{H}_3} \frac{\psi(s)}{P(s)}\,ds$$

equals the double integral

$$\int_{Q_3} \frac{\psi(s)}{P(s)} \, ds$$

for the same reasons as before. If we next replace $\psi(s)$ by its expression in terms of $\varphi(x)$, we arrive at

$$\mathscr{E}_3(x) = \int_{\hat{H}_3} \frac{e^{i(x,s)}}{P(s)} \, ds$$

with the estimate

$$|e^{i(x,s)}| \leq C_0 \sigma_1^{-dA} \sigma_2^{-dA} \qquad \left(A = \frac{C}{2} \right)$$

holding on $\hat{H}_3$. This estimate assures the infinite differentiability of $\mathscr{E}_3(x)$ with respect to x_1 and x_2. A similar argument goes for the regions Q_5, Q_7 and Q_9.

Thus for x_1 and x_2 positive, the function $\mathscr{E} = \mathscr{E}_1 + \cdots + \mathscr{E}_9$ is infinitely differentiable. Although the infinite differentiability holds with regard to the space K', the considerations of Sec. 2.6 imply that it also holds in the usual sense.

Consider now an arbitrary point (x_1, x_2). If x_1 and x_2 are both not equal to zero, then by a rotation (tantamount to changing the signs of the coordinate axes), it is possible to transfer the point to a position where both x_1 and x_2 are positive if not already. Of course the polynomial $P(s)$ is changed by the rotation. But since only the absolute values of σ_j and τ_j appear in the inequality (5), it is preserved for the transformed polynomial $P(s)$. Since the dimensions of $Q_1, \ldots, Q_9$ depend on just the constants C and C_1, these regions are unaffected. The subsequent reasoning is unchanged being based only on the assumption that $|P(s)| \geq R$ and not on the explicit form of $P(s)$.

If one of the variables x_1 or x_2 is equal to zero, a small rotation will make it positive and the previous argument applies. However the rotation should not be so large as to destroy the normal form of the polynomial.

The entire discussion is similar for the case of more than two variables and involves obvious analytical complications concomitant with increasing the number of variables.

Thus Theorem 1 is completely proved.

3.5.3. Theorem 2. *If the manifold* $N(P) = \{s : P(s) = 0\}$ *lies in the region*

$$|\tau| \geq C_2|\sigma|^{1/\beta} - C_3,$$

then a fundamental function $\mathscr{E}(x)$ *for* $\overline{P}\left(i\dfrac{\partial}{\partial x}\right)$ *satisfies the inequality*

$$|D^q\mathscr{E}(x)| \leq CB^q q^{q\beta} \quad (|q| = 0, 1, 2, \ldots)$$

and so belongs to Gevrey's class G_β *outside any neighborhood of the origin.*

Proof. Using the method of proof of the preceding theorem assuming that $x > 0$, we can represent the expression for $\mathscr{E}(x)$ as a finite sum of integrals (9 for $n = 2$) of the form

$$\mathscr{E}_j(x) = \int_{H_j} \frac{e^{i(x,s)}}{P(s)}\, ds. \tag{12}$$

The integrals extend over the parts of Hörmander's staircase deformed as follows. If one of the coordinates say σ_1 varies from a to $+\infty$ in the projection Q_j of H_j, then the horizontal part of the path of integration in the s_1-plane is replaced by the curve

$$a \leq \sigma_1 < \infty, \quad \tau_1 = A\log\sigma_1 - A_1.$$

For definiteness consider the part H_2. We deform the contour once again replacing it by the curve

$$a \leq \sigma_1 < \infty, \quad 2\tau_1 = A_2\sigma_1^{1/\beta} - A_3.$$

We must show that the integral (12) exists along the new contour and has the previous value. It is enough to verify that the integral along the vertical part given by

$$\int_{A\log\sigma_1 - A_1}^{\frac{1}{2}\left(A_2\sigma_1^{1/\beta} - A_3\right)} \frac{e^{i(x,s)}}{P(s)}\, ds_1$$

approaches 0 as $\sigma_1 \to \infty$. But as before we have $|P(s)| \geq 1$,

$$|e^{ix_1s_1}| = |e^{ix_1(\sigma_1 + i\tau_1)}| = e^{-x_1\tau_1} \leq C\,e^{-Ad\log\sigma_1} = C\sigma_1^{-Ad},$$

and so

$$\left| \frac{\frac{1}{2}\left(A_2\sigma_1^{1/\beta} - A_3\right)}{\int\limits_{A\log\sigma_1 - A_1} \frac{e^{i(x,s)}}{P(s)}\,ds_1 \right| \leq C_1\sigma_1^{1/\beta}\,\sigma_1^{-Ad}.$$

Since A may be taken arbitrarily large, the resulting expression approaches 0 as $\sigma_1 \to \infty$.

On the contour itself we have

$$|e^{ix_1s_1}| \leq C\,e^{-\frac{A_2}{2}d\sigma_1^{1/\beta}},$$

and we obtain for $\dfrac{\partial^{q_1}\mathscr{E}_2(x)}{\partial x_1^{q_1}}$ the estimate

$$\left| \frac{\partial^{q_1}\mathscr{E}_2(x)}{\partial x_1^{q_1}} \right| \leq C_1 \int\limits_a^\infty \sigma_1^{q_1} e^{-a_1\sigma_1^{1/\beta}}\,d\sigma_1 \leq C_1 \int\limits_0^\infty \sigma_1^{q_1} e^{-a_1\sigma_1^{1/\beta}}\,d\sigma_1.$$

The integral can easily be evaluated using the Gamma function after making the substitution $a_1\sigma_1^{1/\beta} = \xi$. We obtain

$$\left| \frac{\partial^{q_1}\mathscr{E}_2(x)}{\partial x_1^{q_1}} \right| \leq C_2\Gamma(q_1\beta + \beta) \leq C_3 B_1^{q_1} q_1^{q_1\beta}.$$

The estimates with respect to different variables are independent of one another and as a result we find that

$$|D^q\mathscr{E}_j(x)| \leq C_4 B_1^{q_1} \dots B_n^{q_n} q_1^{q_1\beta} \dots q_n^{q_n\beta}$$

for $j = 1, \dots, 9$ as asserted.

3.5.4. We now state an algebraic lemma which immediately allows us to simplify and to tie our results together.

LEMMA 2. *If the modulus graph of a polynomial $P(s)$ lies above a curve $\varrho = f(\xi)$ for which $f(\xi) \to \infty$ when $\xi \to \infty$, then there exists a positive exponent γ and coefficient C such that the graph lies above the curve $\varrho = C\xi^\gamma$ for sufficiently large ξ.*

On applying this lemma to the differential operator $P\left(i\dfrac{\partial}{\partial x}\right)$, we arrive at the following theorem.

THEOREM 3. *The equation $P\left(i\dfrac{\partial}{\partial x}\right)u = 0$ is hypoelliptic if and only if whenever*

$P(\sigma + i\tau) = 0$ *and* $|\sigma| \to \infty$ *it follows that* $|\tau| \to \infty$. *Under this condition,* $|\tau| \geq C|\sigma|^\gamma$ *on the surface of zeros of* $P(s)$ *for sufficiently large* $|\sigma|$ *and therefore every solution of the equation belongs to some Gevrey class.*

Thus every hypoelliptic equation is β-hypoelliptic for some $\beta > 0$.

The proof of Lemma 2 is not simple. It depends on the application of an algebraic "decision principle" which has recently been used in connection with various analytical questions.

Let $P_1, \ldots, P_k$ be a certain number of real polynomials of two types of arguments, the main ones $x_1, \ldots, x_n$ and parameters $q_1, \ldots, q_m$. Consider the question: For what real values of the parameters does the system of equations

$$P_j(x_1, \ldots, x_n, q_1, \ldots, q_m) = 0 \quad (j = 1, 2, \ldots, k) \tag{13}$$

have a solution in $x_1, \ldots, x_n$? The following Seidenberg-Tarski theorem holds:

There exists a finite collection of real polynomials $Q_\alpha^{(i)}(q_1, \ldots, q_m)$ $(i = 1, 2, \ldots, M; \alpha = 1, 2, \ldots \alpha_i)$ *with the property that the system* (13) *has a solution in* $x_1, \ldots, x_n$ *if and only if at least one of the* M *systems of inequalities*

$$Q_\alpha^{(i)}(q_1, \ldots, q_m) \geq 0 \quad (\alpha = 1, \ldots, \alpha_i)$$

is satisfied for some fixed i.†

Let us apply the Seidenberg-Tarski theorem to our case. We have the system of equations

$$\begin{aligned} \xi^2 - \sum_{j=1}^n \sigma_j^2 &= 0, \\ \varrho^2 - \sum_{j=1}^n \tau_j^2 &= 0, \\ P(\sigma + i\tau) &= 0. \end{aligned} \right\} \tag{14}$$

The last equation, which is complex, is to be replaced by a pair of real equations in the $2n$ real variables σ_j and τ_j.

---

† For a proof of the Seidenberg-Tarski theorem, see E. A. Gorin, *Asymptotic properties of polynomials and algebraic functions*, Russian Math. Surveys, No. 1, 1961, pp. 93–119 or A. Seidenberg, *A new decision problem for elementary algebra*. Ann. Math., 60 (1954), pp. 365–374.

We take σ_j and τ_j to be the main variables in this system and ξ and ϱ to be parameters. By the Seidenberg-Tarski theorem, there exists a finite collection of polynomials $Q_\alpha^{(i)}(\xi, \varrho)$ such that the system (14) has a solution in the σ_j and τ_j if and only if one of the M systems of inequalities,

$$Q_\alpha^{(i)}(\xi, \varrho) \geq 0, \quad \alpha = 1, ..., \alpha_i,$$

is satisfied. Let $Q_\alpha^{(i)}$ be the region in the first quadrant of the $\xi\varrho$-plane where $Q_\alpha^{(i)}(\xi, \sigma) \geq 0$. The conclusion of the Seidenberg-Tarski theorem may be formulated as follows: A necessary and sufficient condition for the existence of a solution to the system (14) is that the point (ξ, ϱ) belong to the set

$$Q = \bigcup_{i=1}^{M} \prod_{\alpha=1}^{a_i} Q_\alpha^{(i)}.$$

By hypothesis, the modulus graph of $P(s)$ lies above a certain curve

$$\varrho = f(\xi), \quad f(\xi) \to \infty \quad \text{for} \quad \xi \to \infty. \tag{15}$$

On the other hand, it is known from the theory of algebraic curves† that the boundary of each region $Q_\alpha^{(i)}$ has the equation

$$\varrho = A_\alpha^{(i)} \xi^{r_{\alpha_i}} E(\xi), \quad A_\alpha^{(i)} \geq 0,$$

where $E(\xi)$ is a function having the limit 1 when $\xi \to \infty$. Therefore the region Q is also bounded by such curves. But since it lies above the curve (15), its lower boundary is a curve of the form

$$\varrho = A\xi^\gamma E, \quad A > 0, \quad \gamma > 0, \quad E \to 1. \tag{16}$$

Hence if P is a hypoelliptic operator, its modulus graph not only lies above the logarithmic curve (1) but also above some power curve (16).

The supremum of values of γ for which the modulus graph of the surface $P(s) = 0$ lies above the curve (16) is called the *index of hypoellipticity* of the polynomial P. A formula exists for its determination (see Prob. 6 of Sec. 3.6).

Problems

1. Prove that the product of polynomials given by

$$P(s) = P_1(s) \cdots P_r(s)$$

---

† See G. E. Shilov, *Singularities of algebraic curves in the plane*, Uspekhi Mat. Nauk, No. 5, 1950, pp. 180–192 (In Russian).

is hypoelliptic if and only if each factor $P_j(s)$, $j = 1, ..., r$, is hypoelliptic. *Hint.* The surface of zeros of P is the union of the surfaces of zeros of the $P_j(s)$.

2. Given that the operator $P\left(i\dfrac{\partial}{\partial x}\right)$ possesses a fundamental function which is infinitely differentiable outside the ball $U = \{x: |x| \leq R\}$ for fixed R. Prove that P is hypoelliptic.

Hint. Lemma 1 of Sec. 3.4 holds for the operator P when applied to any region $W \subset V$ such that $W - U \subset V$. Hence deduce the inequality

$$|\tau| > A \log |\sigma| - A_1$$

for the points of the surface $P(s) = 0$ for some fixed A. Apply Theorem 3.

3. Given that every solution of $P\left(i\dfrac{\partial}{\partial x}\right)u = 0$ in a region V is infinitely differentiable in a region $W \subset V$ such that $W - U \subset V$, where $U = \{x; |x| \leq R\}$ for fixed R. Prove that P is a hypoelliptic operator.

Hint. Applying the condition to the fundamental function $\mathscr{E}(x)$ of the operator $P\left(i\dfrac{\partial}{\partial x}\right)$, show that it is infinitely differentiable outside a sphere with center at the origin. Use Prob. 2.

4. Suppose that a fundamental function for an operator P is infinitely differentiable in a neighborhood of a closed surface inside of which the origin lies. Show that P is hypoelliptic.

Hint. Using the Poisson formula of Sec. 3.1, show that the condition of Prob. 3 is satisfied.

5. If a fundamental function for the m-th order operator $P\left(\dfrac{\partial}{\partial x}\right)$ has ordinary derivatives up to order m-1 inclusive outside the origin, then P is hypoelliptic (V. V. Grushin).

Hint. Formula (3) of Sec. 3.1 also holds for a solution u in the entire space. Assume $u(x) = e^{-i(x,s_0)}$, where $P(s_0) = 0$, take Fourier transforms, and from the resulting relation

$$\delta(s - s_0) = g(s)\,\delta(s - s_0)$$

conclude that $g(s_0) = 1$. Then using the properties of the Fourier transform of a summable function with compact support, show that the surface $P(s) = 0$ recedes from the real hyperplane when $|s| \to \infty$.

Note. There exist non-hypoelliptic *m*-th order operators $P\left(\dfrac{\partial}{\partial x}\right)$ with a fundamental function having ordinary derivatives up to order $m - 2$ outside the origin (for instance, for the string equation; see Sec. 4.5.6, 2° and Prob. 3 of that section).

6. Let $P(\sigma)$ be a polynomial having just a bounded set of real zeros. Show that $P(\sigma)$ satisfies the estimate

$$|P(\sigma)| \geq C|\sigma|^{\gamma}$$

for sufficiently large $|\sigma|$ for some $C > 0$ and γ (γ may be negative).

Hint. Apply the Seidenberg-Tarski theorem and considerations of Sec. 3.5.4 to the "modulus graph" of the points $\xi = |\sigma|$ and $\varrho = |P(\sigma)|$.

7. Show that the operator $P\left(i\dfrac{\partial}{\partial x}\right)$ has a fundamental function $\mathscr{E}(x) \in S'$ if $P(\sigma)$ has no real zeros.

Hint. By the result of Prob. 6, the division $\dfrac{\psi(\sigma)}{P(\sigma)}$ is possible in the space S.

Note. Actually *every* operator $P\left(i\dfrac{\partial}{\partial x}\right)$ has a fundamental function $\mathscr{E}(x) \in S'$, but the proof of this is not simple.†

3.6. Conditions for Hypoellipticity Based on the Behavior of $P(s)$ in the Real Domain

3.6.1. The conditions for hypoellipticity derived in Secs. 3.4 and 3.5 are in terms of the manifold of complex roots of the polynomial $P(s)$ and are therefore difficult to use in specific situations. In this section, we shall derive conditions for hypoellipticity which make use of the behavior of $P(s)$ in the real domain.

LEMMA 1. *If for any $c > 0$ there is an $A > 0$ such that $P(s)$ has no roots in the domain $|\tau| < c$, $|\sigma| > A$, then*

$$\frac{P(\sigma + \theta)}{P(\sigma)} \to 1 \quad \text{as} \quad |\sigma| \to \infty \tag{1}$$

for any real vector θ.

---

† See L. Hörmander, *On the division of distributions by polynomials*, Ark. Mat., 3 (1958), pp. 555–568.

Proof. It may be assumed that $\theta = (1, 0, ..., 0)$. If $|\sigma| > A + c$ and λ is a point of the surface $P(s) = 0$, then $|\sigma - \lambda| > c$. We can write $P(\sigma)$ in the form

$$P(\sigma) = P_0(\sigma_2, ..., \sigma_n) \prod_{j=1}^{k} (\sigma_1 - \lambda_j(\sigma_2, ..., \sigma_n)).$$

Hence

$$\frac{P(\sigma + \theta)}{P(\sigma)} = \prod_{j=1}^{k} \frac{\sigma_1 + 1 - \lambda_j}{\sigma_1 - \lambda_j} = \prod_{j=1}^{k} \left(1 + \frac{1}{\sigma_1 - \lambda_j}\right) = 1 + R,$$

where

$$|R| < \frac{k}{c} + \frac{k(k-1)}{1\cdot 2}\frac{1}{c^2} + \cdots + \frac{1}{c^k} \leq \frac{k}{c}\left(1 + \frac{1}{c}\right)^{k-1}.$$

It is clear that this quantity can be made arbitrarily small for c sufficiently large. Our lemma is thereby proved.

LEMMA 2. *If* (1) *holds for any real* θ, *then for any* $D^q = \dfrac{\partial^{q_1 + \cdots + q_n}}{\partial\sigma_1^{q_1}...\partial\sigma_n^{q_n}}$ *with* $|q| = \Sigma q_j > 0$,

$$\lim_{|\sigma|\to\infty} \frac{D^q P(\sigma)}{P(\sigma)} = 0.$$

Proof. By Taylor's theorem, we have

$$P(\sigma + \theta) = P(\sigma) + \sum_{|q|>0} \frac{D^q P(\sigma)}{q!} \theta^q,$$

where $\theta = (\theta_1, ..., \theta_n)$ and the sum on the right-hand side is finite. Division yields

$$\frac{P(\sigma + \theta)}{P(\sigma)} = 1 + \sum_{|q|>0} \frac{D^q P(\sigma)}{P(\sigma)} \frac{\theta^q}{q!}$$

and so by the hypothesis,

$$\sum_{|q|>0} \frac{D^q P(\sigma)}{P(\sigma)} \frac{\theta^q}{q!} \to 0$$

as $|\sigma| \to \infty$.

Since the monomials $\dfrac{\theta^q}{q!}$ are linearly independent as functions of $\theta_1, ..., \theta_n$, it follows that $\dfrac{D^q P(\sigma)}{P(\sigma)} \to 0$ for each $q \neq 0$ as $|\sigma| \to \infty$, as required.

COROLLARY. *Under the conditions of Lemma 2,*

$$\lim_{|\sigma| \to \infty} |P(\sigma)| = +\infty.$$

Indeed, among the derivatives of $P(\sigma)$ there is one a constant different from zero. Hence by Lemma 2, $\dfrac{1}{P(\sigma)} \to 0$ and so $|P(\sigma)| \to \infty$.

LEMMA 3. *If for* $q \neq 0$

$$\lim_{|\sigma| \to \infty} \frac{D^q P(\sigma)}{P(\sigma)} = 0, \tag{2}$$

then

$$\lim_{|\sigma| \to \infty} \frac{P(\sigma + \theta)}{P(\sigma)} = 1 \tag{3}$$

uniformly for all $|\theta| < c$ *(complex* θ *included).*
Proof. We have

$$P(\sigma + \theta) = P(\sigma) + \sum_{|q| > 0} D^q P(\sigma) \frac{\theta^q}{q!}.$$

Thus

$$\frac{P(\sigma + \theta)}{P(\sigma)} = 1 + \sum_{|q| > 0} \frac{D^q P(\sigma)}{P(\sigma)} \frac{\theta^q}{q!},$$

and this implies the assertion of the lemma.

COROLLARY. *If* (2) *is satisfied for* $q \neq 0$, *then to any* $c > 0$, *there is an* $A > 0$ *such that* $P(s)$ *has no roots in the region* $|\tau| < c$, $|\sigma| > A$.

Proof. Suppose that for some $c > 0$ there is a sequence of roots $\lambda_\nu = \sigma_\nu + i\tau_\nu$ of $P(s)$ such that $|\tau_\nu| < c$ and $|\sigma_\nu| \to \infty$. The balls of radius c centered at the points $\lambda_1, \ldots, \lambda_\nu, \ldots$ intersect the real subspace R_n in certain real balls. Being non-identically vanishing, $P(s)$ has a point σ'_ν in each of these balls where $P(\sigma'_\nu) \neq 0$. Clearly, $\lambda_\nu = \sigma'_\nu + \theta_\nu$ with $|\theta_\nu| < c$. We thus have

$$\frac{P(\sigma'_\nu + \theta_\nu)}{P(\sigma'_\nu)} = \frac{P(\lambda_\nu)}{P(\sigma'_\nu)} = 0,$$

which contradicts (3).

Lemma 1–3 in conjunction with the results of Secs. 3.4 and 3.5 lead to the following theorem which furnishes necessary and sufficient conditions for $P(s)$ to be hypoelliptic.

THEOREM. *$P(s)$ is a hypoelliptic polynomial if and only if one of the following two equivalent conditions holds:*

(1) $\lim\limits_{|\sigma| \to \infty} \dfrac{P(\sigma + \theta)}{P(\sigma)} \to 1$ *for any vector θ;*

(2) $\lim\limits_{|\sigma| \to \infty} \dfrac{D^q P(\sigma)}{P(\sigma)} = 0$ *for $q \neq 0$.*

The second condition can be weakened still by confining our considerations to just the first derivatives of $P(\sigma)$:

LEMMA 4. *If $\dfrac{D^1 P(\sigma)}{P(\sigma)} \to 0$ as $|\sigma| \to \infty$ $\Big(D^1$ symbolizes any of the derivatives $\dfrac{\partial}{\partial \sigma_j} \Big)$, then $P(s)$ is hypoelliptic.*

Proof. By the mean value theorem,

$$\log \frac{P(\sigma + \theta)}{P(\sigma)} = \log P(\sigma + \theta) - \log P(\sigma) = \frac{\Sigma \theta_j \dfrac{\partial P(\sigma + t\theta)}{\partial \sigma_j}}{P(\sigma + t\theta)},$$

$$0 < t < 1.$$

Letting $|\sigma| \to \infty$ for fixed θ, we conclude that $\log \dfrac{P(\sigma + \theta)}{P(\sigma)} \to 0$. Thus $\dfrac{P(\sigma + \theta)}{P(\sigma)} \to 1$, as required.

3.6.2. We have labeled $P(s)$ elliptic if all solutions of $P\Big(i \dfrac{\partial}{\partial x} \Big) u = 0$ are locally analytic functions. We shall now characterize all elliptic polynomials in terms of their coefficients.

Recall that the locally analytic functions are those functions belonging to Gevrey's class G_1 (Sec. 3.1.5). Applying the results of Secs. 3.4, 3.5, we conclude that

A necessary and sufficient condition for $P(s)$ to be elliptic is that

$$|\tau| \geq C|\sigma| - C_1 \tag{4}$$

for all points $s = \sigma + i\tau$ of the manifold of zeros $N(P)$ of $P(s)$.

A consequence of this is the following explicit definition:

THEOREM. *$P(s)$ is elliptic if and only if its principal part $P_0(s)$ (the collection of all highest degree terms) has just the one real root $s = 0$.*

Proof. Suppose $P_0(\sigma)$ has just one zero at $\sigma = 0$. Let us show that there exists a positive constant C such that for all complex roots of $P(s)$,

$$|\tau| \geq C|\sigma|$$

for sufficiently large $|\sigma|$. Assuming the contrary, we can find a root $s_\nu = \sigma_\nu + i\tau_\nu$ such that $|\tau_\nu| < \dfrac{1}{\nu}|\sigma_\nu|$ for $\nu = 1, 2, \ldots$ with $|\sigma_\nu| \to \infty$. Now set $P(s) = P_0(s) + P_1(s)$ where $P_0(s)$ is the principal part of $P(s)$. We have

$$0 = P(\sigma_\nu + i\tau_\nu) = P_0(\sigma_\nu + i\tau_\nu) + P_1(\sigma_\nu + i\tau_\nu)$$

$$= P_0(\sigma_\nu) + \sum_{0 < |k| \leq m} D^k P_0(\sigma_\nu) \frac{(i\tau_\nu)^k}{k!} + \sum_{|k| \leq m-1} D^k P_1(\sigma_\nu) \frac{(i\tau_\nu)^k}{k!}.$$

Since $P_0(s)$ is a homogeneous polynomial of degree m which does not vanish for real $s = \sigma \neq 0$, it follows that

$$|P_0(\sigma_\nu)| \geq A|\sigma_\nu|^m$$

for some positive A. On the other hand, for large $|\sigma|$,

$$|D^k P_0(\sigma_\nu)| \leq A_k |\sigma_\nu|^{m-|k|},$$

$$|D^k P_1(\sigma_\nu)| \leq B_k |\sigma_\nu|^{m-1-|k|}.$$

As a result, we obtain

$$|P(\sigma_\nu + i\tau_\nu)| \geq A|\sigma_\nu|^m - \sum_{0 < |k| \leq m} A_k |\sigma_\nu|^{m-|k|} \frac{1}{\nu^{|k|}} |\sigma_\nu|^{|k|}$$

$$- \sum_{|k| < m} B_k |\sigma_\nu|^{m-1-|k|} |\sigma_\nu|^{|k|} \to \infty \quad \text{as} \quad |\nu| \to \infty,$$

thus contradicting the assumption that $P(\sigma_\nu + i\tau_\nu) = 0$.

Therefore if $P_0(\sigma)$ has a single root $\sigma = 0$, then $P(s)$ is an elliptic polynomial.

Suppose now that $P_0(\sigma)$ has a real root $\sigma \neq 0$. Since P_0 is homogeneous, the set of its roots comprises straight lines passing through the origin. Performing a linear transformation if necessary, we may assume that $(0, 1, 0, \ldots, 0)$ is a root of $P_0(\sigma)$ while $(1, 0, \ldots, 0)$ is not. Consider the polynomial $Q(s_1, s_2) \equiv P(s_1, s_2, 0, \ldots, 0)$. It can be written in the form

$$Q(s_1, s_2) = \sum_{k=0}^{m} a_k s_1^{m-k} s_2^k + \sum_{k \leq j < m} a_{kj} s_1^{j-k} s_2^k = Q_0(s_1, s_2) + Q_1(s_1, s_2),$$

14*

with

$$Q_0(1, 0) = a_0 \neq 0,$$

$$Q_0(0, 1) = a_m = 0.$$

But then as we saw back in Sec. 3.4.1, the lower boundary of the modulus graph of Q (and hence also of P) has the equation $\varrho = C\xi^\gamma E$ with $\gamma < 1$. Thus the polynomial P is necessarily non-elliptic.

Problems

1. Construct a non-hypoelliptic polynomial $P(s_1, s_2)$ for which

$$\lim_{|\sigma| \to \infty} |P(\sigma_1, \sigma_2)| = \infty. \tag{1}$$

Hint. The surface of zeros of $P(s_1, s_2)$ must not intersect the real plane when $|\sigma| \to \infty$ and also must not recede from it. In fulfilling (1), one can still multiply by a known hypoelliptic polynomial.

Ans. For instance,

$$P(s_1, s_2) = (s_2 - i)(s_2 - is_1) = (s_2^2 - s_1) - is_2(1 + s_1).$$

2. Construct a non-hypoelliptic polynomial $P(s_1, s_2)$ for which

$$\lim_{|\sigma| \to \infty} |P(\sigma_1 + i\tau_1, \sigma_2 + i\tau_2)| = \infty$$

uniformly for $|\tau_1|$ and $|\tau_2| \leq C$ (C is fixed).

Hint. Employ the construction of Prob. 1.

Ans. For instance, $P(s_1, s_2) = (s_2 - 2Ci)(s_2 - is_1)$.

3. Prove that the first order equation in t with one space variable x given by

$$\frac{\partial u}{\partial t} = Q\left(i \frac{\partial}{\partial x}\right) u$$

is hypoelliptic if and only if in the decomposition $Q(\sigma) = Q_1(\sigma) + iQ_2(\sigma)$ the degree of the polynomial Q_2 does not exceed that of the polynomial Q_1.

Hint. Apply Lemma 4. Its conditions become

(1) $\dfrac{1}{(\lambda - Q_2(\sigma))^2 + Q_1^2(\sigma)} \to 0,$ (2) $\dfrac{[Q_1'(\sigma)]^2 + [Q_2'(\sigma)]^2}{(\lambda - Q_2(\sigma))^2 + Q_1^2(\sigma)} \to 0.$

If $Q_2(\sigma)$ is of higher degree than $Q_1(\sigma)$, then (2) will not be satisfied on the curve $\lambda = Q_2(\sigma)$. If the degree m of $Q_1(\sigma)$ is at least 1 and the degree of $Q_2(\sigma)$ does not exceed m, then the numerator of (2) will be of degree at most $2(m - 1)$ while the denominator will be of degree at least $2m$ (relative to σ).

4. Prove that the m-th order equation in t with one space variable x given by

$$\frac{\partial^m u}{\partial t^m} - \sum_{k=0}^{m-1} P_k\left(i\,\frac{\partial}{\partial x}\right) \frac{\partial^k u}{\partial t^k}$$

is hypoelliptic if and only if the leading coefficient A_j in the representation

$$\lambda_j = A_j \sigma^{r_j} E$$

of each root of

$$\lambda^m - \sum_{k=0}^{m-1} P_k(\sigma)\, \lambda^k = 0 \tag{1}$$

at infinity is not pure imaginary.

Hint. Write (1) in the factored form

$$\prod_{j=1}^{m} (\lambda - \lambda_j(\sigma))^{k_j} = 0$$

and use the method of Prob. 3 and Prob. 1 of Sec. 3.5.

5. Obtain for a β-hypoelliptic polynomial $P(s)$ the following refinement of the theorem of Sec. 3.6.1:

(1) $1 - \dfrac{P(\sigma + \theta)}{P(\sigma)} = O(|\sigma|^{-1/\beta}),$

(2) $\dfrac{D^q P(\sigma)}{P(\sigma)} = O(|\sigma|^{-1/\beta}).$

Hint. Refine the respective arguments of Lemma 1–3.

6. Derive the following formula for the index of hypoellipticity (V. P. Palamodov):

$$\beta = - \varlimsup_{|\sigma| \to \infty} \left\{ \frac{\log \dfrac{|\text{grad } P(\sigma)|}{|P(\sigma)|}}{\log |\sigma|} \right\}.$$

Hint. Use Prob. 5.

3.7. Radon's Method

3.7.1. We would now like to delve into the structure of a fundamental function for an operator $P\left(\dfrac{\partial}{\partial x}\right)$ which is not hypoelliptic. It can be stated at the outset that such fundamental functions have singularities (as regards to losing their infinite differentiability) arbitrarily far from the origin (Sec. 3.5, Prob. 4). Moreover, it can be shown that for any region G the equation $P\left(\dfrac{\partial}{\partial x}\right) u = 0$ has a solution with a singularity in precisely this region. By adding such a solution to a fundamental function supposedly infinitely differentiable, we obtain a new fundamental solution now having a singularity in G. Thus the possibilities for constructing "bad" fundamental functions are very great. We are interested in the question of the existence of a fundamental function $\mathscr{E}(x)$ with a minimal set of singularities. We shall show that whenever $P(\sigma_1, \ldots, \sigma_n)$ is a homogeneous polynomial with a sufficiently smooth surface of zeros, there is always a fundamental function for $P\left(\dfrac{\partial}{\partial x}\right)$ whose singularities lie in a certain cone of dimensionality $n - 1$ at the most.

The technique used to obtain these results is *Radon's method* of representing fundamental functions.

Suppose that $P\left(\dfrac{\partial}{\partial x}\right)$ is a given m-th order linear differential operator with constant coefficients.

We wish to solve the equation

$$P\left(\frac{\partial}{\partial x_1}, \ldots, \frac{\partial}{\partial x_n}\right) \mathscr{E}(x_1, \ldots, x_n) = \delta(x_1, \ldots, x_n). \tag{1}$$

We replace the delta-function on the right-hand side by its integral representation over a sphere (Sec. 2.4.6)

$$\delta(x) = \int_\Omega f((\omega, x))\, d\omega,$$

where

$$f(\xi) = \begin{cases} a_n \delta^{(n-1)}(\xi) & \text{for } n \text{ odd}, \\ b_n \xi^{-n} & \text{for } n \text{ even}. \end{cases}$$

Then instead of (1), we consider the equation

$$P\left(\frac{\partial}{\partial x_1}, \ldots, \frac{\partial}{\partial x_n}\right) \mathscr{E}_\omega(x_1, \ldots, x_n) = f((\omega, x)) \tag{2}$$

depending on the parameter ω. It is natural to look for a solution of (2) in the form of a generalized function $\mathscr{E}_\omega(\xi)$ of the argument $(\omega, x) = \xi$ (see Sec. 1.6.2, 4°). It will be recalled that the usual chain rule for differentiation

$$\frac{\partial}{\partial x_j} = \omega_j \frac{d}{d\xi}$$

holds for $f((\omega, x))$ (Sec. 1.7.2, 4°). Thus, equation (2) for $\mathscr{E}_\omega(\xi)$ becomes

$$P\left(\omega_1 \frac{d}{d\xi}, \ldots, \omega_n \frac{d}{d\xi}\right) \mathscr{E}_\omega(\xi) = f(\xi). \tag{3}$$

Equation (3) is now an ordinary differential equation with a parameter ω. Suppose that we have found a solution $\mathscr{E}_\omega(\xi)$ which depends continuously on the parameter ω. Then we assert that

$$\mathscr{E}(x) = \int_\Omega \mathscr{E}_\omega(\xi)\, d\omega \tag{4}$$

furnishes the solution to our stated problem. Indeed, since differentiation is a continuous operation for generalized functions, we have

$$P\left(\frac{\partial}{\partial x}\right) \mathscr{E}(x) = \int_\Omega P\left(\frac{\partial}{\partial x}\right) \mathscr{E}_\omega(\xi)\, d\omega$$

$$= \int_\Omega P\left(\omega_1 \frac{d}{d\xi}, \ldots, \omega_n \frac{d}{d\xi}\right) \mathscr{E}_\omega(\xi)\, d\omega = \int_\Omega f(\xi)\, d\omega = \delta(\xi),$$

as required.

Thus our job is to determine a solution to (3) depending continuously on ω.

Let $P_0\left(\dfrac{\partial}{\partial x}\right)$ be the principal part of the operator $P\left(\dfrac{\partial}{\partial x}\right)$, i.e., all the terms involving m-th order derivatives. We have $P = P_0 + P_1$, where P_1 are the lower order terms, and thus

$$P\left(\omega_1 \frac{d}{d\xi}, \dots, \omega_n \frac{d}{d\xi}\right)$$

$$= P_0\left(\omega_1 \frac{d}{d\xi}, \dots, \omega_n \frac{d}{d\xi}\right) + P_1\left(\omega_1 \frac{d}{d\xi}, \dots, \omega_n \frac{d}{d\xi}\right)$$

$$= P_0(\omega) \frac{d^m}{d\xi^m} + P_1\left(\omega_1 \frac{d}{d\xi}, \dots, \omega_n \frac{d}{d\xi}\right).$$

We see that the coefficient of the highest derivative with respect to ξ is $P_0(\omega)$. If equation (1) is elliptic, then $P_0(\omega)$ does not vanish according to Sec. 3.6.2. In that case, it is not hard to find a solution $\mathscr{E}_\omega(x)$ depending continuously on the parameter ω by the method of Sec. 1.5.3 ff. The situation is more complicated in the general case and we shall postpone its consideration until Sec. 3.7.3.

3.7.2. At this point we shall consider a *homogeneous m*-th degree elliptic polynomial. The polynomial P_1 is now absent and equation (3) reduces to

$$P_0(\omega) \frac{d^m \mathscr{E}_\omega(\xi)}{d\xi^m} = f(\xi), \quad P_0(\omega) \neq 0. \tag{5}$$

Let us examine the two cases of even and odd n. If n is odd, $f(\xi) = a_n \delta^{(n-1)}(\xi)$ and (5) is easily solved explicitly to obtain

$$\mathscr{E}_\omega(\xi) = \begin{cases} \dfrac{a_n}{P_0(\omega)} \delta^{(n-1-m)}(\xi), & \text{if } m \leq n-1, \\[3mm] \dfrac{a_{nm}}{P_0(\omega)} \xi_+^{m-n}, & \text{if } m \geq n. \end{cases}$$

For even n, we have $f(\xi) = b_n \xi^{-n}$ and so

$$\mathscr{E}_\omega(\xi) = \begin{cases} \dfrac{b_{nm}}{P_0(\omega)} \xi^{-n+m} & \text{if } m \leq n-1, \\[3mm] \dfrac{b_{nm}}{P_0(\omega)} \xi^{m-n} \log |\xi| & \text{if } m \geq n. \end{cases}$$

The solutions found obviously depend continuously on ω. Therefore by (4), we can write

$$
\mathscr{E}(x) = \begin{cases} a_n \displaystyle\int_\Omega \frac{\delta^{(n-1-m)}(\xi)}{P_0(\omega)}\, d\omega & \text{for } n \text{ odd}, \quad m \leqq n-1, \\[3mm] a_{nm} \displaystyle\int_\Omega \frac{\xi_+^{m-n}}{P_0(\omega)}\, d\omega & \text{for } n \text{ odd}, \quad m \geqq n, \\[3mm] b_{nm} \displaystyle\int_\Omega \frac{\xi^{-n+m}}{P_0(\omega)}\, d\omega & \text{for } n \text{ even}, \quad m \leqq n-1, \\[3mm] b_{nm} \displaystyle\int_\Omega \frac{\xi^{m-n} \log |\xi|}{P_0(\omega)}\, d\omega & \text{for } n \text{ even}, \quad m \geqq n. \end{cases} \tag{6}
$$

The resulting expressions for $\mathscr{E}(x)$ are infinitely differentiable with respect to x for $x \neq 0$. This follows apart from general considerations of elliptic equations from our discussion of Sec. 2.3.7.

3.7.3. Let us examine the case of a non-elliptic polynomial. We still take P to be homogeneous so that $P \equiv P_0$. In this case, $P_0(\omega)$ vanishes at certain points of the sphere Ω. We suppose further that at least one of the derivatives of P does not vanish at the points ω_0 where $P_0(\omega) = 0$. In that event, $P_0(\omega_1, ..., \omega_n) = 0$ can be solved for the corresponding variable in the neighborhood of ω_0. This makes it possible to introduce a new coordinate system in the neighborhood of ω_0 with $P_0(\omega)$ itself one of the coordinates.

We shall again consider our required fundamental function to be of the form (6), i.e.,

$$
\int_\Omega \frac{f(\xi)\, d\omega}{P_0(\omega)}
$$

but interpreted as a Cauchy principal value,

$$
I(x) = \lim_{\varepsilon \to 0} \int_{|P_0| \geqq \varepsilon} \frac{f(\xi)\, d\omega}{P_0(\omega)} = \lim I_\varepsilon(x). \tag{7}
$$

Let us first show that the limit exists in K'. We need to recall certain information concerning the Cauchy principal value of integrals of ordinary functions. Suppose we have

$$\int_{\Omega} \frac{g(\omega)\, d\omega}{P_0(\omega)} ,\tag{8}$$

where $g(\omega)$ is infinitely differentiable and $P_0(\omega)$ satisfies the above-mentioned conditions.

We wish to show that the limit of

$$\int_{|P_0| \geq \varepsilon} \frac{g(\omega)\, d\omega}{P_0(\omega)}$$

exists when $\varepsilon \to 0$. To this end, we introduce in the region $|P_0(\omega)| \leq \varepsilon_0$, $\varepsilon \leq \varepsilon_0$, the coordinates $\theta = P_0(\omega)$ and ω'. Then accordingly $g(\omega)\, d\omega = G(\theta, \omega')\, d\theta\, d\omega'$ with $G(\theta, \omega')$ an infinitely differentiable function.

In particular, we can write

$$G(\theta, \omega') = G(0, \omega') + \theta G_1(\theta, \omega'),$$

where $G_1(\theta, \omega')$ is a bounded continuous function. We then have

$$\int_{\varepsilon \leq |\theta| \leq \varepsilon_0} \frac{G(0, \omega')\, d\theta}{\theta} = G(0, \omega') \int_{\varepsilon \leq |\theta| \leq \varepsilon_0} \frac{d\theta}{\theta} = 0,$$

and

$$\int_{\varepsilon \leq |\theta| \leq \varepsilon_0} \frac{\theta G_1(\theta, \omega')\, d\theta}{\theta} = \int_{\varepsilon \leq |\theta| \leq \varepsilon_0} G_1(\theta, \omega')\, d\theta \to \int_{|\theta| \leq \varepsilon_0} G_1(\theta, \omega')\, d\theta.$$

This implies the existence of

$$\lim_{\varepsilon \to 0} \int_{\varepsilon \leq |P_0| \leq \varepsilon_0} \frac{g(\omega)\, d\omega}{P_0(\omega)} = \lim_{\varepsilon \to 0} \int_{\omega'} \int_{\varepsilon \leq |\theta| \leq \varepsilon_0} \frac{G(\theta, \omega')\, d\theta}{\theta}\, d\omega'$$

$$= \int_{\omega'} \int_{|\theta| \leq \varepsilon_0} \frac{G(\theta, \omega') - G(0, \omega')}{\theta}\, d\theta\, d\omega'.$$

This result summed with the integral over the region $|P_0(\omega)| \geq \varepsilon_0$ is our required Cauchy principal value of (8).

Applying $I_\varepsilon(x)$ of (7) to a test function $\varphi(x)$, we obtain

$$(I_\varepsilon(x), \varphi(x)) = \int_{|P_0| \geq \varepsilon} (f(\xi), \varphi(x)) \frac{d\omega}{P_0(\omega)} .$$

The function of ω

$$(f(\xi), \varphi(x)) = (f((\omega, x)), \omega(x))$$

has been shown in Sec. 2.2.7 to be infinitely differentiable with respect to ω. By our above discussion,

$$(I(x), \varphi) = \lim_{\varepsilon \to 0} \int_{|P| \geq \varepsilon} \frac{(f(\xi), \varphi) \, d\omega}{P_0(\omega)}$$

exists. Owing to the completeness of K', the limit (7) is again a continuous functional, which is what we needed.

Let us show that the functional

$$\mathscr{E}(x) = \int_\Omega \frac{f(\xi) \, d\omega}{P_0(\omega)} \qquad (9)$$

defined as in the above is a solution of the equation

$$P_0\left(\frac{\partial}{\partial x}\right) \mathscr{E}(x) = \delta(x).$$

First of all, we have

$$P_0\left(\frac{\partial}{\partial x}\right) \mathscr{E}(x) = \int_\Omega \frac{P_0\left(\frac{\partial}{\partial x}\right) f(\xi) \, d\omega}{P_0(\omega)} ,$$

and

$$P_0\left(\frac{\partial}{\partial x}\right) f(\xi) = P_0\left(\omega_1 \frac{d}{d\xi}, \dots, \omega_n \frac{d}{d\xi}\right) f(\xi) = P_0(\omega) \frac{d^m f}{d\xi^m} .$$

Hence by definition of the functional $f(\xi)$,

$$P_0\left(\frac{\partial}{\partial x}\right) \mathscr{E}(x) = \int_\Omega \frac{d^m f}{d\xi^m} \, d\omega = \delta(x).$$

Thus the functional (9) is the required fundamental solution.

3.7.4. We now show that $\mathscr{E}(x)$ is everywhere infinitely differentiable with respect to x except for a certain cone of dimensionality less than n.

We shall call a direction x *special* if the equator $(\omega, x) = 0$ is tangent (on the sphere $|\omega| = 1$) to the surface $P_0(\omega) = 0$ and *ordinary* if either the equator $(\omega, x) = 0$ and $P_0(\omega) = 0$ have no point in common or else at each intersection they have different tangent planes (of dimensionality $n - 2$ of course).

The special directions form a cone which can be constructed as follows. Take any point ω_0 on the curve $P_0(\omega) = 0$ and draw a great circle through this point tangent to the curve. Let the equation of the circle be $(\omega, x_0) = 0$.

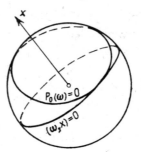

FIG. 12. A special direction

Then x_0 is a special direction. The dimensionality of the cone of special directions is clearly equal to that of the cone determined by the curve $P_0(\omega) = 0$ and so is less than the dimensionality of the whole space.

We wish to show that $\mathscr{E}(x)$ *is infinitely differentiable outside the cone of special directions.*

Consider at first the Cauchy principal value of the integral

$$F(x) = \int_{\Omega} \frac{f((\omega, x)) \, g(\omega)}{P_0(\omega)} \, d\omega,$$

when $f(\xi)$ is infinitely differentiable for all $\xi \neq 0$ and ordinary for all ξ. We can write $g(\omega) = g_1(\omega) + g_2(\omega) + g_3(\omega)$, where g_1, g_2, and g_3 are infinitely differentiable functions such that

$$g_1(\omega) = 0 \quad \text{for} \quad |P_0(\omega)| < \frac{\varepsilon}{2},$$

$$g_2(\omega) = 0 \quad \text{for} \quad |P_0(\omega)| \geq \varepsilon \quad \text{or} \quad |(\omega, x)| < \frac{\varepsilon}{2},$$

$$g_3(\omega) = 0 \quad \text{for} \quad |P_0(\omega)| \geq \varepsilon \quad \text{or} \quad |(\omega, x)| > \varepsilon.$$

Correspondingly,

$$F(x) = F_1(x) + F_2(x) + F_3(x),$$

where

$$F_1(x) = \int\limits_{|P_0| \geq \frac{\varepsilon}{2}} \frac{f((\omega, x))\, g_1(\omega)}{P_0(\omega)}\, d\omega,$$

$$F_2(x) = \int\limits_{\substack{|P_0| \leq \varepsilon \\ |\xi| \geq \varepsilon/2}} \int \frac{f((\omega, x))\, g_2(\omega)}{P_0(\omega)}\, d\omega,$$

$$F_3(x) = \int\limits_{\substack{|P_0| \leq \varepsilon \\ |\xi| \leq \varepsilon}} \int \frac{f((\omega, x))\, g_3(\omega)}{P_0(\omega)}\, d\omega.$$

$F_1(x)$ is infinitely differentiable for $x \neq 0$ by the results of Sec. 2.3.7.
  Now introduce new variables in the integral

$$F_2(x) = \int\limits_{|\xi| \geq \varepsilon/2} \int\limits_{|P_0| \leq \varepsilon} \frac{f((\omega, x))\, g_2(\omega)}{P_0(\omega)}\, d\omega$$

one of which is $\theta = P_0(\omega)$. Denote the aggregate of all others by ω'. Then $f((\omega, x))\, g_2(\omega)\, d\omega$ can be written as $g(\theta, \omega', x)\, d\theta\, d\omega'$ where $g(\theta, \omega', x)$ is an infinitely differentiable function of all its arguments. By definition of Cauchy principal value, we have

$$F_2(x) = \int\limits_{\omega'} \int\limits_{|\theta| \leq \varepsilon} \frac{g(\theta, \omega', x) - g(0, \omega', x)}{\theta}\, d\theta\, d\omega'$$

$$= \int\limits_{\omega'} \int\limits_{|\theta| \leq \varepsilon} g_1(\theta, \omega', x)\, d\omega'\, d\theta,$$

$g_1(\theta, \omega' x)$ again being an infinitely differentiable function of its arguments. But then $F_2(x)$ is also infinitely differentiable, as required.

Finally consider the integral

$$F_3(x) = \int_{|P_0| \leqq \varepsilon} \int_{|\xi| \leqq \varepsilon} \frac{f((\omega, x)) g_3(\omega)}{P_0(\omega)} \, d\omega .$$

We shall show that $F_3(x)$ is infinitely differentiable for non-special x. Let x be a non-null fixed vector which is not special. Using this assumption we introduce in the region $|P_0(\omega)| \leqq \varepsilon$, $|\xi| \leqq \varepsilon$, a coordinate system consisting of $\xi = (\omega, x)$, $\theta = P_0(\omega)$, and ω' the collection of all remaining coordinates. We can write

$$\frac{f((\omega, x)) g_3(\omega) \, d\omega}{P_0(\omega)} = f(\xi) \frac{G(\xi, \omega', \theta; x)}{\theta} \, d\theta \, d\omega' \, d\xi ,$$

where $G(\xi, \omega', \theta; x)$ is an infinitely differentiable function of its arguments. According to the definition of Cauchy principal value,

$$\int_\Omega \frac{f((\omega, x)) g_3(x)}{P_0(\omega)} \, d\omega$$

$$= \int_{\omega'} \int_{|\xi| \leqq \varepsilon} \int_{|\theta| \leqq \varepsilon} f(\xi) \frac{G(\xi, \omega', \theta; x) - G(\xi, \omega', 0; x)}{\theta} \, d\theta \, d\omega' \, d\xi .$$

Integrating in this with respect to θ and ω', we obtain a function

$$K(x, \xi) = \int_{|\theta| \leqq \varepsilon} \int_{\omega'} \frac{G(\xi, \omega', \theta; x) - G(\xi, \omega', 0; x)}{\theta} \, d\omega' \, d\theta ,$$

which again is infinitely differentiable with respect to ξ and x and vanishes for $|\xi| \geqq \varepsilon$. Summing up, we have

$$\int_\Omega \frac{f((\omega, x)) g_3(\omega)}{P_0(\omega)} \, d\omega = \int_{|\xi| \leqq \varepsilon} f(\xi) \, K(x, \xi) \, d\xi = (f(\xi), K(x, \xi)) \qquad (10)$$

and so $\mathscr{E}(x)$ is infinitely differentiable. Now employing the same technique as in Sec. 2.3.7, we can prove that the expression (10) is infinitely differentiable in x for $x \neq 0$ when $f(\xi)$ is any functional of interest to us and not merely an ordinary function. With this, the proof of the infinite differentiability of $\mathscr{E}(x)$ outside the cone of special directions is completed.

Problems

1. Using Radon's method, obtain a fundamental function for the iterated Laplacian Δ^m (Sec. 1.7.2, 9°).

Hint. $P_0(\omega) \equiv 1$. Use Prob. 11 of Sec. 2.4.

2. Using Radon's method, obtain a fundamental function for the operator $\Delta + k^2$ for $n = 3$ (Sec. 1.7, Prob. 7).

Hint. The integral of $e_+^{ik(\omega, x)}$ over Ω can be evaluated by expanding in a series with the help of the formulas of Sec. 2.4.6.

3. Let $P(\sigma)$ be an m-th degree homogeneous polynomial with a smooth surface of zeros (as in Sec. 3.7.4). Then a solution $u(x)$ of $P\left(\dfrac{\partial}{\partial x}\right)u = 0$ which is infinitely differentiable in a neighborhood of the intersection of the cone of special directions with a fixed arbitrary closed surface containing the origin inside is also infinitely differentiable in a neighborhood of the origin.

Hint. Use the Poisson formula of Sec. 3.1.

Chapter 4

Equations in a Half-Space

4.1. Well-Posed Boundary Value Problems

We shall be considering differential equations in which one independent variable has been singled out. It will be denoted by t and will (as a rule) vary over the interval $0 \leqq t < \infty$. There are many classical problems in mathematical physics that lead to equations of this type. Generally one considers equations with the highest t-derivative solved for such as

$$\frac{\partial^m u(t, x)}{\partial t^m} = \sum_{k=0}^{m-1} p_k\left(i\frac{\partial}{\partial x}\right)\frac{\partial^k u(t, x)}{\partial t^k}, \quad x = (x_1, \ldots, x_n), \tag{1}$$

where the $p_k\left(i\dfrac{\partial}{\partial x}\right)$ are certain polynomials in $i\dfrac{\partial}{\partial x_1}, \ldots, i\dfrac{\partial}{\partial x_n}$ of maximum degree say p. A solution of (1) is required to satisfy *initial conditions* imposed as a rule on the functions $u(0, x), \dfrac{\partial u(0, x)}{\partial t}, \ldots, \dfrac{\partial^{m-1} u(0, x)}{\partial t^{m-1}}$. In some instances a desirable behavior for the solution $u(t, x)$ is specified for $t \to \infty$. All such data defines a *boundary value problem* for equation (1).† In addition to this, one specifies the class of functions of x (possibly generalized) to which the solution $u(t, x)$ is to belong for each fixed t.

A boundary value problem is considered to be *well-posed in a class R* if for any admissible set of boundary data it has a solution in R which is unique and depends continuously on the data. The last condition requires R to have a certain topology in which the operator mapping an admissible initial function $u(x)$ into a solution $u(t, x)$ is continuous in R.

Let us give a few classical examples of well-posed boundary value problems involving one spatial variable.

---

† When t stands for time, it is customary to call it an initial value problem.

1°. For the heat equation

$$\frac{\partial u(t,x)}{\partial t} = \frac{\partial^2 u(t,x)}{\partial x^2}$$

the *Cauchy problem* is well-posed: To find a solution taking on an assigned initial value $u(0, x)$. The class R may be chosen to be all bounded functions, for example.

2°. The *Cauchy problem* is also well-posed for the wave equation

$$\frac{\partial^2 u(t,x)}{\partial t^2} = \frac{\partial^2 u(t,x)}{\partial x^2}.$$

In this case, one has to assign $u(0, x)$ and $\dfrac{\partial u(0, x)}{\partial t}$. The class R may be taken to be all generalized functions.

3°. For Laplace's equation

$$\frac{\partial^2 u(t,x)}{\partial t^2} = -\frac{\partial^2 u(t,x)}{\partial x^2}$$

the *Dirichlet problem* ($u(0, x)$ prescribed) plus the condition $u(t, x) = O(t^h)$ as $t \to \infty$ is well-posed. The class R may be taken here to be the collection $\mathscr{H}$ of square integrable functions (in x) and their derivatives of all orders (in K').

We shall be indicating what boundary value problems are well-posed for any equation of the form (1). In so doing, we shall take R to be the same class for every equation, namely, the collection $\mathscr{H}$ of all square integrable functions $f(x) = f(x_1, ..., x_n)$ defined in real n-dimensional Euclidean space $R_n = R_n(x)$ plus their derivatives of all orders (in K'). (In many instances, though, it will be possible to broaden R.) However, as we shall see below, the definitive result can also be formulated in classical terms without recourse to generalized functions. The solution of the corresponding boundary value problem will be an ordinary function if we require certain additional smoothness properties of the initial data.

Relative to the Fourier transform, the dual of $\mathscr{H}$ is the space H of all square integrable functions $g(\sigma) = g(\sigma_1, ..., \sigma_n)$ defined in real n-dimensional Euclidean space $R_n = R_n(\sigma)$ plus their products with polynomials. In particular, H includes all functions $g(\sigma)$ of at most power growth at infinity. The reason for this is that every such function divided by some

15 Shilov

polynomial of sufficiently high degree, for example, of the form $(1 + \sigma_1^2 + \cdots + \sigma_n^2)^q$ or $(1 + i\sigma_1)^m \cdots (1 + i\sigma_n)^m$ is square integrable. It is also obvious that H admits multiplication by any locally bounded function of at most power growth at infinity.

The growth of $g(\sigma)$ at infinity is intimately connected with the order of singularity of the function $f(x)$ of which it is the Fourier transform. Multiplying $g(\sigma)$ by a polynomial is equivalent to differentiating $f(x) = F^{-1}[g]$ an appropriate number of times. This for example makes it possible to write down a useful formula involving orders of singularity. Namely, if $P_m(\sigma)$ is an m-th degree polynomial, then

$$s\{F^{-1}[P_m(\sigma)\, g(\sigma)]\} \leqq m + s\{F^{-1}[g(\sigma)]\} = m + s(f).$$

Moreover, if $g(\sigma) \in L_2$, then so does $f(x)$ and $s(f) \leqq 0$. Thus if $g_0(\sigma)$ becomes square integrable after division by an m-th degree polynomial $P_m(\sigma)$, then the order of the $f_0(x)$ corresponding to it does not exceed m.

The space $\mathscr{H}$ is not closed in K', and so a more natural mode of convergence in $\mathscr{H}$ would be one under which it is complete. It is more convenient for that purpose to consider the space H. This space can be expressed in the form of a union

$$H = \bigcup_{q=0}^{\infty} H^{(q)},$$

where $H^{(q)}$ is a Hilbert space with the inner product

$$(g_1(\sigma), g_2(\sigma))_q = \int_{R_n} \frac{\overline{g_1}(\sigma)\, g_2(\sigma)\, d\sigma}{(1 + |\sigma|^2)^q}.$$

We shall say that a sequence $g_\nu(\sigma) \in H$ *converges to zero in H* if for some q (and hence for all sufficiently large q) the norm

$$(g_\nu, g_\nu)_q = \|g_\nu\|_q^2$$

is defined for each $g_\nu(\sigma)$ and $\|g_\nu\|_q \to 0$ as $\nu \to \infty$. In accordance with this, a sequence $f_\nu(x) \in \mathscr{H}$ *converges to zero in $\mathscr{H}$* if there exists a q such that $f_\nu = (1 - \varDelta)^q \hat{f_\nu}, \hat{f_\nu} \in L_2$, and $\hat{f_\nu}(x)$ converges to zero in L_2.

A linear operator B in H (or $\mathscr{H}$) is said to be continuous if whenever $g_\nu(\sigma) \to 0$ in H ($f_\nu(x) \to 0$ in $\mathscr{H}$), $Bg_\nu \to 0$ in H ($Bf_\nu \to 0$ in $\mathscr{H}$). Thus for example, multiplying by a function $G(\sigma)$ such that

$$|G(\sigma)|^2 \leqq C(1 + |\sigma|^2)^m$$

is a continuous operation in H. For suppose we have

$$\|g_\nu(\sigma)\|_q^2 = \int\limits_{R_n} \frac{|g_\nu(\sigma)|^2 \, d\sigma}{(1 + |\sigma|^2)^q} \to 0$$

for $q \geq q_0$. Then for $q \geq q_0 + m$,

$$\|G(\sigma) \, g_\nu(\sigma)\|_q^2 \leq C \int\limits_{R_n} \frac{(1 + |\sigma|^2)^m |g_\nu(\sigma)|^2}{(1 + |\sigma|^2)^q} \, d\sigma = C \int\limits_{R_n} \frac{|g_\nu(\sigma)|^2 \, d\sigma}{(1 + |\sigma|^2)^{q-m}} \to 0,$$

as required.

A sequence $g_\nu(\sigma) \in H$ is *bounded* in H if every $g_\nu(\sigma)$ belongs to the same $H^{(q)}$ and is bounded in its norm.

A sequence $g_\nu(\sigma)$ *increases in H no faster than a (numerical) function* $\pi(\nu)$ if the sequence $\dfrac{1}{\pi(\nu)} g_\nu(\sigma)$ is bounded in H.

Similar definitions can of course be formulated for a family $g_t(\sigma)$ with a continuously varying parameter t. With the help of the inverse Fourier transform, the definitions can be transfered to $\mathcal{H}$.

Consider now equation (1) assuming that its solution is to be found in the class of functions $u(t, x)$ which belong to $\mathcal{H}$ together with the derivatives $\dfrac{\partial u(t, x)}{\partial t}, \ldots, \dfrac{\partial^m u(t, x)}{\partial t^m}$ for $t \geq 0$. Along with (1), consider the algebraic equation

$$\lambda^m = \sum_{k=0}^{m-1} p_k(\sigma) \lambda^k$$

obtained from (1) by replacing $\dfrac{\partial}{\partial t}$ by λ and $i \dfrac{\partial}{\partial x}$ by σ. The equation (counting multiplicities) has m roots $\lambda_0(\sigma), \ldots, \lambda_{m-1}(\sigma)$. For each σ, we number them so that

$$\operatorname{Re} \lambda_0(\sigma) \leq \operatorname{Re} \lambda_1(\sigma) \leq \cdots \leq \operatorname{Re} \lambda_{m-1}(\sigma).$$

Let A_{j-1} be the set of $\sigma \in R_n$ for which

$$\operatorname{Re} \lambda_0(\sigma) \leq \cdots \leq \operatorname{Re} \lambda_{j-1}(\sigma) \leq 0 \quad (j = 1, 2, \ldots, m).$$

Clearly, $A_0 \supset A_1 \supset \cdots \supset A_{m-1}$. All such sets are to be considered to within a set of measure zero. The following basic theorem specifies the well-posed problem for equation (1):

BASIC THEOREM. *Suppose that on each set A_j there is given a function $v_j(\sigma)$*
15*

which is extendable to all of $R_n(\sigma)$ as an element of H (we shall write $v_j(\sigma) \in H(A_j)$). Then equation (1) has a solution $u(t, x)$ for which the Fourier transform of $\dfrac{\partial^{J-1} u(0, x)}{\partial t^{J-1}}$ coincides with $v_{j-1}(\sigma)$ on A_{j-1} ($j = 1, 2, ..., m$).

$u(t, x)$ belongs to $\mathscr{H}$ for $t \geq 0$ and as $t \to \infty$ increases in $\mathscr{H}$ together with its t-derivatives up to order $m - 1$ no faster than a power of t. It is unique in the class of all solutions of (1) belonging to $\mathscr{H}$ which increase there together with their t-derivatives up to order $m - 1$ no faster than a power of t as $t \to \infty$ and satisfy the same initial conditions. In addition, it depends continuously on the given data $v_j(\sigma)$ in the topology of $\mathscr{H}$ when the $v_j(\sigma)$ vary continuously in the topology of H.

The proof of this theorem will be given in Sec. 4.4. At this point, we shall exemplify it by several specific equations.

1°. The heat equation

$$\frac{\partial u}{\partial t} = \frac{\partial^2 u}{\partial x^2}.$$

Here $\lambda_0 = -\sigma^2$ and the set $A_0 = \{\sigma: -\sigma^2 \leq 0\}$ coincides with the entire σ-axis. The well-posed problem consists in the assignment of $v_0(\sigma) = F[u(0, x)]$ for all σ. But by virtue of the uniqueness of Fourier transforms, this is the same as assigning $u(0, x)$. Thus the well-posed problem consists in the assignment of $u(0, x)$.

2°. The wave equation

$$\frac{\partial^2 u}{\partial t^2} = \frac{\partial^2 u}{\partial x^2}.$$

Here $\lambda^2 = -\sigma^2$ so that $\lambda_{0,1} = \pm i|\sigma|$. The sets $A_0 = \{\sigma: \text{Re } \lambda_0 \leq 0\}$ and $A_1 = \{\sigma: \text{Re } \lambda_1 \leq 0\}$ both coincide with the entire σ-axis. The well-posed problem consists in assigning $v_0(\sigma) = F[u(0, x)]$ and $v_1(\sigma) = F\left[\dfrac{\partial u(0, x)}{\partial t}\right]$ for all σ, which is equivalent to assigning $u(0, x)$ and $\dfrac{\partial u(0, x)}{\partial t}$ themselves.

3°. Laplace's equation

$$\frac{\partial^2 u}{\partial t^2} = -\frac{\partial^2 u}{\partial x^2}.$$

Here $\lambda^2 = \sigma^2$ and $\lambda_{0,1} = \pm|\sigma|$. The set $A_0 = \{\sigma: -|\sigma| \leq 0\}$ is the entire

σ-axis, while the set $A_1 = \{\sigma: |\sigma| \le 0\}$ consists of a single point. In other words, it is equivalent to the empty set. One should assign $v_0(\sigma) = F[u(0, x)]$ for all σ, which is equivalent to assigning $u(0, x)$ itself. We thus obtain the Dirichlet problem.

Let us now examine two non-classical equations.

4°. The backward heat equation $\dfrac{\partial u}{\partial t} = -\dfrac{\partial^2 u}{\partial t^2}$. In this case, $\lambda_0 = \sigma^2$ and $A_0 = \emptyset$. The class of all solutions $u(t, x) \in \mathscr{H}$ contains just one which increases in $\mathscr{H}$ no faster than a power of t, namely, $u(t, x) \equiv 0$.

5°. The ultrahyperbolic equation

$$\frac{\partial^2 u}{\partial t^2} = \frac{\partial^2 u}{\partial x_1^2} + \frac{\partial^2 u}{\partial x_2^2} - \frac{\partial^2 u}{\partial x_3^2}.$$

Here

$$\lambda_{0, 1} = \pm \sqrt{-\sigma_1^2 - \sigma_2^2 + \sigma_3^2}.$$

Re $\lambda_0(\sigma)$ is non-positive everywhere and Re $\lambda_1(\sigma)$ is non-positive in the region $A_1 = \{\sigma_1^2 + \sigma_2^2 \ge \sigma_3^2\}$. The well-posed problem amounts to assigning $v_0(\sigma)$ everywhere, i.e., to assigning $u(0, x)$ in $\mathscr{H}$, and $v_1(\sigma)$ in the region A_1, i.e., to assigning

$$\iiint\limits_{R_3} \frac{\partial u(0, x)}{\partial t} e^{i(x, \sigma)} dx_1 \, dx_2 \, dx_3$$

for $\sigma \in A_1$.

Problem

1. Determine the well-posed boundary value problems for the equations

(a) $\dfrac{\partial^4 u}{\partial t^4} = \dfrac{\partial^4 u}{\partial x^4}$; (b) $\dfrac{\partial^4 u}{\partial t^4} = -\dfrac{\partial^4 u}{\partial x^4}$; (c) $\dfrac{\partial u}{\partial t} = -\dfrac{\partial^2 u}{\partial x^2} - k^2 u.$

Ans. One should assign:

(a) $u(0, x), \dfrac{\partial u(0, x)}{\partial t}, \dfrac{\partial^2 u(0, x)}{\partial t^2}$; (b) $u(0, x), \dfrac{\partial u(0, x)}{\partial t}$;

(c) $\displaystyle\int\limits_{-\infty}^{\infty} u(0, x) e^{ix\sigma} dx$ on the set $A_0 = \{\sigma: |\sigma| \le k\}.$

4.2. Subsidiary Information

In order to prove the basic theorem, we shall need some information and estimates from interpolation theory.

4.2.1. Let $f(\lambda)$ be an analytic function in a convex domain G, and let $\lambda_0, ..., \lambda_{m-1}$ be distinct fixed points in G. We wish to construct a polynomial $R(\lambda)$ having the same values as $f(\lambda)$ at these points, namely, $f_0 = f(\lambda_0), ..., f_{m-1} = f(\lambda_{m-1})$. Such a polynomial can be constructed using Newton's formula

$$R(\lambda) = b_0 + b_1(\lambda - \lambda_0) + b_2(\lambda - \lambda_0)(\lambda - \lambda_1) + \cdots$$
$$+ b_{m-1}(\lambda - \lambda_0) \cdots (\lambda - \lambda_{m-2}). \qquad (1)$$

The m unknown coefficients $b_0, ..., b_{m-1}$ are determined from the m conditions $R(\lambda_0) = f_0, ..., R(\lambda_{m-1}) = f_{m-1}$. Substituting the values $\lambda_0, ..., \lambda_{m-1}$ in turn in (1) for λ, we arrive at the system of equations

$$\left.\begin{aligned}
f_0 &= b_0, \\
f_1 &= b_0 + b_1(\lambda_1 - \lambda_0), \\
f_2 &= b_0 + b_1(\lambda_2 - \lambda_0) + b_2(\lambda_2 - \lambda_0)(\lambda_2 - \lambda_1), \\
&\cdot \cdot \cdot \cdot \cdot \cdot \cdot \cdot \cdot \cdot \cdot \cdot \cdot \cdot \cdot \cdot \cdot \cdot \\
f_{m-1} &= b_0 + b_1(\lambda_{m-1} - \lambda_0) + b_2(\lambda_{m-1} - \lambda_0)(\lambda_{m-1} - \lambda_1) + \cdots \\
&\quad + b_{m-1}(\lambda_{m-1} - \lambda_0) \cdots (\lambda_{m-1} - \lambda_{m-2}).
\end{aligned}\right\} \quad (2)$$

This system allows one to find $b_0, ..., b_{m-1}$ sequentially. To simplify and symmetrize formulae, we introduce the notation

$$[f_j] = f_j,$$

$$[f_{0k}] = \frac{[f_k] - [f_0]}{\lambda_k - \lambda_0},$$

$$[f_{01l}] = \frac{[f_{0l}] - [f_{01}]}{\lambda_l - \lambda_1},$$

$$\cdot \cdot \cdot \cdot \cdot \cdot \cdot \cdot \cdot \cdot$$

$$[f_{01...kp}] = \frac{[f_{0...k-1,p}] - [f_{0...k-1,k}]}{\lambda_p - \lambda_k}.$$

The first equation of (2) yields $b_0 = f_0 = [f_0]$. The second yields
$b_1 = \dfrac{f_1 - f_0}{\lambda_1 - \lambda_0} = \dfrac{[f_1] - [f_0]}{\lambda_1 - \lambda_0} = [f_{01}]$. Let us verify inductively that
$b_k = [f_{01...k}]$ for $k = 0, 1, ..., m - 1$. Suppose that it is true for $j < k$.
from the $(k + 1)$-st equation of (2), we find that

$$b_k = \frac{[f_k] - [f_0] - [f_{01}](\lambda_k - \lambda_0) - \cdots - [f_{01...k-1}](\lambda_k - \lambda_0) \cdots (\lambda_k - \lambda_{k-1})}{(\lambda_k - \lambda_0)(\lambda_k - \lambda_1) \cdots (\lambda_k - \lambda_{k-1})}$$

$$= \frac{[f_k] - [f_0]}{(\lambda_k - \lambda_0) \cdots (\lambda_k - \lambda_{k-1})} - \frac{[f_{01}]}{(\lambda_k - \lambda_1) \cdots (\lambda_k - \lambda_{k-1})} - \cdots$$

$$- \frac{[f_{01...k-1}]}{\lambda_k - \lambda_{k-1}} = \frac{[f_{0k}] - [f_{01}]}{(\lambda_k - \lambda_1) \cdots (\lambda_k - \lambda_{k-1})} - \cdots - \frac{[f_{01...k-1}]}{\lambda_k - \lambda_{k-1}}$$

$$= \frac{[f_{01k}] - [f_{012}]}{(\lambda_k - \lambda_2) \cdots (\lambda_k - \lambda_{k-1})} - \cdots - \frac{[f_{01...k-1}]}{\lambda_k - \lambda_{k-1}} = \cdots = [f_{01...k}].$$

Thus Newton's interpolation polynomial has the form

$$R(\lambda) = [f_0] + [f_{01}](\lambda - \lambda_0) + [f_{012}](\lambda - \lambda_0)(\lambda - \lambda_1) + \cdots$$
$$+ [f_{01...m-1}](\lambda - \lambda_0) \cdots (\lambda - \lambda_{m-2}).$$

The coefficients b_k can be expressed in terms of the derivatives of $f(\lambda)$. To this end, consider the quantities

$$u_j = f(\lambda_j),$$
$$u_{0k} = \int_0^1 f'[\lambda_0 + t_k(\lambda_k - \lambda_0)]\, dt_k,$$
$$u_{011} = \int_0^1 \int_0^{t_1} f''[\lambda_0 + t_1(\lambda_1 - \lambda_0) + t_l(\lambda_l - \lambda_1)]\, dt_1\, dt_l,$$
$$\cdots\cdots\cdots\cdots\cdots\cdots\cdots\cdots\cdots\cdots$$
$$u_{01...kp} = \int_0^1 \int_0^{t_1} \cdots \int_0^{t_k} f^{(k+1)}[\lambda_0 + t_1(\lambda_1 - \lambda_0)$$
$$+ t_2(\lambda_2 - \lambda_1) + \cdots + t_k(\lambda_k - \lambda_{k-1})$$
$$+ t_p(\lambda_p - \lambda_k)]\, dt_1 \ldots dt_k\, dt_p.$$

$$(3)$$

In order to establish a relationship among the $u_{01...kp}$, we integrate with respect to t_p in the expression for $u_{01...kp}$. This yields

$$u_{01...kp} = \int_0^1 \int_0^{t_1} \cdots \int_0^{t_{k-1}} \frac{1}{\lambda_p - \lambda_k} f^{(k)}[\lambda_0 + t_1(\lambda_1 - \lambda_0) + \cdots$$

$$+ t_k(\lambda_k - \lambda_{k-1}) + t_p(\lambda_p - \lambda_k)] \Big|_{t_p=0}^{t_k} dt_1 \ldots dt_k$$

$$= \frac{1}{\lambda_p - \lambda_k} \left\{ \int_0^1 \cdots \int_0^{t_{k-1}} f^{(k)}[\lambda_0 + t_1(\lambda_1 - \lambda_0) + \cdots \right.$$

$$+ t_k(\lambda_p - \lambda_{k-1})] dt_1 \ldots dt_k - \int_0^1 \cdots \int_0^{t_{k-1}} f^{(k)}[\lambda_0 + t_1(\lambda_1 - \lambda_0) + \cdots$$

$$\left. + t_k(\lambda_k - \lambda_{k-1})] dt_1 \cdots dt_k \right\} = \frac{u_{01...k-1,p} - u_{01...k-1,k}}{\lambda_p - \lambda_k}. \qquad (4)$$

Since $u_0 = f_0 = [f_0]$, on applying formula (4) for $u_{01}, u_{02}, \ldots$, we conclude that $u_{01} = [f_{01}], \ldots, u_{01...m-1} = [f_{01...m-1}]$.

We see that the coefficients in Newton's interpolation polynomial can also be evaluated using (3), namely,

$$b_k = [f_{01...k}] = u_{01...k} = \int_0^1 \int_0^{t_1} \cdots \int_0^{t_{k-1}} f^{(k)}[\lambda_0 + t_1(\lambda_1 - \lambda_0) + \cdots$$

$$+ t_k(\lambda_k - \lambda_{k-1})] dt_1 \ldots dt_k. \qquad (5)$$

From this we can derive a useful estimate for the coefficients. The argument of the function $f^{(k)}$ in (5) may be written as

$$\lambda = \lambda_0(1 - t_1) + \lambda_1(t_1 - t_2) + \cdots + \lambda_{k-1}(t_{k-1} - t_k) + \lambda_k t_k. \qquad (6)$$

The sum of the coefficients of $\lambda_0, \ldots, \lambda_k$ is

$$(1 - t_1) + (t_1 - t_2) + \cdots + (t_{k-1} - t_k) + t_k = 1.$$

Therefore as the center of mass of non-negative masses $1 - t_1$, $t_1 - t_2, \ldots$ $t_{k-1} - t_k, t_k$ situated at the points $\lambda_0, \lambda_1, \ldots, \lambda_{k-1}$, and λ_k respectively, the number λ will lie within the smallest convex polygon Q_k containing $\lambda_0, \ldots, \lambda_k$. Suppose that

$$M_k = \max_{\lambda \in Q_k} |f^{(k)}(\lambda)|.$$

Estimating the absolute value of the integral (5), we obtain

$$|b_k| = |u_{01...k}| \leqq \frac{M_k}{k!}. \tag{7}$$

4.2.2. Let $P = \|p_{jk}\|$ be an m-th order matrix. We wish to obtain an estimate of the norm of the matrix e^{tP} for $t \geqq 0$.

Let $\lambda_0, \lambda_2, ..., \lambda_{m-1}$ be the eigenvalues of P (i.e., the roots of the equation $\det (P - \lambda I) = 0$). When $\lambda_0, ..., \lambda_{m-1}$ are all distinct, the matrix e^{tP} is known to be representable in the form $R(P)$, where $R(\lambda)$ is (any) polynomial for which

$$R(\lambda_0) = e^{t\lambda_0}, ..., R(\lambda_{m-1}) = e^{t\lambda_{m-1}}.$$

$R(\lambda)$ may be taken in the form of Newton's interpolation polynomial

$$R(\lambda) = b_0 + b_1(\lambda - \lambda_0) + b_2(\lambda - \lambda_0)(\lambda - \lambda_1) + \cdots$$
$$+ b_{m-1}(\lambda - \lambda_0) \cdots (\lambda - \lambda_{m-2}),$$

in which

$$|b_k| \leqq \frac{1}{k!} \max_{\lambda \in Q_k} \left| \frac{d^k e^{t\lambda}}{d\lambda^k} \right| = \max_{\lambda \in Q_k} \frac{t^k}{k!} e^{t \operatorname{Re} \lambda}. \tag{8}$$

Let $\Lambda_k = \max \{\operatorname{Re} \lambda_0, ..., \operatorname{Re} \lambda_k\}$. The estimate (8) becomes

$$|b_k| \leqq \frac{t^k}{k!} e^{t\Lambda_k} \leqq \frac{t^k}{k!} e^{t\Lambda_{m-1}}.$$

Now using

$$e^{tP} = R(P) = b_0 I + b_1(P - \lambda_0 I) + \cdots$$
$$+ b_{m-1}(P - \lambda_0 I) \cdots (P - \lambda_{m-2} I),$$

we estimate the norm of e^{tP} obtaining

$$\|e^{tP}\| \leqq e^{t\Lambda_{m-1}}$$
$$\times \left(1 + t\|P - \lambda_0 I\| + \cdots + \frac{t^{m-1}}{(m-1)!} \|P - \lambda_0 I\| ... \|P - \lambda_{m-2} I\|\right).$$

Suppose that e_j is the eigenvector corresponding to λ_j. Then we have

$$|\lambda_j| = \frac{\|Pe_j\|}{\|e_j\|} \leqq \sup_{\|e\|=1} \|Pe\| = \|P\|$$

and therefore

$$\|P - \lambda_j I\| \leqq 2\|P\|.$$

Thus

$$\| e^{tP} \| \leq e^{t \Lambda_{m-1}} \left(1 + 2t \| P \| + \cdots + \frac{(2t)^{m-1}}{(m-1)!} \| P \|^{m-1} \right). \qquad (9)$$

This estimate holds for every operator P with distinct eigenvalues. Since an arbitrary operator P with multiple eigenvalues is the limit in norm of operators with distinct eigenvalues, the estimate (9) is also valid for general P.

4.2.3. Consider m analytic functions $f^0(\lambda), \dots, f^{m-1}(\lambda)$ and m complex numbers $\lambda_0, \lambda_1, \dots, \lambda_{m-1}$. The determinant

$$W\{f^k(\lambda_j)\} = \begin{vmatrix} f^0(\lambda_0) & f^0(\lambda_1) & \cdots & f^0(\lambda_{m-1}) \\ f^1(\lambda_0) & f^1(\lambda_1) & \cdots & f^1(\lambda_{m-1}) \\ \cdots & \cdots & \cdots & \cdots \\ f^{m-1}(\lambda_0) & f^{m-1}(\lambda_1) & \cdots & f^{m-1}(\lambda_{m-1}) \end{vmatrix} \qquad (10)$$

is called the generalized Vandermonde determinant. Let us transform it as follows. From the j-th column ($j > 1$) substract the first column and divide by $\lambda_{j-1} - \lambda_0$. This yields

$$W\{f^k(\lambda_j)\} = (\lambda_1 - \lambda_0) \dots (\lambda_{m-1} - \lambda_0) \begin{vmatrix} [f^0_0] & [f^0_{01}] & \cdots & [f^0_{0,m-1}] \\ [f^1_0] & [f^1_{01}] & \cdots & [f^1_{0,m-1}] \\ \cdots & \cdots & \cdots & \cdots \\ [f^{m-1}_0] & [f^{m-1}_{01}] & \cdots & [f^{m-1}_{0,m-1}] \end{vmatrix}.$$

Then from the k-th column ($k > 2$) subtract the second and divide by $\lambda_{k-1} - \lambda_1$. This yields

$$W\{f^k(\lambda_j)\} = (\lambda_1 - \lambda_0) \dots (\lambda_{m-1} - \lambda_0)(\lambda_2 - \lambda_1) \dots (\lambda_{m-1} - \lambda_1)$$

$$\times \begin{vmatrix} [f^0_0] & [f^0_{01}] & [f^0_{012}] & \cdots & [f^0_{01,m-1}] \\ [f^1_0] & [f^1_{01}] & [f^1_{012}] & \cdots & [f^1_{01,m-1}] \\ \cdots & \cdots & \cdots & \cdots & \cdots \\ [f^{m-1}_0] & [f^{m-1}_{01}] & [f^{m-1}_{012}] & \cdots & [f^{m-1}_{01,m-1}] \end{vmatrix}.$$

Proceeding in this way, we finally arrive at

$$W\{f^k(\lambda_j)\} = \prod_{0 \leq j < k \leq m-1} (\lambda_k - \lambda_j) \begin{vmatrix} [f^0_0] & [f^0_{01}] & \cdots [f^0_{01\dots,m-1}] \\ [f^1_0] & [f^1_{01}] & \cdots [f^1_{01\dots,m-1}] \\ \cdots & \cdots & \cdots \\ [f^{m-1}_0] & [f^{m-1}_{01}] & \cdots [f^{m-1}_{01\dots,m-1}] \end{vmatrix}. \quad (11)$$

Now set $f^0 = 1, f^1 = \lambda, ..., f^{m-1} = \lambda^{m-1}$. Then $(f^k)^{(k)} = k!, (f^k)^{(k+1)} = 0$ and by (5),

$$[f^k_{01...j}] = \int_0^1 \int_0^{t_1} \cdots \int_0^{t_{j-1}} (f^k)^{(j)}[\lambda_0 + t_1(\lambda_1 - \lambda_0) + \cdots$$
$$+ t_j(\lambda_j - \lambda_{j-1})] \, dt_1 \, .. \, dt_j = \begin{cases} 0 & \text{for } j > k, \\ 1 & \text{for } j = k. \end{cases}$$

For $j < k$, $[f^k_{01...j}]$ is a homogeneous polynomial in $\lambda_0, ..., \lambda_j$ of degree $k - j$. Thus we obtain the value of the Vandermonde determinant:

$$W\{\lambda_j^k\} = \prod_{0 \leq j < k < m} (\lambda_k - \lambda_j) \begin{vmatrix} 1 & 0 & ... & 0 \\ \cdot & 1 & ... & 0 \\ \cdot & \cdot & \cdot & \cdot \\ \cdot & & ... & 1 \end{vmatrix} = \prod_{0 \leq j < k < m} (\lambda_k - \lambda_j).$$

But if $f^0, f^1, ..., f^{m-1}$ are arbitrary polynomials in λ, then we have

$$W\{f^k(\lambda_j)\} = \prod_{j < k} (\lambda_k - \lambda_j) \begin{vmatrix} [f^0_0] & [f^0_{01}] & \cdots & [f^0_{01...m-1}] \\ \cdot & \cdot & \cdot & \cdot \cdot \cdot \cdot \cdot \cdot \cdot \cdot \cdot \\ [f^{m-1}_0] & [f^{m-1}_0] & \cdots & [f^{m-1}_{01...m-1}] \end{vmatrix}$$
$$= \prod_{j < k} (\lambda_k - \lambda_j) R(\lambda_0, ..., \lambda_{m-1}), \tag{12}$$

where $R(\lambda_0, ..., \lambda_{m-1})$ is a polynomial of the indicated arguments.

Our next operation involving formula (11) is to write it as

$$\frac{W\{f^k(\lambda_j)\}}{\prod_{j < k} (\lambda_k - \lambda_j)} = \begin{vmatrix} [f^0_0] & [f^0_{01}] & \cdots & [f^0_{01...m-1}] \\ \cdot & \cdot & \cdot & \cdot \cdot \cdot \cdot \cdot \cdot \cdot \cdot \cdot \\ [f^{m-1}_0] & [f^{m-1}_{01}] & \cdots & [f^{m-1}_{01...m-1}] \end{vmatrix} \tag{13}$$

and to attempt to pass to the limit letting $\lambda_1 \to \lambda_0, ..., \lambda_{m-1} \to \lambda_0$. From (5) it follows that

$$\lim_{\substack{\lambda_1 \to \lambda_0 \\ \cdot \cdot \cdot \cdot \\ \lambda_k \to \lambda_0}} [f_{01...k}] = f^{(k)}(\lambda_0) \frac{1}{k!}.$$

Thus the limit of the right-hand side of (13) exists. Hence it follows that the left-hand side has a limit its value being

$$
\lim_{\substack{\lambda_1 \to \lambda_0 \\ \cdots \cdots \\ \lambda_{m-1} \to \lambda_0}} \frac{W\{f^k(\lambda_j)\}}{\prod_{j<k}(\lambda_k - \lambda_j)}
$$

$$
= \begin{vmatrix}
f^0(\lambda_0) & \dfrac{1}{1!}f^{0\prime}(\lambda_0) & \cdots & \dfrac{1}{(m-1)!}f^{0(m-1)}(\lambda_0) \\
\cdots & \cdots & \cdots & \cdots \\
f^{m-1}(\lambda_0) & \dfrac{1}{1!}f^{m-1\prime}(\lambda_0) & \cdots & \dfrac{1}{(m-1)!}f^{m-1(m-1)}(\lambda_0)
\end{vmatrix}.
$$

The determinant (10) can also be transformed in a slightly more complicated way. We can divide up the columns in (10) into several groups numbered say from 0 to $k-1$, from k to $l-1$, ..., and from p to $m-1$. The subtraction and division is done just within each group. We find analogously that

$$
\frac{W\{f^k(\lambda_j)\}}{\displaystyle\prod_{0 \le i < j < k}(\lambda_j - \lambda_i) \prod_{k \le i_k < j_k < l}(\lambda_{j_k} - \lambda_{i_k}) \cdots \prod_{p \le i_p < j_p < m}(\lambda_{j_p} - \lambda_{i_p})}
$$

$$
= \begin{vmatrix}
[f_0^0]\cdots[f_{01\ldots k-1}^0] & [f_k^0]\cdots[f_{k,k+1\ldots l-1}^0] & \cdots & [f_p^0]\cdots[f_{p,p+1\ldots m-1}^0] \\
\cdots & \cdots & \cdots & \cdots \\
[f_0^{m-1}]\cdots[f_{01\ldots k-1}^{m-1}] & [f_k^{m-1}]\cdots[f_{k,k+1\ldots l-1}^{m-1}] & \cdots & [f_p^{m-1}]\cdots[f_{p,p+1\ldots m-1}^{m-1}]
\end{vmatrix}.
$$

If we pass to the limit in this letting $\lambda_j \to \lambda_0$, $\lambda_{j_k} \to \lambda_k$, ..., $\lambda_{j_p} \to \lambda_p$, we obtain

$$
\lim \frac{W\{f^k(\lambda_j)\}}{\displaystyle\prod_{0 \le i < j < k}(\lambda_j - \lambda_i) \prod_{k \le i_k < j_k < l}(\lambda_{j_k} - \lambda_{i_k}) \cdots \prod_{p \le i_p < j_p < m}(\lambda_{j_p} - \lambda_{i_p})}
$$

$$
= \begin{vmatrix}
f^0(\lambda_0) & \dfrac{f^{0\prime}(\lambda_0)}{1!} & \cdots & \dfrac{f^{0(k-1)}(\lambda_0)}{(k-1)!} & \cdots & f^0(\lambda_p) & \cdots & \dfrac{f^{0(m-p)}(\lambda_p)}{(m-p)!} \\
\cdots & \cdots & \cdots & \cdots & \cdots & \cdots & \cdots & \cdots \\
f^{m-1}(\lambda_0) & \dfrac{f^{m-1\prime}(\lambda_0)}{1!} & \cdots & \dfrac{f^{m-1(k-1)}(\lambda_0)}{(k-1)!} & \cdots & f^{m-1}(\lambda_p) & \cdots & \dfrac{f^{m-1(m-p)}(\lambda_p)}{(m-p)!}
\end{vmatrix}.
$$

In the case of the Vandermonde determinant, this result reduces to

$$\lim \frac{\displaystyle\prod_{0 \le i < j < m} (\lambda_j - \lambda_i)}{\displaystyle\prod_{0 \le i < j < k} (\lambda_j - \lambda_i) \prod_{k \le i_k < j_k < l} (\lambda_{j_k} - \lambda_{i_k}) \cdots \prod_{p \le i_p < j_p < m} (\lambda_{j_p} - \lambda_{i_p})}$$

$$= \begin{vmatrix} 1 & 0 & 0 & \cdots 1 & 0 & 0 & \cdots \\ \lambda_0 & 1 & 0 & \cdots \lambda_p & 1 & 0 & \cdots \\ \lambda_0^2 & 2\lambda_0 & 1 & \cdots \lambda_p^2 & 2\lambda_p & 1 & \cdots \\ \cdot & \cdot & \cdot & \cdot & \cdot & \cdot & \cdot \\ \lambda_0^{m-1} & (m-1)\lambda_0^{m-2} & \binom{m-1}{2}\lambda_0^{m-3} & \cdots \lambda_p^{m-1} & (m-1)\lambda_p^{m-2} & \binom{m-1}{2}\lambda_p^{m-3} & \cdots \end{vmatrix}.$$

If in particular the numbers $\lambda_0, \lambda_k, \ldots, \lambda_p$ are distinct, then the determinant on the right-hand side will differ from zero. The reason for this is that after cancellation is performed on the left-hand side, the numerator will contain only differences of the form $\lambda_j - \lambda_i$ with λ_i and λ_j belonging to distinct groups. The limit of each such difference is the difference of the first elements of their respective groups, in other words, a non-zero quantity. The matrix

$$\begin{Vmatrix} 1 & 0 & 0 & \cdots 1 & 0 & 0 & \cdots \\ \lambda_0 & 1 & 0 & \cdots \lambda_p & 1 & 0 & \cdots \\ \lambda_0^2 & 2\lambda_0 & 1 & \cdots \lambda_p^2 & 2\lambda_p & 1 & \cdots \\ \cdot & \cdot & \cdot & \cdot & \cdot & \cdot & \cdot \\ \lambda_0^{m-1} & (m-1)\lambda_0^{m-2} & \binom{m-1}{2}\lambda_0^{m-3} & \cdots \lambda_p^{m-1} & (m-1)\lambda_p^{m-2} & \binom{m-1}{2}\lambda_p^{m-3} & \cdots \end{Vmatrix} \tag{14}$$

is called the generalized Vandermonde maxtrix and plays an important role in the sequel. Not only is its determinant non-vanishing but also any minor formed from its first s rows and s columns since it has the exact same structure as the entire determinant.

4.2.4. We shall give one more expression for the coefficients $b_0, \ldots, b_{m-1}$ in Newton's interpolation polynomial to be used later on in Sec. 4.8. It is apparent from (2) that b_k can be expressed as a linear combination of $f_0, \ldots, f_{k-1}$. Indeed,

$$b_k = \sum_{j=0}^{k} \frac{f_j}{\displaystyle\prod_{i=0}^{k}{}' (\lambda_j - \lambda_i)}, \tag{15}$$

wherein the prime on the product symbol means that the factor for $i = j$ is to be omitted. To prove this, we put $c_0 = b_0 = f_0$ and

$$c_{0...k, l} = \sum_{j = 0, 1, ..., k, l} \frac{f_j}{{\prod_{i = 0, 1, ..., k, l}}' (\lambda_j - \lambda_i)}. \tag{16}$$

We shall show that the quantities defined by (16) satisfy the same recursion relations as $[f_{01...kl}]$, namely,

$$c_{0...kl} = \frac{c_{0...k-1, l} - c_{0...k-1, k}}{\lambda_l - \lambda_k}. \tag{17}$$

And since $c_0 = b_0 = f_0 = [f_0]$, the relation $c_{0...k} = [f_{0...k}] = b_k$ will turn out to be valid together with (17). We have: (See the next page for the appropriate computations)

The expression

$$F(\lambda_0, ..., \lambda_{r-1}) = \sum_{j = 0, ..., r-1} \frac{F(\lambda_j)}{{\prod_{i = 0, ..., r-1}}' (\lambda_j - \lambda_i)}$$

is called the *divided difference* of the function $F(\lambda)$ at the points $\lambda_0, ..., \lambda_{r-1}$. For $r = 1$, the divided difference is $F(\lambda_0)$ itself. For $r > 1$, it satisfies the relation

$$F(\lambda_0, ..., \lambda_{r-1}) = \int_0^1 \int_0^{t_1} \cdots \int_0^{t_{r-2}} F^{(r-1)}[\lambda_0 + t_1(\lambda_1 - \lambda_0) + \cdots$$

$$+ t_{r-1}(\lambda_{r-1} - \lambda_{r-2})] \, dt_1 ... \, dt_{r-1}, \tag{18}$$

which is a consequence of (5). An estimate for $F(\lambda_0, ..., \lambda_{r-1})$ corresponding to (7) is given by

$$|F(\lambda_0, ..., \lambda_{r-1})| \leqq \frac{1}{(r-1)!} \max_{\lambda \in Q_{r-1}} |F^{(r-1)}(\lambda)|. \tag{19}$$

4.2.5. Later on we shall need an estimate for the compound divided difference defined by

$$F(\lambda_0, ..., \lambda_{r-1}; \lambda_r, ..., \lambda_{m-1})$$

$$\equiv \sum_{j = 0, ..., r-1} \frac{1}{{\prod_{i = 0, ..., r-1}}' (\lambda_j - \lambda_i)} \sum_{s = j, r, ..., m-1} \frac{F(\lambda_j, \lambda_s)}{{\prod_{k = j, r, ..., m-1}}' (\lambda_s - \lambda_k)}. \tag{20}$$

$$\frac{c_{0...k-1,\,l} - c_{0...k-1,\,k}}{\lambda_l - \lambda_k}$$

$$= \frac{\displaystyle\sum_{j=0,...,k-1,\,l} \frac{f_j}{\displaystyle\prod'_{i=0,...,k-1,\,l}(\lambda_j - \lambda_i)} - \sum_{j=0,...,k-1,\,k} \frac{f_j}{\displaystyle\prod'_{i=0,...,k-1,\,k}(\lambda_j - \lambda_i)}}{\lambda_l - \lambda_k}$$

$$= \frac{\displaystyle\sum_{j=0,...,k-1} \frac{f_j}{\displaystyle\prod'_{i=0,...,k-1}(\lambda_j - \lambda_i)}\left(\frac{1}{\lambda_j - \lambda_l} - \frac{1}{\lambda_j - \lambda_k}\right) + \frac{f_l}{\displaystyle\prod'_{i=0,...,k-1,\,l}(\lambda_l - \lambda_i)} - \frac{f_k}{\displaystyle\prod'_{i=0,...,k-1,\,k}(\lambda_k - \lambda_i)}}{\lambda_l - \lambda_k}$$

$$= \sum_{j=0,...,k-1} \frac{f_j}{\displaystyle\prod'_{i=0,...,k-1}(\lambda_j - \lambda_i)\,(\lambda_j - \lambda_l)\,(\lambda_j - \lambda_k)} + \frac{f_l}{\displaystyle\prod'_{i=0,...,k-1,\,l}(\lambda_l - \lambda_i)\,(\lambda_l - \lambda_k)}$$

$$- \frac{f_k}{\displaystyle\prod'_{i=0,...,k-1,\,k}(\lambda_k - \lambda_i)\,(\lambda_l - \lambda_k)} = \sum_{j=0,...,k-1,\,k,\,l} \frac{1}{\displaystyle\prod'_{i=0,...,k-1,\,k,\,l}(\lambda_j - \lambda_i)} = c_{0...k\,l},$$

as required.

The function $F(\lambda, \zeta)$ is analytic with respect to ζ in the convex polygon spanned by the points $\lambda_r, \ldots, \lambda_{m-1}$ and with respect to λ in the convex polygon spanned by $\lambda_0, \ldots, \lambda_{r-1}$. The expression (20) can be obtained as follows. Form

$$\Phi(\lambda) \equiv \Phi(\lambda; \lambda, \lambda_r, \ldots, \lambda_{m-1}), \tag{21}$$

the divided difference for $F(\lambda, \zeta)$ with respect to its second argument at the points $\lambda, \lambda_r, \ldots, \lambda_{m-1}$ keeping its first argument λ fixed. Then form the divided difference again at the points $\lambda = \lambda_0, \ldots, \lambda_{r-1}$. According to formula (5),

$$F(\lambda_0, \ldots, \lambda_{r-1}; \lambda_r, \ldots, \lambda_{m-1}) = \int_0^1 \int_0^{t_1} \cdots \int_0^{t_{r-2}} \Phi^{(r-1)}[\lambda_0 + t_1(\lambda_1 - \lambda_0) + \cdots$$
$$+ t_{r-1}(\lambda_{r-1} - \lambda_{r-2})] \, dt_1 \ldots dt_{r-1}. \tag{22}$$

Let us determine $\Phi^{(r-1)}(\lambda)$ once again resorting to (5). We have

$$\Phi(\lambda) = \int_0^1 \int_0^{t_r} \cdots \int_0^{t_{m-2}} \frac{\partial^{m-r}}{\partial \zeta^{m-r}} F[\lambda; \lambda + t_r(\lambda_r - \lambda) + \cdots$$
$$+ t_{m-1}(\lambda_{m-1} - \lambda_{m-2})] \, dt_r \ldots dt_{m-1}. \tag{23}$$

Differentiating (23) with respect to λ, we obtain

$$\Phi^{(r-1)}(\lambda) = \int_0^1 \int_0^{t_r} \cdots \int_0^{t_{m-2}} \frac{\partial^{m-r}}{\partial \zeta^{m-r}} \left[\frac{\partial}{\partial \lambda} + (1 - t_r) \frac{\partial}{\partial \zeta} \right]^{r-1}$$
$$\times F[\lambda; \lambda + t_r(\lambda_r - \lambda) + \cdots + t_{m-1}(\lambda_{m-1} - \lambda_{m-2})] \, dt_r \ldots dt_{m-1}. \tag{24}$$

Substituting (24) in (22), we find

$$F(\lambda_0, \ldots, \lambda_{r-1}; \lambda_r, \ldots, \lambda_{m-1})$$
$$= \int_0^1 \int_0^{t_1} \cdots \int_0^{t_{r-2}} \int_0^1 \int_0^{t_r} \cdots \int_0^{t_{m-2}} \frac{\partial^{m-r}}{\partial \zeta^{m-r}} \left[\frac{\partial}{\partial \lambda} + (1 - t_r) \frac{\partial}{\partial \zeta} \right]^{r-1}$$
$$\times F[\bar{\lambda}; \bar{\lambda} + t_r(\lambda_r - \bar{\lambda}) + \cdots + t_{m-1}(\lambda_{m-1} - \lambda_{m-2})] \, dt_1 \ldots dt_{m-1}, \tag{25}$$

where $\bar{\lambda} = \lambda_0 + t_1(\lambda_1 - \lambda_0) + \cdots + t_{r-1}(\lambda_{r-1} - \lambda_{r-2})$ lies in the convex polygon spanned by $\lambda_0, \dots, \lambda_{r-1}$. Equation (25) leads to the estimate

$$|F(\lambda_0, \dots, \lambda_{r-1}; \lambda_r, \dots, \lambda_{m-1})| \leq C \max \left| \frac{\partial^{m-1} F(\lambda, \zeta)}{\partial \lambda^k \partial \zeta^{m-1-k}} \right|, \qquad (26)$$

where $C \leq 2^{r-1}$ and the maximum is taken over all $k \leq m - 1$ and all points of the polygon spanned by the points $\zeta = \bar{\lambda}, \lambda_r, \dots, \lambda_{m-1}, \bar{\lambda}$ being any point of the polygon spanned by $\lambda_0, \dots, \lambda_{r-1}$.

4.3. Ordinary Differential Equations and Systems

4.3.1. Consider the system of ordinary differential equations

$$\left.\begin{aligned}
\frac{dv_0(t)}{dt} &= p_{00} v_0(t) + \cdots + p_{0, \, m-1} v_{m-1}(t), \\
\cdots \cdots \cdots \cdots \cdots \cdots \cdots \cdots \cdots \cdots \\
\frac{dv_{m-1}(t)}{dt} &= p_{m-1, \, 0} v_0(t) + \cdots + p_{m-1, \, m-1} v_{m-1}(t),
\end{aligned}\right\} \qquad (1)$$

where $p_{00}, \dots, p_{m-1, \, m-1}$ are (complex) constants. It is known that a solution $v(t) = \{v_0(t), \dots, v_{m-1}(t)\}$ is uniquely determined for all $t \geq 0$ by the assignment of $v(0) = \{v_0(0), \dots, v_{m-1}(0)\}$. The vector $v(0)$ will be said to define a *well-posed initial value problem* if the solution $v(t)$ has the behavior

$$|v(t)| \equiv \sqrt{\sum_{j=0}^{m-1} v_j^2(t)} = O(t^h)$$

as $t \to \infty$. In other words the solution increases no faster than a power of t at infinity.

The question we raise is: What conditions must $v(0)$ satisfy in order that it define a well-posed problem?

To answer this, suppose that P is an operator defined in the m-dimensional space Q of vectors $\xi = (\xi_0, \dots, \xi_{m-1})$ by the formulas

$$(P\xi)_0 = p_{00} \xi_0 + \cdots + p_{0, \, m-1} \xi_{m-1},$$

$$\cdots \cdots \cdots \cdots \cdots \cdots \cdots \cdots \cdots$$

$$(P\xi)_{m-1} = p_{m-1, \, 0} \xi_0 + \cdots + p_{m-1, \, m-1} \xi_{m-1}.$$

Let $\lambda_0, \dots, \lambda_{m-1}$ be the eigenvalues of P (among which there may be equal ones), and let $e_0, \dots, e_{m-1}$ be a basis for the space Q comprising the corresponding eigenvectors (and root vectors in case of multiplicities). The λ_j are assumed to be numbered according to increasing real parts:

$$\text{Re } \lambda_0 \leqq \text{Re } \lambda_1 \leqq \cdots \leqq \text{Re } \lambda_{m-1}.$$

Let r be determined by the relation

$$\text{Re } \lambda_{r-1} \leqq 0 < \text{Re } \lambda_r. \tag{2}$$

In other words, r is the number of eigenvalues of P (counting multiplicities) with a non-positive real part. In particular, $r = 0$ if $\text{Re } \lambda_j > 0$ for all j and $r = m$ if $\text{Re } \lambda_j \leqq 0$ for all j.

Q may be decomposed into a direct sum of subspaces Q^- and Q^+ each invariant under the operator P, the first being generated by the vectors $e_0, ..., e_{r-1}$ and the second by the vectors $e_r, ..., e_{m-1}$. We can now formulate the following theorem:

THEOREM 1. *The vector $v(0)$ defines a well-posed problem for the system* (1) *if and only if it belongs to the subspace Q^-.*

Proof. The solution to (1) with initial value $v(0)$ can be written as

$$v(t) = e^{tP}v(0).$$

Let $v(0)$ belong to Q^-. Then the solution $v(t)$ is also in Q^- for all $t \geqq 0$ since the invariance of Q^- under P implies its invariance under e^{tP}. We can thus confine our attention to Q^-. The Jordan form of the matrix P in Q^- will in general be quasi-diagonal with blocks of the form

$$
\begin{Vmatrix}
\lambda_j & 1 & 0 & \cdots & 0 \\
0 & \lambda_j & 1 & \cdots & 0 \\
0 & 0 & \lambda_j & \cdots & 0 \\
\cdot & \cdot & \cdot & \cdot & \cdot & \cdot \\
0 & 0 & 0 & \cdots & \lambda_j
\end{Vmatrix}
\tag{3}
$$

along its diagonal.

As is known, if $f(\lambda)$ is any sufficiently smooth function, then a knowledge of the Jordan form of P makes it possible to write down the matrix $f(P)$ at once in the same basis. Namely, $f(P)$ is also quasi-diagonal and is obtained by replacing each block of the form (3) by†

$$\left\|\begin{array}{ccccc} f(\lambda_j) & f'(\lambda_j) & \dfrac{1}{2!}f''(\lambda_j) & \cdots & \cdots \\ 0 & f(\lambda_j) & f'(\lambda_j) & \cdots & \cdots \\ 0 & 0 & f(\lambda_j) & \cdots & \cdots \\ \cdot & \cdot & \cdot & \cdot & \cdot \\ 0 & 0 & 0 & \cdots & f(\lambda_j) \end{array}\right\|.$$

In our case $f(\lambda) = e^{t\lambda}$ and the corresponding block is given by

$$\left\|\begin{array}{ccccc} e^{t\lambda_j} & te^{t\lambda_j} & \dfrac{t^2}{2}e^{t\lambda_j} & \cdots & \cdots \\ 0 & e^{t\lambda_j} & te^{t\lambda_j} & \cdots & \cdots \\ 0 & 0 & e^{t\lambda_j} & \cdots & \cdots \\ \cdot & \cdot & \cdot & \cdot & \cdot \\ 0 & 0 & 0 & \cdots & e^{t\lambda_j} \end{array}\right\|.$$

We point out that the numbers $e^{t\lambda_j}$ do not exceed 1 in absolute value since Re $\lambda_j \leq 0$ in the subspace Q^-.

Thus each basis vector $e_0, \ldots, e_{r-1}$ of Q^- is transformed under e^{tP} into a vector of the form

$$e^{tP}e_k = \sum_{j=0}^{k} c_{jk}t^{k-j}e^{t\lambda_k}e_j.$$

The c_{jk} are fixed constants of which c_{kk} is definitely non-vanishing (always having the value 1).

Hence for any vector $v(0) = \sum_{k=0}^{r-1} \xi_k e_k \in Q^-$, we have

$$v(t) = e^{tP}v(0) = \sum_k \xi_k e^{tP}e_k = \sum_k \sum_j c_j t^{k-j}e^{t\lambda_k}e_j$$

and so

$$\|v(t)\| \leq C(1+t)^{m-1}.$$

Thus if $v(0) \in Q$, the problem is well-posed.

† See F. R. Gantmacher, *The Theory of Matrices*, Chelsea, New York, 1957, Chapt. V, Sec. 1, pp. 98–99.

16*

Now suppose that $v(0)$ has a non-null component in Q^+. In other words, $v(0) = \sum\limits_{k=0}^{m-1} \xi_k e_k$ and at least one $\xi_r, \ldots, \xi_{m-1}$ is different from zero. Let ξ_p be the last non-vanishing component of $v(0)$. As is apparent from the foregoing, the vector e_p is transformed by e^{tP} into a vector of the form

$$\sum_{j=0}^{p} c_{jp} t^{p-j} e^{t\lambda_p} e_j, \quad c_{pp} \neq 0,$$

where this time $\operatorname{Re} \lambda_p > 0$. The preceding vectors $e_s, s < p$, are transformed under e^{tP} into similar expressions in which however there appear terms not containing e_p. Thus the vector $e^{tP} v(0)$ has a p-th component $t^{j_p} e^{t\lambda_p} e_p$ which is exponentially increasing with increasing t. This implies that the norm of $e^{tP} v(0)$ is itself exponentially increasing so that the problem is not well-posed. The theorem is proved.

4.3.2. Consider now a single m-th order equation

$$\frac{d^m v(t)}{dt^m} = \sum_{k=0}^{m-1} p_k \frac{d^k v(t)}{dt^k} \tag{4}$$

with complex coefficients $p_0, \ldots, p_{m-1}$. Its solution $v(t)$ is uniquely determined by the assignment of $v(0), \ldots, v^{(m-1)}(0)$. A problem is called well-posed if its initial data is such that it has a unique solution $v(t)$ which with its derivatives $v'(t), \ldots, v^{(m-1)}(t)$ increases no faster than a power of t as $t \to \infty$. To characterize the well-posed problems, we construct a first order system equivalent to (4) as follows:

$$\left. \begin{aligned} \frac{dv_0(t)}{dt} &= v_1, \\[1em] \frac{dv_1(t)}{dt} &= v_2, \\[1em] \cdots \cdots \cdots \cdots \cdots \cdots \cdots \cdots \cdots \cdots \\[1em] \frac{dv_{m-2}(t)}{dt} &= v_{m-1}, \\[1em] \frac{dv_{m-1}(t)}{dt} &= p_0 v_0 + p_1 v_1 + p_2 v_2 + \cdots + p_{m-1} v_{m-1}. \end{aligned} \right\} \tag{5}$$

The system is obtained by introducing the new unknowns $v_0(t) = v(t)$, $v_1(t) = v'(t), \ldots, v_{m-1}(t) = v^{(m-1)}(t)$. The matrix P corresponding to (5) has the special form

$$P = \begin{Vmatrix} 0 & 1 & 0 & \cdots & 0 \\ 0 & 0 & 1 & \cdots & 0 \\ \cdot & \cdot & \cdot & \cdots & \cdot \\ 0 & 0 & 0 & \cdots & 1 \\ p_0 & p_1 & p_2 & \cdots & p_{m-1} \end{Vmatrix}. \qquad (6)$$

In order to apply the general scheme of Sec. 4.3.1 to the present case, we must determine the eigenvalues and eigenvectors of P in the m-dimensional space Q.

P operates on a vector $\xi = (\xi_0, \xi_1, ..., \xi_{m-1})$ through the formulas

$$(P\xi)_0 = \xi_1,$$
$$(P\xi)_1 = \xi_2,$$
$$\cdots\cdots\cdots\cdots\cdots\cdots\cdots$$
$$(P\xi)_{m-2} = \xi_{m-1},$$
$$(P\xi)_{m-1} = p_0\xi_0 + p_1\xi_1 + p_2\xi_2 + \cdots + p_{m-1}\xi_{m-1}.$$

Let ξ be an eigenvector corresponding to the eigenvalue λ. Then we have

$$\xi_1 = \lambda\xi_0,$$
$$\xi_2 = \lambda\xi_1,$$
$$\cdots\cdots\cdots\cdots\cdots\cdots\cdots$$
$$\xi_{m-1} = \lambda\xi_{m-2},$$
$$p_0\xi_0 + p_1\xi_1 + p_2\xi_2 + \cdots + p_{m-1}\xi_{m-1} = \lambda\xi_{m-1}.$$

Setting $\xi_0 = 1$, we find sequentially that

$$\xi_0 = 1, \quad \xi_1 = \lambda, \quad \xi_2 = \lambda^2, ..., \xi_{m-1} = \lambda^{m-1}$$

and

$$\lambda^m = \sum_{k=0}^{m-1} p_k\lambda^k. \qquad (7)$$

Thus each eigenvalue λ of P is a root of (7). The eigenvector for λ is collinear with the vector

$$(1, \lambda, \lambda^2, ..., \lambda^{m-1})$$

and is thereby uniquely determined to within collinearity (which means that if λ is a multiple eigenvalue, then only root vectors will occur).

As before, let r be the number of eigenvalues of P (counting multiplicities) whose real parts are non-positive. The next theorem furnishes conditions for the initial value problem for (4) to be well-posed:

THEOREM 2. *Any assignment of the data* $v(0), ..., v^{(r-1)}(0)$ *defines a well-posed problem for equation* (4).

Proof. Having the application of Theorem 1 in mind, we must first find a basis for the subspace Q^-. If the eigenvalues $\lambda_0, ..., \lambda_{r-1}$ are all simple, the columns of the matrix

$$
\begin{Vmatrix}
1 & 1 & \dots & 1 \\
\lambda_0 & \lambda_1 & \dots & \lambda_{r-1} \\
\lambda_0^2 & \lambda_1^2 & \dots & \lambda_{r-1}^2 \\
\cdot & \cdot & \cdot & \cdot \\
\lambda_0^{m-1} & \lambda_1^{m-1} & \dots & \lambda_{r-1}^{m-1}
\end{Vmatrix}
$$

will be a basis. If there are multiple eigenvalues among $\lambda_0, ..., \lambda_{r-1}$, the basis matrix will have a more complicated form. Suppose that λ_0 has multiplicity k. Then as is apparent from the Jordan form of P, the basis vectors $e_0, ..., e_{k-1}$ will be related by the equations

$$
\left.
\begin{aligned}
Pe_0 &= \lambda_0 e_0, \\
Pe_1 &= e_0 + \lambda_0 e_1, \\
&\cdot \cdot \cdot \cdot \cdot \cdot \cdot \cdot \cdot \cdot \cdot \cdot \cdot \\
Pe_{k-1} &= e_{k-2} + \lambda_0 e_{k-1}.
\end{aligned}
\right\} \tag{8}
$$

The vectors $e_0, ..., e_{k-1}$ are not uniquely determined by the equations (8). But any set $e_0, ..., e_{k-1}$ satisfying (8) can be taken as the basis for the corresponding invariant subspace of P (the collection of all solutions of the equation $(P - \lambda_0 I)^k \xi = 0$). We shall construct one such set. To that end, consider the first k columns of the matrix (14) of Sec. 4.2., namely,

$$
\begin{Vmatrix}
1 & 0 & 0 & \dots & 0 \\
\lambda_0 & 1 & 0 & \dots & 0 \\
\lambda_0^2 & 2\lambda_0 & 1 & \dots & 0 \\
\cdot & \cdot & \cdot & \cdot & \cdot \\
\lambda_0^{m-1} & (m-1)\lambda_0^{m-2} & \binom{m-1}{2}\lambda_0^{m-3} & \dots & \binom{m-1}{k-1}\lambda_0^{m-k}
\end{Vmatrix}. \tag{9}
$$

They define k vectors in the space Q. Denote them by $e_0, e_1, ..., e_{k-1}$. We wish to show that they satisfy (8).

As we already know, (9) is the limit of the matrix

$$\left\| \begin{array}{cccc} [f_0^0] & [f_{01}^0] & \cdots & [f_{01...k-1}^0] \\ [f_0^1] & [f_{01}^1] & \cdots & [f_{01...k-1}^1] \\ \cdot \cdot \cdot & \cdot \cdot \cdot & \cdot \cdot \cdot & \cdot \cdot \cdot \\ [f_0^{m-1}] & [f_{01}^{m-1}] & \cdots & [f_{01...k-1}^{m-1}] \end{array} \right\|, \qquad (10)$$

with $f^0 = 1, f^1 = \lambda, ..., f^{m-1} = \lambda^{m-1}$. More precisely, (9) results from (10) if we let $\lambda_1 \to \lambda_0, ..., \lambda_{k-1} \to \lambda_0$ assuming that $\lambda_0, \lambda_1, ..., \lambda_{k-1}$ are all distinct. Let $f_{01...js}$ denote a vector with components $[f_{01...js}^0], ..., f_{01...js}^{m-1}]$. We have $Pf_j = \lambda_j f_j$ $(j = 0, 1, ..., k - 1)$. Further,

$$Pf_{0s} = P\frac{f_s - f_0}{\lambda_s - \lambda_0} = \frac{Pf_s - Pf_0}{\lambda_s - \lambda_0} = \frac{\lambda_s f_s - \lambda_0 f_0}{\lambda_s - \lambda_0} = f_0 + \lambda_s f_{0s}.$$

Let us show that

$$Pf_{01...js} = f_{01...j} + \lambda_s f_{01...js} \quad (s > j).$$

Assume that it is true for a fewer number of subscripts, i.e.,

$$Pf_{01...j-1,s} = f_{01...j-1} + \lambda_s f_{01...j-1,s}, \quad s > j - 1.$$

Then

$$Pf_{01...js} = P\frac{f_{01...j-1,s} - f_{01...j-1,j}}{\lambda_s - \lambda_j} = \frac{Pf_{01...j-1,s} - Pf_{01,...j-1,j}}{\lambda_s - \lambda_j}$$

$$= \frac{f_{01...j-1} + \lambda_s f_{01...j-1,s} - f_{01...j-1} - \lambda_j f_{01...j-1,j}}{\lambda_s - \lambda_j}$$

$$= \frac{\lambda_s f_{01...j-1,s} - \lambda_j f_{01...j-1,j}}{\lambda_s - \lambda_j} = f_{01...j-1,j} + \lambda_s f_{01...j-1,js}.$$

In particular we have

$$Pf_{01...j-1,j} = f_{01...j-1} + \lambda_j f_{01...j-1,j} \quad (j = 1, 2, ..., k - 1).$$

Letting $\lambda_1 \to \lambda_0, ..., \lambda_k \to \lambda_0$ in this relation, we obtain

$$Pe_j = e_{j-1} + \lambda_0 e_j \quad (j = 1, 2, ..., k - 1),$$

as required.

Thus in the general case a basis matrix for Q^- is given by the generalized Vandermonde matrix

$$E = \begin{Vmatrix} 1 & 0 & 0 & \cdots & 0 & \cdots \\ \lambda_0 & 1 & 0 & \cdots & 0 & \cdots \\ \lambda_0^2 & 2\lambda_0 & 1 & \cdots & 0 & \cdots \\ \cdots\cdots\cdots\cdots\cdots\cdots\cdots\cdots\cdots\cdots & & & & & \\ \lambda_0^{m-1} & (m-1)\lambda_0^{m-2} & \binom{m-1}{2}\lambda_0^{m-3} & \cdots & \binom{m-1}{k-1}\lambda_0^{m-k} & \cdots \end{Vmatrix}.$$

$$\underbrace{\qquad\qquad\qquad\qquad\qquad\qquad\qquad\qquad\qquad\qquad}_{r}$$

We now prove the following lemma:

LEMMA. *If P is an operator defined by the matrix* (6), *then given any numbers* $\xi_0, ..., \xi_{r-1}$ *it is always possible to uniquely determine numbers* $\xi_r, ..., \xi_{m-1}$ *so that the vector* $\xi = (\xi_0, ..., \xi_{m-1})$ *lies in the subspace* Q^-.

Proof. Let $c = (c_0, ..., c_{r-1})$ be an unknown vector. We wish to show that the system of equations

$$\xi = Ec \tag{11}$$

is consistent. The first r equations of this system have a determinant coming from the first r rows of the matrix E. By the remark at the end of Sec. 4.2.3 it is different from zero. Since $\xi_0, ..., \xi_{r-1}$ are known quantities, the unknowns $c_0, ..., c_{r-1}$ are uniquely determined. The next $m - r$ equations of the system (11) then uniquely determine $\xi_r, ..., \xi_{m-1}$, as required.

For the sequel, we shall derive formulas expressing $\xi_r, ..., \xi_{m-1}$ in terms of $\xi_0, ..., \xi_{r-1}$. If the eigenvalues $\lambda_0, ..., \lambda_{r-1}$ are distinct, then

$$\begin{vmatrix} \xi_0 & 1 & 1 & \cdots & 1 \\ \xi_1 & \lambda_0 & \lambda_1 & \cdots & \lambda_{r-1} \\ \cdots & \cdots & \cdots & \cdots & \cdots \\ \xi_{r-1} & \lambda_0^{r-1} & \lambda_1^{r-1} & \cdots & \lambda_{r-1}^{r-1} \\ \xi_s & \lambda_0^s & \lambda_1^s & \cdots & \lambda_{r-1}^s \end{vmatrix} = 0$$

for any $s \geqq r$. This is due to the fact that the rank of the augmented matrix of the system (11) remains equal to r. If we expand the determinant according to the first column, we obtain

$$\xi_s = \sum_{j=0}^{r-1} \xi_j \frac{W\{f_{ij}^k\}}{W\{\lambda_i^k\}},$$

where $\{f_{ij}^0, f_{ij}^1, ..., f_{ij}^{r-1}\} \equiv \{1, \lambda_i, ..., \lambda_i^{j-1}, \lambda_i^s, \lambda_i^{j+1}, ..., \lambda_i^{r-1}\}$. As shown in Sec. 4.2.3, the coefficient of ξ_j is a homogeneous polynomial in $\lambda_0, ..., \lambda_{r-1}$ of degree $s - j$. In summary, we have established that ξ_s can be expressed (for $s \geq r$) in terms of $\xi_0, ..., \xi_{r-1}$ by means of

$$\xi_s = \sum_{j=0}^{r-1} \xi_j R_{js}(\lambda_0, ..., \lambda_{r-1}), \tag{12}$$

where

$$R_{js}(\lambda_0, ..., \lambda_{r-1}) = \frac{W\{f_{ij}^k\}}{W\{\lambda_i^k\}} \tag{13}$$

is a homogeneous polynomial in $\lambda_0, ..., \lambda_{r-1}$ of degree $s - j$.

Let us show that this formula remains valid for the case of multiple roots. We factor out the expression

$$\prod_{0 \leq i < j < k} (\lambda_j - \lambda_i) ... \prod_{p \leq i_p < j_p < m} (\lambda_{j_p} - \lambda_{i_p})$$

from the numerator and denominator of the fraction in (13) and cancel. This has no affect on the polynomial R_{js}. We then pass to the limit on the right-hand side letting $\lambda_1, ..., \lambda_{k-1} \to \lambda_0$ and so on. As a result, the numerator and denominator of each such fraction turn into the necessary minors making up the coefficients of $\xi_0, ..., \xi_{r-1}$ in the expression for ξ_s. As for the left-hand side, it is simply the value of the polynomial R_{js} for $\lambda_0 = \cdots = \lambda_{k-1}, \lambda_k = \lambda_{k+1} = \cdots = \lambda_{l-1}$, etc. Thus (12) remains valid for the case of multiple roots.

4.3.3. We again consider the system

$$\frac{dv_0(t)}{dr} = p_{00}v_0(t) + \cdots + p_{0,m-1}v_{m-1}(t),$$

$$\cdots \cdots \cdots \cdots \cdots \cdots \cdots \cdots \cdots \cdots \cdots \tag{1}$$

$$\frac{dv_{m-1}(t)}{dt} = p_{m-1,0}v_0(t) + \cdots + p_{m-1,m-1}v_{m-1}(t).$$

We wish to pose a more general initial value problem for it.

Consider the system of equations

$$\left. \begin{aligned} a_{00}\xi_0 + a_{01}\xi_1 + \cdots + a_{0,m-1}\xi_{m-1} &= b_0, \\ \cdots \cdots \cdots \cdots \cdots \cdots \cdots \cdots \cdots \\ a_{q0}\xi_0 + a_{q1}\xi_1 + \cdots + a_{q,m-1}\xi_{m-1} &= b_q. \end{aligned} \right\} \tag{14}$$

The question raised is this. What should the matrix $A = \|a_{jk}\|$ be so that for any constants $b_0, \ldots, b_q$ there exist a unique vector $v(0)$ guaranteeing a well-posed initial value problem for (1) with data $v_0 = \xi_0, \ldots, v_{m-1} = \xi_{m-1}$ satisfying (14) This question is now easy to answer.

Let E denote the portion of the Jordan matrix for the operator P comprising the columns corresponding to the basis vectors $e_0, \ldots, e_{r-1}$ of the subspace Q^-. E has r columns and m rows. The condition that a vector ξ lies in Q^- or equivalently is a linear combination of $e_0, \ldots, e_{r-1}$ may be written as

$$\xi = Ec, \tag{15}$$

where $c = (c_0, \ldots, c_{r-1})$ is a vector uniquely determined in terms of ξ. On the other hand, the system (14) may be written as

$$A\xi = b. \tag{16}$$

Let A satisfy the stated condition. That is, for any b there is a unique vector ξ satisfying (15) and (16). Then the system of equations

$$AEc = b \tag{17}$$

has a unique solution for any b. The maxtrix AE clearly has $q + 1$ rows and r columns. The fact that (17) has a unique solution means that $q = r - 1$ and that the now square matrix AE is invertible. In other words,

$$\det AE \neq 0. \tag{18}$$

Let us show that condition (18) is not only necessary but also sufficient for the existence and uniqueness of the required vector $\xi = v(0)$. Indeed let (18) be satisfied. Then from any prescribed b, we can determine

$$c = (AE)^{-1}b \tag{19}$$

uniquely and then from c the vector

$$\xi = Ec = E(AE)^{-1} b \in Q^-. \tag{20}$$

It follows from (20) that $A\xi = b$ and so (14) has been satisfied. The uniqueness of ξ results from (20) and the uniqueness of c.

In summary, we have:

THEOREM 3. *The system of equations* (14) *defines a well-posed problem for system* (1) *for any vector b if and only if* (18) *holds.*

4.3.4. The condition $v(t) = O(t^h)$ figuring in all the results of this section can be replaced by the growth condition

$$v(t) = O(t^h e^{\alpha t}), \tag{21}$$

where α is a fixed real constant.

All of the results of this section carry over to the case (21) under the same general formulation except for one difference. The number r is to be determined from the relation

$$\operatorname{Re} \lambda_{r-1} \leqq \alpha < \operatorname{Re} \lambda_r \tag{22}$$

rather than relation (2). The proof here proceeds just as before. However, it is possible to reduce case (21) to the previous case by introducing a new unknown vector function $v_\alpha(t) = v(t) e^{-\alpha t}$. It is easy to verify that the system which $v_\alpha(t)$ must satisfy has eigenvalues $\lambda_0 - \alpha, ..., \lambda_{m-1} - \alpha$. Thus the condition essential to applying the results of Secs. 4.3.1–3 is that $\operatorname{Re} \lambda_j - \alpha \leqq 0$. This guarantees that $v_\alpha(t)$ will increase no faster than $O(t^h)$. But this means that the condition $\operatorname{Re} \lambda_j \leqq \alpha$ guarantees that $v(t)$ will increase no faster than $O(t^h e^{\alpha t})$.

4.4. Partial Differential Equations

4.4.1. Consider the system of partial differential equations

$$\frac{\partial u_j(t, x)}{\partial t} = \sum_{k=0}^{m-1} p_{jk}\left(i \frac{\partial}{\partial x}\right) u_k(t, x) \quad (j = 0, 1, ..., m - 1), \tag{1}$$

where p_{jk} is a polynomial in $i \dfrac{\partial}{\partial x_1}, ..., i \dfrac{\partial}{\partial x_n}$ of at most p-th degree. Letting $u(t, x)$ be a vector function with components $u_0(t, x), ..., u_{m-1}(t, x)$

and $P\left(i\dfrac{\partial}{\partial x}\right)$ the $m \times m$ matrix $\left\|p_{jk}\left(i\dfrac{\partial}{\partial x}\right)\right\|$, we can write (1) in the vector form

$$\frac{\partial u(t, x)}{\partial t} = P\left(i\frac{\partial}{\partial x}\right)u(t, x). \tag{2}$$

$u(t, x)$ is supposed to be in $\mathscr{H}$ for each $t \geqq 0$ (i.e., each component $u_j(t, x)$ is in $\mathscr{H}$).

Differentiation with respect to t is understood as differentiation with respect to a parameter in the topology of $\mathscr{H}$.

With each system (1) (or (2)) is associated an initial condition $u(0, x) = u_0(x)$ and a growth condition at infinity of the form

$$u(t, x) = O(t^h).$$

A problem is well-posed if there exists a unique solution to (1) satisfying all the stated conditions and depending continuously on the admissible initial data $u(0, x)$ (under convergence in $\mathscr{H}$).

We shall now specify what problem is well-posed for (1). To this end, we take Fourier transforms with respect to x thus obtaining the dual system in H,

$$\frac{dv_j(t, \sigma)}{dt} = \sum_{k=0}^{m-1} p_{jk}(\sigma) v_k(t, \sigma) \quad (j = 0, 1, ..., m-1). \tag{3}$$

The system (3) is a system of ordinary differential equations with constant coefficients for each fixed value of the parameter σ. Thus we may apply the results of Sec. 4.3.1 to answer our question concerning a well-posed problem.

In accordance with these results, we have to consider the m-dimensional space Q of vectors $\xi = (\xi_0, ..., \xi_{m-1})$ in which $P = P(\sigma)$ is an operator defined by

$$(P(\sigma)\xi)_0 = p_{00}(\sigma)\,\xi_0 + \cdots + p_{0, m-1}(\sigma)\,\xi_{m-1},$$

$$\cdots \cdots \cdots \cdots \cdots \cdots \cdots \cdots \cdots \cdots \cdots \cdots$$

$$(P(\sigma)\,\xi)_{m-1} = p_{m-1, 0}(\sigma)\,\xi_0 + \cdots + p_{m-1, m-1}(\sigma)\,\xi_{m-1}.$$

We must then decompose Q into a direct sum of subspaces $Q^-(\sigma)$ and $Q^+(\sigma)$ invariant under $P(\sigma)$. $Q^-(\sigma)$ is generated by the eigenvectors and root vectors corresponding to the eigenvalues $\lambda(\sigma)$ of $P(\sigma)$ such that Re $\lambda(\sigma) \leq 0$ and $Q^+(\sigma)$ by the remaining eigenvectors and root vectors of $P(\sigma)$. The dimension $r = r(\sigma)$ of $Q^-(\sigma)$ also depends on σ. For fixed σ, the well-posed problem for (3) is to take $v(0, \sigma) \in Q^-(\sigma)$. It turns out that the fulfillment of this condition for all σ (up to a set of measure zero) defines a well-posed problem for the system (3) in H and thereby a well-posed problem for the system (1) in $\mathscr{H}$:

THEOREM 1. *If the Fourier transform $v_0(\sigma)$ of the vector function $u_0(x)$ in $\mathscr{H}$ belongs to $Q^-(\sigma)$ for almost every σ, then the system (2) has a unique solution $u(t, x)$ in $\mathscr{H}$ for $t \geq 0$ equaling $u_0(x)$ at $t = 0$. The solution increases in $\mathscr{H}$ no faster than a power of t as $t \to \infty$. However if $v_0(\sigma)$ does not satisfy the stated condition (i.e., it has a non-null component in $Q^+(\sigma)$ on a set of positive measure), then the solution of (2) with initial data $u_0(x)$, even if one should exist in $\mathscr{H}$, will always increase there faster than any power of t.*

Proof. Let us begin by constructing the required solution. The solution of (3) with initial data $v_0(\sigma)$ is given by

$$v(t, \sigma) = e^{tP(\sigma)} v_0(\sigma).$$

Since $v_0(\sigma) \in Q^-(\sigma)$, according to Sec. 4.3.1 $v(t, \sigma)$ will increase no faster than a power of t as $t \to \infty$ for fixed σ. We must show that as an element of H, $v(t, \sigma)$ increases there no faster than a power of t. The method of Sec. 4.3.1 which was good for fixed σ, involves estimation constants whose dependence on σ is difficult to visualize. Therefore in estimating the norm of $e^{tP(\sigma)}$ we shall use inequality (9) of Sec. 4.2 in which the dependence of the constants on σ is easily taken into consideration. In the present case, the inequality has the form

$$\|e^{tP(\sigma)}\| \leq e^{t\Lambda(\sigma)} \left[1 + 2t\|P(\sigma)\| + \cdots + \frac{(2t)^{m-1}}{(m-1)!} \|P(\sigma)\|^{m-1} \right],$$

where $\Lambda(\sigma) = \max \text{Re } \lambda_j(\sigma)$ and only the roots $\lambda_j(\sigma)$ corresponding to the invariant subspace $Q^-(\sigma)$ to which $v(0, \sigma)$ belongs need of course be considered. Thus we may regard $\Lambda(\sigma) \leq 0$ and $e^{t\Lambda(\sigma)} \leq 1$. Now we know † that the square of the norm of a matrix in an orthonormal basis

† See G. E. Shilov, *Introduction to the Theory of Linear Spaces*, Prentice-Hall, Inc., New Jersey, 1961, Sec. 53, pp. 145–146.

does not exceed the sum of the squares of its elements. Therefore

$$\|P(\sigma)\|^2 \leqq \sum\sum |p_{jk}(\sigma)|^2 \leqq C^2(1 + |\sigma|^2)^p,$$

and so

$$\|P(\sigma)\| \leqq C(1 + |\sigma|)^p.$$

Finally we have

$$\|e^{tP(\sigma)}\| \leqq 1 + 2tC(1 + |\sigma|)^p + \cdots + \frac{(2tC)^{m-1}}{(m-1)!}(1 + |\sigma|)^{p(m-1)}. \tag{4}$$

For sufficiently large t we can restrict ourselves to the leading term obtaining

$$\|e^{tP(\sigma)}\| \leqq C_1 t^{m-1}(1 + |\sigma|)^{p(m-1)}. \tag{5}$$

The estimates (4) and (5) are simultaneously estimates for each element of the matrix $e^{tP(\sigma)}$. By hypothesis, each component of $v(0, \sigma)$ belongs to H. The components of $v(t, \sigma)$ are sums of products of the elements of $e^{tP(\sigma)}$ and components of $v(0, \sigma)$. Therefore

$$\|v(t, \sigma)\| \leqq C_2 t^{m-1}(1 + |\sigma|)^{p(m-1)} \|v(0, \sigma)\|. \tag{6}$$

But the space H admits multiplication by a function with power growth and therefore $v(t, \sigma) \in H$ for each fixed t. The estimate (6) shows that $t^{1-m}\|v(t, \sigma)\|$ is bounded in H and so

$$v(t, \sigma) = O(t^{m-1})$$

in H. Hence

$$u(t, x) = O(t^{m-1})$$

in $\mathscr{H}$. The uniqueness of the solution $v(t, \sigma)$ found follows from the uniqueness of the solution for each σ. The continuous dependence of $v(t, \sigma)$ on $v(0, \sigma)$ and therefore of $u(t, x)$ on $u(0, x)$ is apparent from the estimate (6).

REMARK. According to the observation made in Sec. 4.1 the following estimates hold for the order of singularity of the solution:

$$s(v(t, \sigma)) \leqq p(m - 1) + s(v(0, \sigma)),$$

$$s(u(t, x)) \leqq p(m - 1) + s(u(0, x)).$$

Thus if $u(0, x)$ is a sufficiently smooth function so that $s(u(0, x)) \leqq -p(m - 1)$, then the solution $u(t, x)$ will also be an ordinary function. But if $s(u(0, x)) > -p(m - 1)$, $u(t, x)$ will not be an ordinary function in general. The wave equation (Sec. 4.7.1, 3°) serves as an actual example of this behavior.

We proceed now to prove the necessity of the condition. Suppose that on a set G of positive measure the vector function $v_0(\sigma)$ has a non-null component $v_0^+(\sigma)$ in the subspace $Q^+(\sigma)$. We may regard

$$\operatorname{Re} \lambda_k(\sigma) > c > 0$$

on G for some k. By Sec. 4.3.1, we have for each such σ that

$$\lim_{t \to \infty} e^{-ct}\|v(t, \sigma)\|_\sigma = \infty. \tag{7}$$

By Egorov's theorem, the limit relation (7) holds uniformly on a set $G_0 \subset G$ of positive measure. But then in any space $H^{(q)}$ with $q > n/2$,

$$\|v(t, \sigma)\|_q^2 = \int_{R_n} \frac{\|v(t, \sigma)\|_\sigma^2 \, d\sigma}{(1 + |\sigma|^2)^q} \geq \int_{G_0} \frac{e^{2ct} d\sigma}{(1 + |\sigma|^2)^q} = e^{2ct} C_q$$

for sufficiently large t. Thus the corresponding solution $u(t, x)$ in $\mathscr{H}$, if it exists, increases faster than an exponential as $t \to \infty$. The theorem is proved.

4.4.2. Consider now the single equation

$$\frac{\partial^m u(t, x)}{\partial t^m} = \sum_{k=0}^{m-1} p_k\left(i\,\frac{\partial}{\partial x}\right) \frac{\partial^k u(t, x)}{\partial t^k}. \tag{8}$$

Assuming that the solution $u(t, x)$ and its derivatives $\dfrac{\partial^k u(t, x)}{\partial t^k}$ $(k = 0, 1, ..., m)$ belong to $\mathscr{H}$, we take Fourier transforms in (8). This yields the ordinary differential equation

$$\frac{d^m v(t, \sigma)}{dt^m} = \sum_{k=0}^{m-1} p_k(\sigma) \frac{d^k v(t, \sigma)}{dt^k} \tag{9}$$

with parameter σ. For each fixed σ, (9) is an equation with constant coefficients and we can apply the results of Sec. 4.3.2 to construct a well-posed problem. Namely, let $r = r(\sigma)$ be the number of roots of the equation

$$\lambda^m(\sigma) = \sum_{k=0}^{m-1} p_k(\sigma)\, \lambda^k(\sigma)$$

which for given σ have non-positive real parts. Then the well-posed problem (again for given σ) consists in assigning $v(0, \sigma), \dfrac{dv(0, \sigma)}{dt}, \ldots,$ $\dfrac{d^{r-1}v(0, \sigma)}{dt^{r-1}}$.

It turns out that the imposition of this condition for all σ (up to a set of measure zero) constitutes a well-posed problem for (9) in H and hence a well-posed problem for (8). (It should be mentioned that the number r of assignable initial conditions depends on σ.) It is this fact which makes up the content of the basic theorem of Sec. 4.1. We restate it here as

THEOREM 2. *Let G_r be the set of $\sigma \in R_n$ for which exactly r of the quantities* Re $\lambda_0(\sigma), \ldots,$ Re $\lambda_{m-1}(\sigma)$ *are non-positive. Let $v_0(\sigma), \ldots, v_{r-1}(\sigma)$ be given functions on each G_r extendable to all of $R_n(\sigma)$ (for instance with value identically zero) as elements of H. Then equation (8) has a solution $u(t, x)$ such that the Fourier transform of* $\dfrac{\partial^{k-1}u(0, x)}{\partial t^{k-1}}$ *coincides on each G_r with* $v_{k-1}(\sigma)$ $(k = 1, \ldots, r)$. $u(t, x)$ *belongs to $\mathscr{H}$ for $t \geqq 0$ and increases there no faster than a power of t as $t \to \infty$ together with all its t-derivatives up to order $m - 1$. It is unique in the class of all solutions of (8) belonging to $\mathscr{H}$ which increase there no faster than a power of t together with all their t-derivatives up to order $m - 1$. Moreover, it depends continuously on the data $v_k(\sigma)$ in the topology of $\mathscr{H}$ when the $v_k(\sigma)$ vary continuously in the topology of their corresponding spaces $H(G_r)$.*

Proof. Equation (9) is equivalent to the system

$$
\left.
\begin{aligned}
\frac{dv_0(t, \sigma)}{dt} &= v_1(t, \sigma), \\[2mm]
\frac{dv_1(t, \sigma)}{dt} &= v_2(t, \sigma), \\
&\cdots\cdots\cdots\cdots\cdots\cdots \\
\frac{dv_{m-2}(t, \sigma)}{dt} &= v_{m-1}(t, \sigma), \\[2mm]
\frac{dv_{m-1}(t, \sigma)}{dt} &= p_0(\sigma)\, v_0(t, \sigma) + p_1(\sigma)\, v_1(t, 0) + \cdots + p_{m-1}(\sigma) v_{m-1}(t, \sigma).
\end{aligned}
\right\} \quad (10)
$$

By Theorem 1 an initial vector

$$v(\sigma) = (v_0(0, \sigma), ..., v_{m-1}(0, \sigma))$$

defines a well-posed problem for (10) if and only if for almost each σ it belongs to the corresponding space $Q^-(\sigma)$.

Let $v_j(0, \sigma)$ be given on G_j ($j = 0, 1, ..., r - 1$) and extendable (with value identically zero) to an element of H. By the lemma of Sec. 4.3.2 for each $\sigma \in G_r$ we can tack on quantities $v_r(\sigma), ..., v_{m-1}(\sigma)$ to $v_0(\sigma), ..., v_{r-1}(\sigma)$ so that the resulting vector $v(\sigma) = (v_0(\sigma), ..., v_{m-1}(\sigma))$ lies in $Q^-(\sigma)$. Let us show that as a function of σ it belongs to H. By virtue of (12) of Sec. 4.3, $v(\sigma)$ is uniquely determined through the formula

$$v_s(\sigma) = \sum_{j=0}^{r-1} v_j(\sigma) R_{js}(\lambda_0(\sigma), ..., \lambda_{r-1}(\sigma)),$$

where R_{js} is a fixed homogeneous polynomial of its arguments of degree $s - j$. Each eigenvalue $\lambda_j(\sigma)$ satisfies the estimate (cf. Sec. 4.2.2).

$$|\lambda_j(\sigma)| \leq \|P(\sigma)\| \leq C(1 + |\sigma|)^p$$

and so

$$|R_{js}(\lambda_0(\sigma), ..., \lambda_{r-1}(\sigma))| \leq C_1(1 + |\sigma|)^{p(s-j)} \leq C_1(1 + |\sigma|)^{p(m-1)}. \quad (11)$$

With multiplication by functions of power growth admissible in H, we thus see that each $v_s(\sigma)$ is in H whenever all $v_j(\sigma)$ are ($j = 0, ..., r - 1$). Since $v(\sigma) \in Q^-(\sigma)$ by construction, Theorem 1 implies that $v(t, \sigma)$ is in H and satisfies the estimate $v(t, \sigma) = O(t^{m-1})$.

The uniqueness of the solution follows from the uniqueness of the vector $v(0, \sigma)$ generated from the given data $v_0(0, \sigma), ..., v_{r-1}(0, \sigma)$. Its continuous dependence on $v_0(0, \sigma), ..., v_{r-1}(0, \sigma)$ follows from the continuous dependence of the succeeding functions $v_s(0, \sigma)$ on the data and from Theorem 1. This completes the proof of the theorem.

As a consequence of the remark in Theorem 1 it is possible to make certain assertions about the existence of a solution in classical terms. Namely, if the initial data $\dfrac{\partial^k u(0, x)}{\partial t^k}$ satisfy the conditions of Theorem 2 and are moreover ordinary functions having square integrable derivatives up to a certain fixed order q, then the solution $u(t, x)$ will be an ordinary function. The number q is determined by the degree with respect to σ of the polynomials on the right-hand side of (11), i.e. by the quantity

17 Shilov

$p(m - 1)$ with $p(m - 1)$ still added to it, as we indicated in the remark in Theorem 1 and thus $q = 2p(m - 1)$. This estimate, of course, actually overstates the number of derivatives needed.

4.4.3. Our next theorem shows that the conditions set forth in Theorem 2 are necessary and sufficient for the correct formulation of a problem when the Fourier transforms of the initial data

$$u(0, x), \quad \frac{\partial u(0, x)}{\partial t}, \quad ..., \quad \frac{\partial^{m-1} u(0, x)}{\partial t^{m-1}}$$

are prescribed on any nested family of sets in $R_n(\sigma)$.

THEOREM 3. *If the Fourier transform $v_j(\sigma)$ of $\dfrac{\partial^j u(0, x)}{\partial t^j}$ is prescribed on G'_j $(j = 0, ..., m - 1)$, where $G'_0 \supset G'_1 \supset \cdots \supset G'_{m-1}$, and G'_j and the corresponding G_j are disjoint (up to a set of measure zero), then the problem is not well-posed. More precisely, if the measure of $G'_j - G_j \cap G'_j$ is positive for at least one $j = j_0$, then a solution does not exist in general .If G'_j is a subset of G_j $(j = 0, ..., m - 1)$ and $G_j - G'_j$ has positive measure for at least one j, then a solution is not unique.*

Proof. Consider the first case: the measure of $G'_j - G_j \cap G'_j$ is positive. Suppose $\sigma \in G'_j - G_j \cap G'_j$. Since $\sigma \notin G_j$, the subspace $Q^-(\sigma)$ for this σ has dimension less than j. Therefore $v_0(\sigma), ..., v_{j-1}(\sigma)$ may be assigned so that whatever values $v_j(\sigma), ..., v_{m-1}(\sigma)$ have, the vector $v(\sigma) = (v_0(\sigma), ..., v_{m-1}(\sigma))$ will not be in $Q^-(\sigma)$. This construction can be done for every $\sigma \in G'_j - G_j G'_j$ and moreover in such a way that $v_i(\sigma)$ is continuous and has compact support $(0 \leq i < j)$. By Theorem 1, the solution for each problem with corresponding initial data, if it exists, increases in $\mathscr{H}$ faster than any power of t (perhaps not $u(t, x)$ itself but some t-derivative of it of order less than m). Thus, the problem is not well-posed.

In the second case where each G'_j is a subset of G_j and $G_j - G'_j$ has positive measure for some j, we can give $v_{j-1}(\sigma)$ arbitrarily on this difference and in so doing obtain different solutions to the problem. The theorem is thereby completely proved.

4.4.4. Our next goal is to examine a more general boundary value problem for the system

$$\frac{\partial u_j(t, x)}{\partial t} = \sum_{k=0}^{m-1} p_{jk}\left(i \frac{\partial}{\partial x}\right) u_k(t, x) \quad (j = 0, 1, ..., m - 1). \quad (1)$$

Consider the system of equations

$$a_{00}(\sigma)\,v_0(\sigma) + \cdots + a_{0,m-1}(\sigma)\,v_{m-1}(\sigma) = b_0(\sigma),$$
$$\cdots \cdots \cdots \cdots \cdots \cdots \cdots \cdots \cdots \cdots$$
$$a_{q0}(\sigma)\,v_0(\sigma) + \cdots + a_{q,m-1}(\sigma)\,v_{m-1}(\sigma) = b_q(\sigma). \quad (12)$$

Let $b(\sigma) = (b_0(\sigma), ..., b_q(\sigma))$ be any vector (with certain restrictions as to the nature of its dependence on σ to be specified below).

The question raised is this. What should the matrix $A = \|a_{jk}\|$ be so that there should exist a unique solution $u(t, x)$ to (1) taking on the value $u(0, x)$ at $t = 0$ and such that the vector $v(\sigma) = (v_0(\sigma), ..., v_{m-1}(\sigma))$ satisfying (12) is the Fourier transform of $u(0, x)$?

For fixed σ, the problem reduces to the one of Sec. 4.3.3 and so we can apply the corresponding result. Namely, suppose $E = E(\sigma)$ is an $m \times r$ matrix representing a basis for $Q^-(\sigma)$. Then for fixed σ we must have $q = r(\sigma) - 1$ and the square matrix $A(\sigma)\,E(\sigma)$ must have an inverse. If this condition is satisfied, then our desired $v(0, \sigma)$ is given by

$$v(0, \sigma) = E(AE)^{-1}\,b(\sigma).$$

For each σ, $v(0, \sigma)$ belongs to $Q^-(\sigma)$. In order that it define a well-posed problem for (12), it is still necessary to require that it belong to H. Thus we have the added sufficient condition on $b(\sigma)$:

$$E(AE)^{-1}\,b(\sigma) \in H.$$

This then is the additional restriction that has to be imposed on $b(\sigma)$ to assure a well-posed problem.

We have proved the sufficiency of the conditions of the following theorem:

THEOREM 4. *The system* (1) *has a unique solution* $u(t, x)$ *that belongs to* $\mathcal{H}$ *for* $t \geq 0$, *has a Fourier transform at* $t = 0$ *satisfying the conditions* (12), *and increases no faster than a power of t has* $t \to \infty$ *if and only if the matrix* $(AE)^{-1}$ *exists on each set* G_r *for almost every* σ *and* $E(AE)^{-1}\,b$ *belongs to* $H(G_r)$.

To prove the necessity, we argue as follows.

Suppose that AE is singular on a set of positive measure. We can find a non-trivial bounded solution to $AEb(\sigma) = 0$ on the set of positive measure. The function $v(\sigma) = Eb(\sigma)$ belongs to $Q^-(\sigma)$ for each σ and to H on the whole space $R_n(\sigma)$. Thus by our earlier theorem, (1) has a solu-

17*

tion $v(t, \sigma)$ belonging to H for $t \geq 0$ and with growth there no faster than a power of t. For $t = 0$, we have

$$Av(0, \sigma) = Av(\sigma) = 0.$$

Hence, if AE is singular on a set of positive measure, then even if a solution to problem (1), (12) exists in $\mathscr{H}$, it is afortiori not unique. Therefore the problem (1), (12) is not well-posed.

If r is fixed (i.e., independent of σ), then the conditions (12) can be reformulated in terms of the $u_j(0, x)$. We shall assume in addition that each $a_{jk}(\sigma)$ is a polynomial in σ. Let $(B_0(x), ..., B_{r-1}(x))$ be the vector having $(b_0(\sigma), ..., b_{r-1}(\sigma))$ as its Fourier transform. Taking inverse Fourier transforms in (12), we obtain

$$\left.\begin{aligned}
a_{00}\left(i\frac{\partial}{\partial x}\right)u_0(0, x) + \cdots + a_{0,m-1}\left(i\frac{\partial}{\partial x}\right)u_{m-1}(0, x) &= B_0(x), \\
\cdots\cdots\cdots\cdots\cdots\cdots\cdots\cdots\cdots\cdots\cdots\cdots\cdots & \\
a_{r-1,0}\left(i\frac{\partial}{\partial x}\right)u_0(0, x) + \cdots + a_{r-1,m-1}\left(i\frac{\partial}{\partial x}\right)u_{m-1}(0, x) &= B_{r-1}(x).
\end{aligned}\right\} \quad (13)$$

Conditions such as these are termed "oblique derivative" conditions.

4.4.5. Let us discuss some examples.

1°. Laplace's equation $\dfrac{\partial^2 u}{\partial t^2} = -\varDelta u$. Here $m = 2$ and $\lambda_{0,1}(\sigma) = \pm|\sigma|$ so that $r = 1$, $G_0 = R_n$, and $G_1 = \emptyset$, The matrix E has one column of two elements:

$$E = \left\| \begin{matrix} 1 \\ -|\sigma| \end{matrix} \right\|.$$

We wish to examine the Neumann problem. The data is $\dfrac{\partial u}{\partial t} = u_1(x)$ or if we take Fourier transforms, $\dfrac{dv}{dt} = v_1(\sigma)$. The corresponding matrix A is $\|0\ 1\|$ and so

$$AE = \|-|\sigma|\|.$$

We thus see that $(AE)^{-1}$ exists for almost all σ. We further have

$$E(AE)^{-1}v_1 = \left\|\begin{matrix} 1 \\ -|\sigma| \end{matrix}\right\| \left(\begin{matrix} v_1(\sigma) \\ -|\sigma| \end{matrix}\right) = \left\|\begin{matrix} -\dfrac{v_1(\sigma)}{|\sigma|} \\ v_1(\sigma) \end{matrix}\right\|.$$

The function $v_1(\sigma)$ is in H by hypothesis. But $\dfrac{v_1(\sigma)}{|\sigma|}$ must also be in H if the problem is to be well-posed. If we assume in addition that $v_1(\sigma)$ is continuous in a neighborhood of zero and $n \leq 2$, then it follows that $v_1(0) = 0 \left(\text{otherwise } \dfrac{v_1}{|\sigma|} \text{ would not be in } H \right)$. This means that

$$\int_{R_n} u_1(x)\, dx = 0.$$

Thus if the data $u_1(x) = \dfrac{\partial u(0, x)}{\partial t}$ is integrable, then its integral over $R_n(x)$ must equal zero when $n \leq 2$. This is the classical criterion making the Neumann problem well-posed.†

2°. Consider the same equation but with initial data now the "oblique derivative"

$$\frac{\partial^2 u(0, x)}{\partial t\, \partial X} = U(x), \qquad \frac{\partial}{\partial X} = \sum_{j=1}^{n} \gamma_j \frac{\partial}{\partial x_j}.$$

Taking Fourier transforms, we obtain

$$-il \frac{\partial v}{\partial t} = V(\sigma), \qquad l = \sum_{j=1}^{n} \gamma_j \sigma_j.$$

Hence

$$A = \| 0 \quad -il \|, \qquad AE = \| il|\sigma| \|,$$

$$(AE)^{-1} = \left\| -i\,\frac{1}{l|\sigma|} \right\|, \qquad E(AE)^{-1} V(\sigma) = \left\| \begin{array}{c} \dfrac{-iV}{l(\sigma)\,|\sigma|} \\[2ex] \dfrac{iV}{l(\sigma)} \end{array} \right\|.$$

The present problem is well-posed provided that $\dfrac{V(\sigma)}{l(\sigma)\,|\sigma|} \in H$. This means particularly that $V(\sigma)$ has to vanish if $l = (\gamma, \sigma) = 0$. A requisite condition on $U(x)$ is therefore

$$\int_{R_n} U(x)\, e^{i(x, \sigma)}\, dx = 0 \quad \text{if} \quad (\gamma, \sigma) = 0.$$

---

† See P. Garabedian, *Partial Differential Equations*, John Wiley, New York, 1964, Chapt. 7, Sec. 1, p. 230.

4.4.6. The condition $u(t, x) = O(t^h)$ in $\mathscr{H}$ may be replaced

$$u(t, x) = O(t^h e^{\alpha t}) \quad \text{in} \quad \mathscr{H} \tag{14}$$

for some fixed real α. All of the results of this section carry over to the case (14) under the same general set-up except for one difference. The number r is to be determined from the relation

$$\operatorname{Re} \lambda_{r-1}(\sigma) \leqq \alpha < \operatorname{Re} \lambda_r(\sigma)$$

instead of the previous one. The proof merely involves making the substitution

$$u_\alpha(t, x) = e^{-\alpha t} u(t, x)$$

and using the considerations of Sec. 4.3.4.

Related to this are the systems of equations that are called *proper in the sense of Petrovsky*. So labeled is any system such as (1) for which

$$\operatorname{Re} \lambda_j(\sigma) \leqq C,$$

i.e., the real parts of all the eigenvalues are uniformly bounded. If a system is proper in the sense of Petrovsky, then by Theorem 1 the *Cauchy problem* for it with data

$$u(0, x) = \{u_0(0, x), \ldots, u_{m-1}(0, x)\}$$

possesses a solution with the following properties. It is in $\mathscr{H}$ for each $t \geqq 0$ and increases there no faster than $t^h e^{\alpha t}$ as $t \to \infty$; it is unique in the class of all $u(t, x)$ belonging to $\mathscr{H}$ for $t \geqq 0$ that satisfy this growth condition; it takes on the value $u(0, x)$ at $t = 0$.

4.5. Fundamental Solutions of Regular Boundary Value Problems

4.5.1. We have seen that multiplication by a locally bounded function $V(\sigma)$ of at most power growth at infinity is a well-defined and continuous operation in H. We wish to ascertain its dual in $\mathscr{H}$. If $G(x)$ is the inverse Fourier transform of $V(\sigma)$, then it is natural to define the desired operation as the convolution with $G(x)$. But this convolution is something new as compared to the convolution considered in Sec. 2.5 since neither the functional $G(x)$ nor the generalized function $f \in \mathscr{H}$ with which the convolution is taken need have compact support. We proceed to clarify its meaning.

$V(\sigma)$ can be expressed as the product of some bounded integrable function $V_0(\sigma)$ and say the q-th power of the polynomial $1 + |\sigma|^2$. Correspondingly, its inverse Fourier transform $G(x)$ can be represented by

$$G(x) = (1 - \Delta)^q \, G_0(x),$$

where $G_0(x)$ is a continuous function that approaches 0 as $|x| \to \infty$. Further, any arbitrary function $g(\sigma) \in H$ is the product of a polynomial $Q(\sigma)$ and a square integrable function $g_0(\sigma)$; the corresponding generalized function $f \in \mathscr{H}$ comes from applying the operator $Q\left(i \dfrac{\partial}{\partial x}\right)$ to the square integrable function $f_0(x) = F^{-1}[g_0(\sigma)]$. We have

$$F^{-1}[V(\sigma) \, g(\sigma)] = F^{-1}[(1 + |\sigma|^2)^q \, Q(\sigma) \, V_0(\sigma) \, g_0(\sigma)]$$

$$= (1 - \Delta)^q \, Q\left(i \frac{\partial}{\partial x}\right) F^{-1}[V_0(\sigma) \, g_0(\sigma)].$$

Thus our task has been reduced to determining $F^{-1}[V_0(\sigma) \, g_0(\sigma)]$. We have already termed this operation the convolution of $G_0(x) = F^{-1}[V_0(\sigma)]$ and $f_0(x) = F^{-1}[g_0(\sigma)]$, and we now must define it explicity. The customary convolution

$$\int_{R_n} G_0(x - \xi) f_0(\xi) \, d\xi$$

does not exist in general in the present case. However, it does exist in the mean, i.e., as the limit of

$$[G_0 * f_0]_N \, (x) = \int_{|\xi| \leq N} G_0(x - \xi) f_0(\xi) \, d\xi$$

in the $L_2(R_n)$ metric. Since $G_0(x)$ is bounded and $f_0(x)$ is integrable on $[-N, N]$ this last expression is meaningful. It represents the customary convolution of $G_0(x)$ and the square integrable function $f_0^N(x)$ equaling $f_0(x)$ for $|x| \leq N$ and 0 for $|x| > N$. Let $g_0^N(\sigma)$ denote the Fourier transform of $f_0^N(x)$. Since the sequence $f_0^N(x)$ converges to $f_0(x)$ in the mean as $N \to \infty$, so does the sequence $g_0^N(\sigma)$ converge to $g_0(\sigma)$. The sequence $V_0(\sigma) \, g_0^N(\sigma)$ also converges to $V_0(\sigma) \, g_0(\sigma)$ in the mean owing to the boundedness of $V_0(\sigma)$. Hence it follows that $[G_0 * f_0]_N(x) = F^{-1}[V_0(\sigma) \, g_0^N(\sigma)]$

has the limit in the mean $F^{-1}[V_0(\sigma) g_0(\sigma)]$. Thus

$$F^{-1}[V_0(\sigma) g_0(\sigma)] = \lim_{N\to\infty} \{F^{-1}[V_0] * F^{-1}[g_0^N]\}$$

$$= \lim_{N\to\infty} G_0(x) * f_0^N(x) = \lim_{N\to\infty} \int_{|\xi|\leq N} G_0(x - \xi) f_0(\xi)\, d\xi$$

in the mean.

Therefore if $G(x)$ is a functional whose Fourier transform $V(\sigma)$ is a locally bounded function of at most power growth at infinity, then the convolution of $G(x)$ and any functional $f \in \mathscr{H}$ is defined by

$$G(x) * f(x) = (1 - \Delta)^q Q\left(i\frac{\partial}{\partial x}\right) \lim_{N\to\infty} \int_{|x|\leq N} G_0(x - \xi) f_0(\xi)\, d\xi,$$

where $(1 - \Delta)^q G_0(x) = G(x)$ and $Q\left(i\dfrac{\partial}{\partial x}\right) f_0(x) = f(x)$. The operation is unique (ie., it does not depend on the choice of G_0 and f_0) and is continuous in $\mathscr{H}$.

Any generalized function $G(x)$ of the stated form will henceforth be termed a *convolutor* in $\mathscr{H}$. We point out that if G_1 and G_2 are two convolutors in $\mathscr{H}$, then so is $G_1 * G_2$, and for any $f \in \mathscr{H}$, $G_1 * (G_2 * f) = (G_1 * G_2) * f$.

The last two assertions are equivalent to the corresponding properties of the functions of σ.

4.5.2. The following differentiation formula holds:

$$\frac{\partial}{\partial x_j}(G * f) = \frac{\partial G}{\partial x_j} * f = G * \frac{\partial f}{\partial x_j}. \tag{1}$$

Indeed, the string of equations (1) is equivalent to

$$i\sigma_j(V(\sigma) g(\sigma)) = (i\sigma_j V(\sigma)) g(\sigma) = V(\sigma) (i\sigma_j g(\sigma)),$$

and the latter are obviously valid.

As we have already stated, $G * f$ is a continuous operation in $\mathscr{H}$. That is, whenever $f_\nu \to f$ in $\mathscr{H}$, then $G * f_\nu \to G * f$ in $\mathscr{H}$. We now indicate a condition for G_ν to converge to G from which it will follow that $G_\nu * f \to G * f$. (The analogous condition in Sec. 2.5 was that the functionals G_ν had uniform compact support.) We shall say that a sequence of convolutors $G_\nu(x)$ *converges properly* to a convolutor $G(x)$ if their Fourier transforms $V_\nu(\sigma)$ have uniform bounded growth, i.e., for all ν

$$|V_\nu(\sigma)| \leqq C(1 + |\sigma|^2)^q$$

for some C and q independent of ν, and $V_\nu(\sigma)$ converges uniformly to $V(\sigma)$ in each block of R_n. Let us show that

If $G_\nu(x)$ converges properly to $G(x)$, then $G_\nu * f$ converges to $G * f$ in $\mathscr{H}$.

It is enough to verify that $V_\nu(\sigma) g(\sigma) \to V(\sigma) g(\sigma)$ in H. We have $V_\nu(\sigma) = (1 + |\sigma|^2)^q v_\nu(\sigma)$, $V(\sigma) = (1 + |\sigma|^2)^q v(\sigma)$, and $g(\sigma) = Q(\sigma) g_0(\sigma)$. Here $g_0(\sigma) \in L_2(R_n)$, each $v_\nu(\sigma)$ is bounded by constant C, and the sequence $v_\nu(\sigma)$ converges uniformly to $v(\sigma)$ in each block.

Hence $g_0(\sigma) v_\nu(\sigma) \to g_0(\sigma) v(\sigma)$ in L_2. But then $V_\nu(\sigma) g(\sigma) = (1 + |\sigma|^2)^q \times Q(\sigma) g_0(\sigma) v_\nu(\sigma) \to (1 + |\sigma|^2)^q Q(\sigma) g_0(\sigma) v(\sigma) = V(\sigma) g(\sigma)$ in H, as required.

Proper convergence is defined in a similar way when ν is a continuously varying parameter say $\nu = t \geqq 0$. In this case also, the proper convergence of $G(t)$ to G implies that $G(t) * f$ converges to $G * f$ for any $f \in \mathscr{H}$.

A function $G(t)$ is said to be *properly differentiable* if $\dfrac{G(t + \Delta t) - G(t)}{\Delta t}$ converges properly to some function $G'(t)$ as $\Delta t \to 0$ which is called the derivative of G with respect to t. Just as before we have

$$\frac{G(t + \Delta t) - G(t)}{\Delta t} * f \to G'(t) * f \quad \text{in} \quad \mathscr{H},$$

or

$$\frac{d}{dt}(G(t) * f) = G'(t) * f.$$

4.5.3. The equations and systems considered in Sec. 4.4 will be labeled *regular* if the number r of eigenvalues $\lambda_j(\sigma)$ that have non-positive real parts (or are bounded from above) is fixed for almost all σ. Well-posed problems for regular equations can be specified in terms of assigning the initial data $u_0(0, x), ..., u_{r-1}(0, x)$. Examples of regular equations are the heat equation, wave equation, Laplace's equation as well as

$$\frac{\partial^4 u}{\partial t^4} = \frac{\partial^4 u}{\partial x^4} \qquad (m = 4,\ r = 3),$$

and

$$\frac{\partial^4 u}{\partial t^4} = -\frac{\partial^4 u}{\partial x^4} \qquad (m = 4,\ r = 2).$$

A regular equation (system) for which $r = m$ is proper in the sense of Petrovsky (Sec. 4.4.6.). The well-posed problem for such an equation is of course the Cauchy problem with the m initial data $u_0(0, x), \ldots, u_{m-1}(0, x)$. The well-posed problem for a regular equation is also called a regular boundary value problem and the corresponding r the index of regularity.

Suppose that

$$\frac{\partial^m u(t, x)}{\partial t^m} = \sum_{k=0}^{m-1} p_k \left(i \frac{\partial}{\partial x} \right) \frac{\partial^k u(t, x)}{\partial t^k} \tag{2}$$

is a regular equation with index r. As we know, the regular boundary value problem for it entails giving the r initial data

$$\left. \begin{array}{l} u(0, x) = u_0(x), \\ \cdot \ \cdot \ \cdot \ \cdot \ \cdot \ \cdot \ \cdot \\ \dfrac{\partial^{r-1} u(0, x)}{\partial t^{r-1}} = u_{r-1}(x). \end{array} \right\} \tag{3}$$

The special solution $G(t, x)$ of (2) satisfying the conditions

$$\left. \begin{array}{l} G(0, x) = 0, \\ \dfrac{\partial G(0, x)}{\partial t} = 0, \\ \cdot \ \cdot \ \cdot \ \cdot \ \cdot \\ \dfrac{\partial^{r-2} G(0, x)}{\partial t^{r-2}} = 0, \\ \dfrac{\partial^{r-1} G(0, x)}{\partial t^{r-1}} = \delta(x), \end{array} \right\} \tag{4}$$

is called the *fundamental solution* or *Green's function* for the boundary value problem. Not only does it belong to $\mathscr{H}$ it is also a convolutor in $\mathscr{H}$. And as we shall likewise see below, by knowing the Green's function $G(t, x)$, we shall be able to write down the solution to the general boundary value problem with data (3).

4.5.4. Let us determine the succeeding derivatives of $G(t, x)$ at $t = 0$. Let $V(t, \sigma)$ be the Fourier transform of $G(t, x)$. It satisfies the equation

$$\frac{d^m V(t, \sigma)}{dt^m} = \sum_{k=0}^{m-1} p_k(\sigma) \frac{d^k V(t, \sigma)}{dt^k} \tag{5}$$

and initial conditions

$$V(0, \sigma) = 0,$$

$$\cdots \cdots \cdots$$

$$\frac{d^{r-2}\, V(0, \sigma)}{dt^{r-2}} = 0,$$

$$\frac{d^{r-1}\, V(0, \sigma)}{dt^{r-1}} = 1.$$

V will be called the *resolvent function* for equation (5). According to the general discussion of Secs. 4.3 and 4.4, the succeeding components $v_s(\sigma)$ of the initial vector $v(\sigma)$ ($s \geq r$) of a well-posed problem can be expressed in terms of the preceding ones $v_j(\sigma)$ ($j < r$) by means of

$$v_s(\sigma) = \sum_{j=0}^{r-1} v_j(\sigma)\, R_{js}(\lambda_0(\sigma), \ldots, \lambda_{r-1}(\sigma)).$$

In our case, we thus have

$$\frac{d^s V(0, \sigma)}{dt^s} = R_{r-1, s}(\lambda_0(\sigma), \ldots, \lambda_{r-1}(\sigma)).$$

As a polynomial in the eigenvalues, this function is locally bounded and has no more than power growth at infinity. Now from equation (5) and its successive derivatives, we obtain

$$\frac{d^m V(0, \sigma)}{dt^m} = \sum_{k=0}^{m-1} p_k(\sigma)\, \frac{d^k V(0, \sigma)}{dt^k},$$

$$\frac{d^{m+1} V(0, \sigma)}{dt^{m+1}} = \sum_{k=0}^{m-1} p_k(\sigma)\, \frac{d^{k+1} V(0, \sigma)}{dt^{k+1}},$$

and so forth. We see that the t-derivatives of $V(t, \sigma)$ of any order at $t = 0$ are locally bounded functions having at most power growth at infinity.

Let us show that $G(t, x)$ and its t-derivatives are properly differentiable convolutors. Employing vector notation, we can write

$$W(t, \sigma) = e^{tP(\sigma)} W(0, \sigma).$$

The vector function

$$W(0, \sigma) = \left\{ V(0, \sigma), \frac{dV(0, \sigma)}{dt}, \ldots, \frac{d^{m-1} V(0, \sigma)}{dt^{m-1}} \right\}$$

has fixed growth in σ (power growth at the most). It follows with the help of the estimate (4) of Sec. 4.4 that for bounded t each component of $W(t, \sigma)$ satisfies a uniform power estimate in σ. Thus $G(t, x)$ is a convolutor for fixed arbitrary t. $W(t, \sigma)$ clearly depends continuously on t in each block of R_n. More precisely, whenever $t_\nu \to t_0$, $W(t_\nu, \sigma)$ converges uniformly to $W(t_0, \sigma)$. Taking the uniform σ-power estimate for $W(t, \sigma)$ into consideration, we see that $G(t_\nu, x)$ converges properly to $G(t_0, x)$ as $t_\nu \to t_0$.

To establish that $V(t, \sigma)$ is properly differentiable at $t = 0$, we start from the vector function

$$\frac{1}{\tau}[e^{\tau P(\sigma)} - I] W(0, \sigma). \tag{6}$$

As $\tau \to 0$, it has the limit $P(\sigma)W(0, \sigma) = \dfrac{dW(0, \sigma)}{dt}$ (in the usual sense).

To show that the function (6) converges to $\dfrac{dW(0, \sigma)}{dt}$ properly, we must verify that it has a power estimate in t uniform in σ. We replace $\dfrac{1}{\tau}[e^{\tau P(\sigma)}$ $- I]$ by its corresponding interpolation polynomial. According to Sec. 4.2.1, the coefficients in the interpolation polynomial

$$R(\lambda) = b_0 + b_1(\lambda - \lambda_0) + \cdots + b_{m-1}(\lambda - \lambda_0) \ldots (\lambda - \lambda_{m-2})$$

for any $f(\lambda)$ are estimable by

$$|b_k| \leqq \frac{1}{k!} \max_{\lambda \in Q_k} |f^{(k)}(\lambda)|.$$

Here Q_k is the smallest convex polygon containing $\lambda_0, \ldots, \lambda_k$. In the present case,

$$f(\lambda) = \frac{e^{t\lambda} - 1}{\lambda} = \int_0^t e^{\tau\lambda} \, d\tau.$$

Therefore

$$\max_{\lambda \in Q} |f(\lambda)| \leqq \int_0^t \max_{\lambda \in Q} |e^{\tau\lambda}| \, d\tau \leqq te^{t\Lambda}$$

and

$$\max_{\lambda \in Q} |f^{(k)}(\lambda)| \leqq \max_{\lambda \in Q} t^{k-1}|e^{t\lambda}| = t^{k-1}e^{t\Lambda}$$

for any polygon Q, where $\Lambda = \max_{\lambda \in Q} \operatorname{Re} \lambda$. Hence just as in Sec. 4.2, we have

$$\|R(P)\| \leq |b_0| + 2|b_1| \, \|P\| + \cdots + |b_{m-1}| \, (2\|P\|)^{m-1}$$

$$\leq e^{t\Lambda}\left(t + 2\|P\| + t\frac{(2\|P\|)^2}{2!} + \cdots + t^{m-1}\frac{(2\|P\|)^{m-1}}{(m-1)!}\right), \qquad (7)$$

where $\Lambda = \max(\operatorname{Re} \lambda_0, \ldots, \operatorname{Re} \lambda_{m-1})$. If P is an operator in the subspace $Q^-(\sigma)$, then $e^{t\Lambda} \leq 1$ and (7) furnishes our desired estimate. Further, the derivative of $e^{tP(\sigma)}$ with respect to t is the limit of

$$e^{tP(\sigma)} \frac{e^{\tau P(\sigma)} - I}{\tau}$$

as $\tau \to 0$. We obviously have proper convergence here also. Finally, the k-th derivative of $e^{tP(\sigma)}$ with respect to t is equal to $[P(\sigma)]^k e^{tP(\sigma)}$; it is properly differentiable with respect to t along with $e^{tP(\sigma)}$. Summing up, we have:

The generalized functions $\dfrac{\partial^k G(t, x)}{\partial t^k}$ *are all convolutors in* $\mathscr{H}$ *for* $k \geq r$

which are moreover continuous and properly differentiable.

4.5.5. We now wish to show how the solution to the general problem (2), (3) is determined with the help of the fundamental solution $G(t, x)$. We define the required solution to be

$$u(t, x) = G(t, x) * f_0(x) + \frac{\partial G(t, x)}{\partial t} * f_1(x) + \cdots$$

$$+ \frac{\partial^{r-1} G(t, x)}{\partial t^{r-1}} * f_{r-1}(x), \qquad (8)$$

where $f_0(x), \ldots, f_{r-1}(x)$ are temporarily unknown functions in $\mathscr{H}$.

Our first aim is to prove that the $u(t, x)$ given by (8) satisfies equation (2) for any choice of $f_0, \ldots, f_{r-1}$. It is enough to establish that each term in the sum (8) satisfies the equation.

By virtue of the properties of the convolution, we have

$$\frac{\partial^m}{\partial t^m}\left[\frac{\partial^j G}{\partial t^j} * f_j\right] = \frac{\partial^j}{\partial t^j}\left[\frac{\partial^m G}{\partial t^m}\right] * f_j = \frac{\partial^j}{\partial t^j}\left[\sum_{k=0}^{m-1} P_k\left(i\frac{\partial}{\partial x}\right)\frac{\partial^k G}{\partial t^k}\right] * f_j$$

$$= \sum_{k=0}^{m-1} P_k\left(i\frac{\partial}{\partial x}\right)\frac{\partial^k}{\partial t^k}\left(\frac{\partial^j G}{\partial t^j} * f_j\right),$$

and so $\dfrac{\partial_j G}{\partial t^j} * f_j$ is a solution of equation (2). We next differentiate (8) $r - 1$ times and set $t = 0$. This yields the system of equations

$$u(0, x) = f_{r-1}(x),$$

$$\frac{\partial u(0, x)}{\partial t} = f_{r-2}(x) + \frac{\partial^r G(0, x)}{\partial t^r} * f_{r-1}(x),$$

$$\cdots \cdots \cdots \cdots \cdots \cdots \cdots \cdots \cdots \cdots$$

$$\frac{\partial^{r-2} u(0, x)}{\partial t^{r-2}} = f_1(x) + \frac{\partial^r G(0, x)}{\partial t^r} * f_2(x) + \cdots + \frac{\partial^{2r-3} G(0, x)}{\partial t^{2r-3}} * f_{r-1}(x),$$

$$\frac{\partial^{r-1} u(0, x)}{\partial t^{r-1}} = f_0(x) + \frac{\partial^r G(0, x)}{\partial t^r} * f_1(x) + \cdots + \frac{\partial^{2r-2} G(0, x)}{\partial t^{2r-2}} * f_{r-1}(x),$$

which makes it possible to find the unknown functions $f_{r-1}(x)$ $f_{r-2}(x)$, ..., $f_0(x)$ sequentially.

4.5.6. To illustrate matters, we shall determine the fundamental solutions for the three classical well-posed problems of mathematical physics when $n = 1$ and we shall write out the corresponding solutions satisfying the given data.

1°. The heat equation

$$\frac{\partial u}{\partial t} = \frac{\partial^2 u}{\partial x^2}$$

with $u(0, x)$ given.

Taking Fourier transforms, we obtain

$$\frac{dv}{dt} = -\sigma^2 v$$

with $v(0, \sigma)$ given. The resolvent function $V(t, \sigma)$ is

$$V(t, \sigma) = e^{-t\sigma^2}.$$

Taking its inverse Fourier transform,† we obtain the fundamental solution

$$G(t, x) = \frac{1}{2\sqrt{\pi t}} e^{-\frac{x^2}{4t}}.$$

---

† See H. F. Weinberger, *A First Course in Partial Differential Equations*, Blaisdell, New York, 1965, Sec. 70, p. 328.

The solution $u(t, x)$ to the heat equation with initial value $u(0, x)$ in $\mathscr{H}$ is given by

$$u(t, x) = \frac{1}{2\sqrt{\pi t}}\, e^{-\frac{x^2}{4t}} * u(0, x).$$

When $u(0, x)$ is an ordinary function, this reduces to the classical Poisson integral

$$u(t, x) = \frac{1}{2\sqrt{\pi t}} \int_{-\infty}^{\infty} e^{-\frac{(x-\xi)^2}{4t}}\, u(0, \xi)\, d\xi.$$

2°. The wave equation

$$\frac{\partial^2 u}{\partial t^2} = \frac{\partial^2 u}{\partial x^2}$$

with $u_0(x) = u(0, x)$ and $u_1(x) = \dfrac{\partial(0, x)}{\partial t}$ prescribed.

Taking Fourier transforms, we obtain

$$\frac{d^2 v}{dt^2} = -\sigma^2 v$$

with $v_0(\sigma) = v(0, \sigma)$ and $v_1(\sigma) = \dfrac{dv(0, \sigma)}{dt}$ prescribed. The resolvent function $V(t, \sigma)$ is

$$V(t, \sigma) = \frac{\sin t\sigma}{\sigma}.$$

Since

$$\sigma V(t, \sigma) = \sin t\sigma,$$

$G(t, x)$ satisfies the equation

$$i\, \frac{\partial}{\partial x} G(t, x) = \frac{\delta(x - t) - \delta(x + t)}{2i}.$$

Thus $G(t, x)$ is the indefinite integral of $\frac{1}{2}[\delta(x + t) - \delta(x - t)]$ and so equals $\frac{1}{2}[\theta(x + t) - \theta(x - t)]$ up to an additive constant. The latter function has the value $\frac{1}{2}$ for $|x| < t$ and 0 elsewhere. But a constant function does not belong to $\mathscr{H}$ and so we may say outright that

$$G(t, x) = \tfrac{1}{2}\,[\theta(x + t) - \theta(x - t)].$$

To construct the solution corresponding to the given initial data, we start from the expression

$$u(t, x) = G(t, x) * f_0(x) + \frac{\partial G(t, x)}{\partial t} * f_1(x).$$

In this connection, we obviously have

$$\frac{\partial G(t, x)}{\partial t} = \frac{1}{2} [\delta(x + t) + \delta(x - t)],$$

$$\frac{\partial^2 G(t, x)}{\partial t^2} = \frac{1}{2} [\delta'(x + t) - \delta'(x - t)], \quad \frac{\partial^2 G(0, x)}{\partial t^2} = 0.$$

The functions $f_0(x)$ and $f_1(x)$ are solutions of the system of equation

$$u_0(x) = \qquad\qquad\qquad f_1(x),$$

$$u_1(x) = f_0(x) + \frac{\partial^2 G(0, x)}{\partial t^2} * f_1(x) = f_0(x).$$

Thus

$$u(t, x) = G(t, x) * u_1(x) + \frac{\partial G(t, x)}{\partial t} * u_0(x).$$

If $u_0(x)$ and $u_1(x)$ are ordinary functions, we obtain the classical d'Alembert formula

$$u(t, x) = \frac{1}{2} \int_{x-t}^{x+t} u_1(\xi)\, d\xi + \frac{1}{2} [u_0(x + t) + u_0(x - t)].$$

3°. Laplace's equation

$$\frac{\partial^2 u}{\partial t^2} = - \frac{\partial^2 u}{\partial x^2}$$

with $u_0(x) = u(0, x)$ prescribed.

Taking Fourier transforms, we find that

$$\frac{d^2 v}{dt^2} = \sigma^2 v \qquad\qquad (9)$$

with $v_0(\sigma) = v(0, \sigma)$ prescribed. In considering the general solution to (9), which is

$$v(t, \sigma) = C_1 e^{-t|\sigma|} + C_2 e^{t|\sigma|},$$

we need only retain the first term since we must satisfy the condition $v(t, \sigma) = O(t^h)$. Taking the initial condition into account, we find that the resolvent function is

$$V(t, \sigma) = e^{-t|\sigma|}.$$

Obtaining its inverse Fourier transform, we have

$$G(t, x) = \frac{1}{2\pi} \int_{-\infty}^{\infty} e^{-t|\sigma|} e^{-i\sigma x} \, d\sigma = \frac{1}{2\pi} \left\{ \int_{-\infty}^{0} e^{t\sigma - i\sigma x} \, d\sigma + \int_{0}^{\infty} e^{-t\sigma - i\sigma x} \, d\sigma \right\}$$

$$= \frac{1}{2\pi} \left\{ \frac{1}{t - ix} + \frac{1}{t + ix} \right\} = \frac{1}{\pi} \frac{t}{t^2 + x^2}. \tag{10}$$

The solution $u(t, x)$ to Laplace's equation corresponding to data $u_0(x)$ in $\mathscr{H}$ is given by

$$u(t, x) = \frac{1}{\pi} \frac{t}{t^2 + x^2} * u_0(x).$$

If $u_0(x)$ is an ordinary function, this becomes the classical Poisson formula

$$u(t, x) = \frac{t}{\pi} \int_{-\infty}^{\infty} \frac{u_0(\xi) \, d\xi}{t^2 + (x - \xi)^2}.$$

Problems

1. Let $f(t, x) = f(t, x_1, ..., x_n)$ be a generalized function in K_n' depending continuously on a parameter t. Let $\varphi(t, x) = \varphi(t, x_1, ..., x_n)$ be a test function in $(n + 1)$-dimensional $tx_1...x_n$-space. Set

$$(f(t, x), \varphi(t, x)) = \int_{-\infty}^{\infty} (f(t, x), \varphi(t, x))_x \, dt, \tag{1}$$

where $(\cdot)_x$ means that f operates on φ as a function of x for fixed t. Show that (1) defines a generalized function in K_{n+1}'.

2. Given that $G(t, x)$ is the fundamental solution of the Cauchy problem for

$$\frac{\partial u(t, x)}{\partial t} - P\left(i \frac{\partial}{\partial x}\right) u(t, x) = 0, \quad x = (x_1, ..., x_n).$$

18 Shilov

Using $G(t, x)$, construct a generalized function $\mathscr{E}(t, x)$ in K'_{n+1} as indicated in Prob. 1. Show that this generalized function satisfies the equation

$$\frac{\partial \mathscr{E}(t, x)}{\partial t} - P\left(i \frac{\partial}{\partial x}\right) \mathscr{E}(t, x) = \delta(t, x).$$

i.e., $\mathscr{E}(t, x)$ is a fundamental function for the operator $\dfrac{\partial}{\partial t} - P\left(i \dfrac{\partial}{\partial x}\right)$.

Hint. Prove that

$$\left(\left[\frac{\partial}{\partial t} - P\left(i \frac{\partial}{\partial x}\right)\right] \mathscr{E}(t, x), \varphi(t, x)\right)_{x, t}$$

$$= \int_0^\infty \left(\left[\frac{\partial}{\partial t} - P\left(i \frac{\partial}{\partial x}\right)\right] G(t, x), \varphi(t, x)\right)_x dt$$

$$= \int_0^\infty \left(G(t, x), \left[\frac{\partial}{\partial t} - \bar{P}\left(-i \frac{\partial}{\partial x}\right)\right] \varphi(t, x)\right)_x dt = \varphi(0, 0).$$

Employ the relation

$$\int_0^\infty \left(G(t, x), \frac{\partial \varphi}{\partial t}\right)_x dt = \int_0^\infty \frac{\partial}{\partial t} (G, \varphi)_x \, dt - \int_0^\infty \left(\frac{\partial G}{\partial t}, \varphi\right)_x dt$$

$$= (G(0, x), \varphi(0, x)) - \int_0^\infty \left(\frac{\partial G}{\partial t}, \varphi\right)_x dt = \varphi(0, 0) - \int_0^\infty \left(P\left(i \frac{\partial}{\partial x}\right) G, \varphi\right)_x dt.$$

3. Given that $G(t, x)$ is the fundamental solution of the Cauchy problem for

$$\frac{\partial^m u}{\partial t^m} - \sum_{k=0}^{m-1} \frac{\partial^k}{\partial t^k} P_k\left(i \frac{\partial}{\partial x}\right) u = 0.$$

Show that the generalized function $\mathscr{E}(t, x)$ in K'_{n+1} (Prob. 1) satisfies

$$\frac{\partial^m \mathscr{E}}{\partial t^m} - \sum_{k=0}^{m-1} \frac{\partial^k}{\partial t^k} P_k\left(i \frac{\partial}{\partial x}\right) \mathscr{E} = \delta(t, x). \tag{1}$$

In other words, $\mathscr{E}(t, x)$ is a fundamental function for the operator

$$\frac{\partial^m}{\partial t^m} - \sum_{k=0}^{m-1} \frac{\partial^k}{\partial t^k} P_k\left(i \frac{\partial}{\partial x}\right).$$

Hint. Write (1) as a first order system and repeat the reasoning of Prob. 2 for the vector function.

4. Find the fundamental solution for the iterated (one-dimensional) heat equation given by

$$\left(\frac{\partial}{\partial t} - \frac{\partial^2}{\partial x^2}\right)^m u = 0.$$

Ans.

$$V(t, \sigma) = \frac{t^{m-1}}{(m-1)!} e^{-t\sigma^2}, \quad G(t, x) = \frac{t^{m-1}}{(m-1)!} \frac{1}{2\sqrt{\pi t}} e^{-\frac{x^2}{4t}}.$$

5. Find the fundamental solution for the iterated Laplace equation

$$\left(\frac{\partial^2}{\partial t^2} + \frac{\partial^2}{\partial x^2}\right)^m u = 0.$$

Ans.

$$V(t, \sigma) = \frac{t^{m-1}}{(m-1)!} e^{-t|\sigma|}, \quad G(t, x) = \frac{1}{\pi} \frac{t^{m-1}}{(m-1)!} \frac{t}{t^2 + x^2}.$$

6. Find the fundamental solution for the equation

$$\left(\frac{\partial}{\partial t} - \omega \frac{\partial}{\partial x}\right)^m = 0 \quad (\omega \text{ real}).$$

Ans.

$$V(t, \sigma) = \frac{t^{m-1}}{(m-1)!} e^{i\omega\sigma t}, \quad G(t, x) = \frac{t^{m-1}}{(m-1)!} \delta(x - \omega t).$$

4.6. Formulas for Fundamental Solutions of Regular Equations ($n = 1$)

It is possible to derive transparent practical formulas for the fundamental solutions for regular differential operators that are homogeneous in all arguments. In this section, we shall consider the case of one space variable.

4.6.1. Suppose we are given the equation

$$\frac{\partial^m u(t, x)}{\partial t^m} = \sum_{k=0}^{m-1} a_k \left(i\frac{\partial}{\partial x}\right)^{m-k} \frac{\partial^k u(t, x)}{\partial t^k}. \tag{1}$$

Taking Fourier transforms in (1), we obtain

$$\frac{d^m v(t, \sigma)}{dt^m} = \sum_{k=0}^{m-1} a_k \sigma^{m-k} \frac{d^k v(t, \sigma)}{dt^k}. \tag{2}$$

Its characteristic equation is given by

$$\lambda^m(\sigma) = \sum_{k=0}^{m-1} a_k \sigma^{m-k} \lambda^k(\sigma). \tag{3}$$

Setting $\lambda(\sigma) = \mu(\sigma)\sigma$ and canceling σ^m, we obtain

$$\mu^m(\sigma) = \sum_{k=0}^{m-1} a_k \mu^k(\sigma), \tag{4}$$

obviously showing that $\mu(\sigma)$ really does not depend on σ. We shall assume that the roots $\mu_0, \mu_1, \ldots, \mu_{m-1}$ of (4) are distinct. The general solution to (2) can then be written in the form

$$v(t, \sigma) = \sum_{j=0}^{m-1} C_j(\sigma) e^{t\mu_j \sigma}. \tag{5}$$

As usual, we number the roots μ_j in increasing order of magnitude of their real parts:

$$\operatorname{Re} \mu_0 \leqq \cdots \leqq \operatorname{Re} \mu_{p-1} < \operatorname{Re} \mu_p = \cdots = \operatorname{Re} \mu_{r-1}$$
$$= 0 < \operatorname{Re} \mu_r \leqq \cdots \leqq \operatorname{Re} \mu_{m-1}.$$

Here we have explicitly earmarked a group of roots having zero real parts. Since the solution (5) increases no faster than a power of t as $t \to \infty$, $C_r, \ldots, C_{m-1}$ must vanish for $\sigma > 0$ and $C_0, \ldots, C_{p-1}$ for $\sigma < 0$. For $\sigma > 0$, the space $Q^-(\sigma)$ has dimension r and for $\sigma < 0$ dimension $m - p$. Thus, the condition for (1) to be regular can be written as $m - p = r$.

We point out two important special cases:

1) $r = m$, $p = 0$. In other words,

$$\operatorname{Re} \mu_0 = \cdots = \operatorname{Re} \mu_{m-1} = 0$$

(the equation is hyperbolic). The string equation for which (4) is

$$\mu^2 = -1, \quad \mu_{0,1} = \pm i,$$

is of this type.

2) $r = p$, $m = 2p$ (the equation is elliptic). Laplace's equation is an example of this type since

$$\mu^2 = +1, \quad \mu_{0,1} = \pm 1.$$

Let us examine the general case. It will be recalled that the resolvent function $V(t, \sigma)$ has to satisfy the conditions

$$V(0, \sigma) = \cdots = \frac{d^{r-2}V(0, \sigma)}{dt^{r-2}} = 0, \quad \frac{d^{r-1}V(0, \sigma)}{dt^{r-1}} = 1. \qquad (6)$$

Writing $V(t, \sigma)$ as

$$V(t, \sigma) = \begin{cases} \displaystyle\sum_{j=0}^{r-1} C_j(\sigma)\, e^{t\mu_j\sigma} & (\sigma > 0) \\[2mm] \displaystyle\sum_{s=p}^{m-1} C_s(\sigma)\, e^{t\mu_s\sigma} & (\sigma < 0), \end{cases}$$

we obtain two systems of equations

$$\left.\begin{aligned} C_0(\sigma) + \cdots + C_{r-1}(\sigma) &= 0, \\ \mu_0\sigma C_0(\sigma) + \cdots + \mu_{r-1}\sigma C_{r-1}(\sigma) &= 0, \\ \cdots\cdots\cdots\cdots\cdots\cdots\cdots\cdots \\ (\mu_0\sigma)^{r-1}C_0(\sigma) + \cdots + (\mu_{r-1}\sigma)^{r-1}C_{r-1}(\sigma) &= 1 \end{aligned}\right\} \quad (\sigma > 0), \qquad (7)$$

and

$$\left.\begin{aligned} C_p(\sigma) + \cdots + C_{m-1}(\sigma) &= 0, \\ \mu_p\sigma C_p(\sigma) + \cdots + \mu_{m-1}\sigma C_{m-1}(\sigma) &= 0, \\ \cdots\cdots\cdots\cdots\cdots\cdots\cdots\cdots \\ (\mu_p\sigma)^{r-1}C_p(\sigma) + \cdots + (\mu_{m-1}\sigma)^{r-1}C_{m-1}(\sigma) &= 1 \end{aligned}\right\} \quad (\sigma < 0) \qquad (8)$$

for determining the coefficients $C_j(\sigma)$ and $C_s(\sigma)$. To simplify matters, we set $\sigma^{r-1}C_j(\sigma) = B_j(\sigma)$ for $\sigma > 0$ and $\sigma^{r-1}C_s(\sigma) = A_s(\sigma)$ for $\sigma < 0$. This leads to the following systems of equations for the quantities

$B_j(\sigma)$ and $A_s(\sigma)$:

$$
\left.
\begin{aligned}
B_0 + \cdots + B_{r-1} &= 0, \\
\mu_0 B_0 + \cdots + \mu_{r-1} B_{r-1} &= 0, \\
\cdots \cdots \cdots \cdots \cdots \cdots \cdots \cdots & \\
\mu_0^{r-1} B_0 + \cdots + \mu_{r-1}^{r-1} B_{r-1} &= 1
\end{aligned}
\right\}
\quad (\sigma > 0),
\tag{9}
$$

$$
\left.
\begin{aligned}
A_p + \cdots + A_{m-1} &= 0, \\
\mu_p A_p + \cdots + \mu_{m-1} A_{m-1} &= 0, \\
\cdots \cdots \cdots \cdots \cdots \cdots \cdots \cdots & \\
\mu_p^{r-1} A_p + \cdots + \mu_{m-1}^{r-1} A_{m-1} &= 1
\end{aligned}
\right\}
\quad (\sigma < 0).
\tag{10}
$$

In particular, $B_j(\sigma) = B_j$ and $A_s(\sigma) = A_s$ actually do not depend on σ. The solutions of these systems are

$$
B_j = (-1)^{r+j+1} \frac{W[\mu_0, \ldots, \mu_{j-1}, \mu_{j+1}, \ldots, \mu_{r-1}]}{W[\mu_0, \ldots, \mu_{r-1}]},
$$

$$
A_s = (-1)^{r+s-p+1} \frac{W[\mu_p, \ldots, \mu_{s-1}, \mu_{s+1}, \ldots, \mu_{m-1}]}{W[\mu_p, \ldots, \mu_{m-1}]},
$$

where $W[\ldots]$ stands for the Vandermonde determinant of the indicated arguments.

Altogether we have

$$
V(t, \sigma) =
\begin{cases}
\dfrac{1}{\sigma^{r-1}} \sum_{j=0}^{r-1} B_j e^{\mu_j \sigma t} & (\sigma > 0), \tag{11} \\[2ex]
\dfrac{1}{\sigma^{r-1}} \sum_{s=p}^{m-1} A_s e^{\mu_s \sigma t} & (\sigma < 0), \tag{12}
\end{cases}
$$

where B_j and A_s are the solutions of (9) and (10) respectively.

4.6.2. For a hyperbolic equation, formulas (11) and (12) reduce to a single expression

$$
V(t, \sigma) = \frac{1}{\sigma^{m-1}} \sum_{j=0}^{m-1} B_j e^{\mu_j \sigma t}.
\tag{13}
$$

Since the sum in (13) has an $(m-1)$-st order zero at $\sigma = 0$ (cf. (9)), $V(t, \sigma)$ actually has no singularity there. It can be continued analytically into the complex s-plane ($s = \sigma + i\tau$) as the entire function

$$V(t, s) = \frac{1}{s^{m-1}} \sum_{j=0}^{m-1} B_j e^{\mu_j s t} . \tag{14}$$

For large $|s|$, $V(t, s)$ satisfies the estimate

$$|V(t, s)| \leqq C e^{tM|s|}, \quad M = \max |\mu_j| .$$

That is, $V(t, s)$ is a first order function of type M at the most and is square integrable if $m > 1$. By the Paley-Wiener theorem,[†] its inverse Fourier transform $G(t, x)$ is a square integrable function with compact support in the interval $[-tM, tM]$. Hence, for any (generalized) initial function $u_{m-1}(x)$ with compact support say in $[a, b]$ and $u_j(x) = 0$ ($j = 0, ..., m - 2$), we can conclude that the corresponding solution

$$u(t, x) = G(t, x) * u_{m-1}(x)$$

is carried in $[a - tM, b + tM]$. Thus the support of the solution $u(t, x)$ expands in time at a rate M.

The case $m = 1$ is easy to handle separately. In this case, equation (1) is a first order equation

$$\frac{\partial u}{\partial t} = ib \frac{\partial u}{\partial x} .$$

The transformed equation is

$$\frac{dv}{dt} = b\sigma v, \tag{15}$$

while equation (4) becomes $\mu = b$.

Since by hypothesis Re $\mu = $ Re $b = 0$, we let $ib = a$, a now being real. Thus the original differential equation assumes the form

$$\frac{\partial u}{\partial t} = a \frac{\partial u}{\partial x}$$

with a real.

† See R. Paley and N. Wiener, *Fourier Transforms in the Complex Domain*, Colloq. Publ. Amer. Math. Soc., Providence, 1934, Sec. 6.

The solution $V(t, \sigma)$ of (15) satisfying the initial condition $V(0, \sigma) = 1$ is obviously

$$V(t, \sigma) = e^{-ia\sigma t}.$$

Taking the inverse Fourier transform of $V(t, \sigma)$, we obtain the Green's function $G(t, x)$ which according to Sec. 2.7.5, 6° is

$$G(t, x) = \delta(x + at).$$

The solution to the equation with arbitrary initial data $u_0(x)$ is

$$u(t, x) = G(t, x) * u_0(x) = u_0(x + at).$$

We shall now determine the explicit form of the inverse Fourier transform of $V(t, \sigma)$ for $m > 1$. $V(t, \sigma)$ is the product of the generalized function $\sigma^{1-m} \in Z' \subset K'$ and the infinitely differentiable function $\sum_{j=0}^{m-1} B_j e^{\mu_j \sigma t}$ (cf. Sec. 2.4, (21)). By the convolution theorem, we have

$$F^{-1}\left[\frac{1}{\sigma^{m-1}} \sum_{j=0}^{m-1} B_j e^{\mu_j \sigma t}\right] = F^{-1}\left[\frac{1}{\sigma^{m-1}}\right] * F^{-1}\left[\sum_{j=0}^{m-1} B_j e^{\mu_j \sigma t}\right].$$

Since by Sec. 2.7.5, 4°

$$F\left[\frac{1}{x^{m-1}}\right] = \frac{i^{m-1}\pi}{(m-2)!} \sigma^{m-2} \, \text{sgn} \, \sigma,$$

we obtain using the rule (22) of Sec. 2.7

$$F^{-1}\left[\frac{1}{\sigma^{m-1}}\right] = (-1)^{m-1} \frac{i^{m-1}}{2(m-2)!} x^{m-2} \, \text{sgn} \, x = C_m x^{m-2} \, \text{sgn} \, x.$$

The inverse Fourier transform of $\sum_{j=0}^{m-1} B_j e^{\mu_j \sigma t}$ is

$$\sum_{j=0}^{m-1} B_j \delta(x - \omega_j t), \quad \omega_j = -i\mu_j.$$

In summary then, we have

$$G(t, x) = C_m \sum_{j=0}^{m-1} B_j \delta(x - \omega_j t) * x^{m-2} \, \text{sgn} \, x$$

$$= C_m \sum_{j=0}^{m-1} B_j(x - \omega_j t)^{m-2} \, \text{sgn} \, (x - \omega_j t). \qquad (16)$$

Each of the translates of x^{m-2} sgn x by the amount $\omega_j t$ is continuous and has continuous derivatives up to order $m - 3$ $(m > 2)$. The $(m - 2)$-nd derivative is piecewise constant (equaling $(m-2)!$ sgn$(x - \omega_j t)$). The $(m-1)$-st derivative is a delta-function. Hence it follows that $G(t, x)$ is continuous and possesses continuous derivatives up to order $m - 3$ (for $m > 2$), its $(m-2)$-nd derivative being piecewise constant and its $(m-1)$-st a linear combination of delta-functions. For $x < \omega_0 t$, each sgn$(x - \omega_j t)$ is equal to -1. Removing parentheses and using (9), we see that $G(t, x) = 0$. Similarly, $G(t, x) = 0$ for $x > \omega_{m-1} t$. Between the values $\omega_j t$ and $\omega_{j+1} t$, $G(t, x)$ is a polynomial in x of degree $m - 2$.

4.6.3. Let us next examine the case of a general regular equation. The resolvent function $V(t, \sigma)$ is given by (11) and (12). $V(t, \sigma)$ is clearly infinitely differentiable (and even analytic) outside of the origin.

To clarify the behavior of $V(t, \sigma)$ in a neighborhood of $\sigma = 0$ $(\sigma > 0)$, we expand $e^{\mu_j \sigma t}$ in a power series in σ obtaining

$$\sum_{j=0}^{r-1} B_j e^{\mu_j \sigma t} = \sum_{j=0}^{r-1} \sum_{k=0}^{\infty} B_j \frac{(\mu_j \sigma t)^k}{k!} = \sum_{k=0}^{\infty} \frac{t^k \sigma^k}{k!} \sum_{j=0}^{r-1} B_j \mu_j^k. \tag{17}$$

By means of (9), it follows that the expansion

$$V(t, \sigma) = \frac{t^{r-1}}{(r-1)!} + \sigma \frac{t^r}{r!} \sum_{j=0}^{r-1} B_j \mu_j^r + \sigma^2 \frac{t^{r+1}}{(r+1)!} \sum_{j=0}^{r-1} B_j \mu_j^{r+1} + \cdots \tag{18}$$

is valid for $\sigma > 0$. Similarly, in a neighborhood of $\sigma = 0$ $(\sigma < 0)$, we have

$$V(t, \sigma) = \frac{t^{r-1}}{(r-1)!} + \sigma \frac{t^r}{r!} \sum_{s=p}^{m-1} A_s \mu_s^r + \sigma^2 \frac{t^{r+1}}{(r+1)!} \sum_{s=p}^{m-1} A_s \mu_s^{r+1} + \cdots \tag{19}$$

We see that $V(t, \sigma)$ is continuous at $\sigma = 0$ $(t > 0)$ and if $r \neq m$, has a discontinuous first derivative in general.

Therefore $G(t, x)$ does not have compact support in general. This means that the initial excitation is transmitted instantaneously to the entire x-axis and even if $u_{m-1}(x)$ has compact support, $u(t, x)$ is non-vanishing for all x.

We wish to determine $G(t, x)$. The inverse Fourier transform of $V(t, \sigma)$ multiplied by 2π is the sum of two convergent integrals:

$$2\pi G(t, x) = \int_0^\infty \frac{1}{\sigma^{r-1}} \sum_{j=0}^{r-1} B_j e^{\mu_j \sigma t - i\sigma x} \, d\sigma$$

$$+ \int_{-\infty}^0 \frac{1}{\sigma^{r-1}} \sum_{s=p}^{m-1} A_s e^{\mu_s \sigma t - i\sigma x} \, d\sigma.$$

Substituting $-\sigma$ for σ in the second integral, we obtain

$$2\pi G(t, x) = \int_0^\infty \frac{1}{\sigma^{r-1}} \left\{ \sum_{j=0}^{r-1} B_j e^{\mu_j \sigma t - i\sigma x} + (-1)^{r-1} \sum_{s=p}^{m-1} A_s e^{-\mu_s \sigma t + i\sigma x} \right\} d\sigma.$$

$$(20)$$

The problem is thus reduced to evaluating an integral of the form

$$\int_0^\infty \frac{1}{\xi^{r-1}} \sum_{j=0}^{r-1} C_j e^{-a_j \xi} \, d\xi, \quad \operatorname{Re} a_j \geqq 0,$$

in which the coefficients C_j are such that the sum has an $(r-1)$-st order zero at $\xi = 0$. For $r > 1$, the integral can be evaluated in the following way.

Consider the integral depending on a parameter λ given by

$$I(\lambda) = \int_0^\infty \xi^\lambda \sum_{j=0}^{r-1} C_j e^{-a_j \xi} \, d\xi, \quad \operatorname{Re} a_j \geqq 0.$$

It is convergent for $-r < \operatorname{Re} \lambda < 0$ (and for $-r < \operatorname{Re} \lambda < \infty$ if all $\operatorname{Re} a_j > 0$) and analytic in λ. When $-1 < \operatorname{Re} \lambda < 0$, we can integrate termwise to obtain

$$I(\lambda) = \sum_{j=0}^{r-1} C_j \int_0^\infty \xi^\lambda e^{-a_j \xi} \, d\xi.$$

The integral in this is reduced to

$$\frac{1}{a_j^{\lambda+1}} \int_{L_j} z^\lambda e^{-z} \, dz \qquad (21)$$

by means of the substitution $a_j\xi = z$, where L_j is a ray $z = a_j\xi$, $0 \leq \xi < \infty$, lying in the half-plane Re $z \geq 0$. By Jordan's lemma, the integral

$$\int_{C_R} z^\lambda e^{-z}\, dz \tag{22}$$

along any arc of a circle of radius R in the half-plane Re $z \geq 0$ will tend to zero as $R \to \infty$. Therefore the integral along L_j is equal to the integral along the positive real axis. But the latter has the value $\Gamma(\lambda + 1)$. As a result, for $-1 < \text{Re }\lambda < 0$ we have

$$I(\lambda) = \sum_{j=0}^{r-1} C_j \int_0^\infty \xi^\lambda e^{-a_j\xi}\, d\xi = \sum_{j=0}^{r-1} \frac{C_j}{a_j^{\lambda+1}} \int_{L_j} z^\lambda e^{-z}\, dz = \sum_{j=0}^{r-1} C_j \frac{\Gamma(\lambda + 1)}{a_j^{\lambda+1}}.$$

The right-hand side can be continued analytically to the half-plane Re $\lambda > -r$ and its value at $\lambda = -r + 1$ determined. We have

$$\Gamma(\lambda + 1) = \frac{\Gamma(\lambda + 2)}{\lambda + 1} = \cdots = \frac{\Gamma(\lambda + r)}{(\lambda + 1) \cdots (\lambda + r - 1)}$$

and

$$\sum_{j=0}^{r-1} C_j a_j^{-(\lambda+1)} = (\lambda + r - 1)\frac{d}{d\lambda} \sum_{j=0}^{r-1} C_j a_j^{-(\lambda+1)} \Big|_{\lambda=-r+1} + o(\lambda + r - 1)$$

$$= -(\lambda + r - 1) \sum_{j=0}^{r-1} C_j a_j^{r-2} \log a_j + o(\lambda + r - 1).$$

Thus

$$\int_0^\infty \frac{1}{\xi^{r-1}} \sum_{j=0}^{r-1} C_j e^{-a_j\xi}\, d\xi$$

$$= -\lim_{\lambda \to 1-r} \frac{\Gamma(\lambda + r)}{(\lambda + 1) \cdots (\lambda + r - 2)} \sum_{j=0}^{r-1} C_j a_j^{r-2} \log a_j$$

$$= \frac{(-1)^{r-1}}{(r-2)!} \sum_{j=0}^{r-1} C_j a_j^{r-2} \log a_j.$$

We can now write out the value of the integral in (20) for $r > 1$ by applying the resulting formula. This yields

$$G(t, x) = \frac{(-1)^{r-1}}{2\pi(r-2)!} \left\{ \sum_{j=0}^{r-1} B_j(-\mu_j t + ix)^{r-2} \log(-\mu_j t + ix) \right.$$

$$\left. + (-1)^{r-1} \sum_{s=p}^{m-1} A_s(\mu_s t - ix)^{r-2} \log(\mu_s t - ix) \right\}. \tag{23}$$

When a_j is complex, the value of $\log a_j = \log |a_j| + i \arg a_j$ falls in the strip $|\operatorname{Im} z| \leq \dfrac{\pi}{2}$ since by assumption $\arg a_j$ lies in $\left[-\dfrac{\pi}{2}, \dfrac{\pi}{2}\right]$.

In the first sum, $\operatorname{Re} \mu_j \leq 0$ and so the quantity whose logarithm is being taken falls in the right-hand half-plane. In the second sum, $\operatorname{Re} \mu_s \geq 0$ and the corresponding quantity is also in this half-plane.

We have to handle the case $r = 1$ separately. Since $p \leq r$, we have either $p = 0$ or $p = 1$. Since $m = p + r$, in the first case $m = 1$ and in the second $m = 2$. If $m = 1$ and $p = 0$, then we are dealing with the first order hyperbolic equation discussed above. If $p = 1$ and $m = 2$, then the matter concerns a second order elliptic equation.

The expression (20) becomes

$$G(t, x) = \frac{1}{2\pi} \int_0^\infty [e^{\mu_0 \sigma t - i\sigma x} + e^{-\mu_1 \sigma t + i\sigma x}] \, d\sigma$$

and can be evaluated at once to yield

$$G(t, x) = \frac{1}{2\pi} \left[\frac{1}{ix - \mu_0 t} + \frac{1}{\mu_1 t - ix} \right] = \frac{1}{2\pi} \frac{(\mu_1 - \mu_0) t}{(ix - \mu_0 t)(\mu_1 t - ix)}. \tag{24}$$

For the simplest case of the Laplacian, we have $\mu_{0,1} = \mp 1$ and (24) reduces to (10) of Sec. 4.5.

Problems

1. Derive a formula for the fundamental solution of a homogeneous hyperbolic equation using the general formula (23).

Hint. If the logarithm is defined as in the text and A is real, then $\log iA - \log(-iA) = \pi \operatorname{sgn} A$.

2. Find the fundamental solutions for the equations

a) $\dfrac{\partial^4 u}{\partial t^4} = \dfrac{\partial^4 u}{\partial x^4}$,

b) $\dfrac{\partial^4 u}{\partial t^4} + \dfrac{\partial^4 u}{\partial x^4} = 0$.

Ans.

a) $G(t, x) = \dfrac{1}{2\pi}\left\{ t \log \sqrt{x^2 + t^2} - x \arctan \dfrac{x}{t} - \dfrac{1}{2}(t + x) \log |t + x| \right.$

$$\left. - \dfrac{1}{2}(t - x) \log |t - x| + \dfrac{1}{8}\left[|t + x| + |t - x| \right] \right\}$$

(S. A. Timoshkina);

b) $G(t, x) = \dfrac{1}{\pi\sqrt{2}}\left[\arctan\left(1 - \dfrac{x\sqrt{2}}{t} \right) + \arctan\left(1 + \dfrac{x\sqrt{2}}{t} \right) \right]$

(W. I. Averbukh).

3. Construct the fundamental solution for

$$\left(\dfrac{\partial^2}{\partial t^2} - \dfrac{\partial^2}{\partial x^2} \right)^m u(t, x) = 0.$$

Hint. To compute $V(t, \sigma)$, take Laplace transforms with respect to t and determine the inverse transform by means of contour integration and residues.†

Ans.

$$G(t, x) = \dfrac{(-1)^{m-1}}{4^m(m - 1)!}\left\{ \operatorname{sgn}(x - t) \sum_{j=0}^{m-1} \dfrac{(2t)^j(x - t)^{2m-2-j}}{(m - 1 - j)!\,j!} \right.$$

$$\left. + (-1)^m \operatorname{sgn}(x + t) \sum_{j=0}^{m-1} \dfrac{(-2t)^j(x + t)^{2m-2-j}}{(m - 1 - j)!\,j!} \right\}$$

(V. E. Kondrashov).

4. Derive an expression for the fundamental solution of a homogeneous hyperbolic equation in the case of multiple roots.

Ans.

$$G(t, x) = \sum_{k=1}^{l} \sum_{j=0}^{r_k-1} \dfrac{t^j b_{kj}}{j!} \dfrac{(x + \omega_k t)^{r_k-2-j} \operatorname{sgn}(x + \omega_k t)}{2 i^{r_k-1-j}(r_k - 2 - j)!},$$

† See J. Dettman, *Applied Complex Variables*, Macmillan, New York, 1965, Chapt. 9.

where r_k is the multiplicity of the root $\mu_k = i\omega_k$ and the coefficients b_{kj} satisfy the equations

$$\sum_{k=1}^{r_k} \sum_{j=0}^{\min(r_k-1,\, s)} b_{kj}\mu_k^{s-j}\binom{s}{j} = \begin{cases} 0, & \text{if } 0 \le s \le r_k - 2, \\ 1, & \text{if } s = r_k - 1 \end{cases}$$

(V. E. Kondrashov).

5. Derive an expression for the fundamental solution of a regular homogeneous equation in the case of multiple roots.

Ans.

$$G(t, x) = \frac{(-1)^{r-1}}{2\pi}$$

$$\times \left\{ \sum_{k=1}^{l} \sum_{j=0}^{r_k-1} \frac{(-1)^j t^j b_{kj}}{j!(r_k - 2 - j)!}(-\mu_j t + ix)^{r-2-j}[\log(-\mu_j t + ix) + \alpha_j] \right.$$

$$\left. + \sum_{k=1}^{l} \sum_{j=0}^{r_k-1} \frac{t^j b_{kj}}{j!(r_k - 2 - j)!}(\mu_j t - ix)^{r-2-j}[\log(\mu_j t - ix) + a_j] \right\},$$

where

$$a_j = \sum_{n=1}^{r_k - 2 - j} \frac{1}{n}$$

and r_k and b_{kj} are the same as in Prob. 4 (V. E. Kondrashov).

4.7. Fundamental Solutions of Regular Equations ($n > 1$)

4.7.1. Consider the differential equation in several space variables given by

$$\frac{\partial^m u(t, x)}{\partial t^m} = \sum_{k=0}^{m-1} p_k\left(i\frac{\partial}{\partial x}\right)\frac{\partial^k u(t, x)}{\partial t^k}, \qquad x = (x_1, ..., x_n). \tag{1}$$

Taking Fourier transforms in (1), we arrive at the equation

$$\frac{d^m v(t, \sigma)}{dt^m} = \sum_{k=0}^{m-1} p_k(\sigma)\frac{d^k v(t, \sigma)}{dt^k}, \qquad \sigma = (\sigma_1, ..., \sigma_n). \tag{2}$$

Suppose that (1) is regular. Then among the roots of

$$\lambda^m(\sigma) = \sum_{k=0}^{m-1} p_k(\sigma)\, \lambda^k(\sigma) \tag{3}$$

there are exactly r which for (almost) every σ have a non-positive real part. If they are all simple, then the resolvent function is

$$V(t, \sigma) = \sum_{j=0}^{r-1} C_j e^{t\lambda_j(\sigma)}, \tag{4}$$

where the coefficients $C_j = C_j(\sigma)$ satisfy the system of equations

$$C_0 + \cdots + C_{r-1} = 0,$$

$$\cdots\cdots\cdots\cdots\cdots\cdots\cdots$$

$$\lambda_0^{r-2} C_0 + \cdots + \lambda_{r-1}^{r-2} C_{r-1} = 0,$$

$$\lambda_0^{r-1} C_0 + \cdots + \lambda_{r-1}^{r-1} C_{r-1} = 1.$$

After finding $C_0, \ldots, C_{r-1}$ and substituting their values in (4), we must take inverse Fourier transforms. This procedure is not always technically feasible and so we shall proceed in a different way in the general case. However in simple cases it does achieve the objective and to illustrate it we shall look at a few classical examples.

1°. The heat equation

$$\frac{\partial u(t, x)}{\partial t} = \Delta u \quad \left(\Delta = \sum_{j=1}^{n} \frac{\partial^2}{\partial x_j^2} \right). \tag{5}$$

The transformed equation is given by

$$\frac{dv(t, \sigma)}{dt} = -\varrho^2 v \quad \left(\varrho^2 = \sum_{j=1}^{n} \sigma_j^2 \right)$$

and has the resolvent function

$$V(t, \sigma) = e^{-t\varrho^2} = \prod_{j=1}^{n} e^{-t\sigma_j^2}.$$

Since the resolvent function is a product of n factors each depending on just one of the variables, we can take inverse Fourier transforms with respect to each one separately and multiply the results together. Inasmuch as (Sec. 4.5.6, 1°)

$$F^{-1}[e^{-t\sigma_j^2}] = \frac{1}{\sqrt{4\pi t}} e^{-\frac{x_j^2}{4t}},$$

the Green's function for (5) is

$$G(t, x) = \prod_{j=1}^{n} \frac{1}{\sqrt{4\pi t}} e^{-\frac{x_j^2}{4t}} = \frac{1}{(4\pi t)^{n/2}} e^{-\frac{|x|^2}{4t}}, \quad |x|^2 = \sum_{j=1}^{n} x_j^2.$$

2°. Laplace's equation

$$\frac{\partial^2 u(t, x)}{\partial t^2} = -\Delta u.$$

The transformed equation

$$\frac{d^2 v(t, \sigma)}{dt^2} = \varrho^2 v$$

has the resolvent function

$$V(t, \sigma) = e^{-t\rho}.$$

The Green's function is given by the inverse Fourier transform

$$G(t, x) = \frac{1}{(2\pi)^n} \int_{R_n} e^{-t\rho} e^{-i(x,\sigma)} \, d\sigma.$$

By formula (25) of Sec. 2.8, we have

$$G(t, x) = \frac{\Omega_{n-1}}{2\pi^n} \Gamma\left(\frac{n-1}{2}\right) \Gamma\left(\frac{n+1}{2}\right) \frac{t}{(t^2 + |x|^2)^{\frac{n+1}{2}}}.$$

3°. The wave equation

$$\frac{\partial^2 u(t, x)}{\partial t^2} = \Delta u.$$

The resolvent function for the transformed equation

$$\frac{d^2 v(t, \sigma)}{dt^2} = -\varrho^2 v$$

is

$$V(t, \sigma) = \frac{\sin t\varrho}{\varrho}.$$

The next step is to determine its inverse Fourier transform. Recalling formula (27) of Sec. 2.8, we can write down the answer at once for the case $n = 3$:

$$G(t, x) = \frac{\delta(r - t)}{4\pi t}.$$

This means that if the assigned data is

$$u_0(x) = u(0, x) = 0, \quad u_1(x) = \frac{\partial u(0, x)}{\partial t},$$

then the solution $u(t, x)$ is given by

$$u(t, x) = G(t, x) * u_1(x) = \frac{\delta(r - t)}{4\pi t} * u_1(x)$$

$$= \frac{1}{4\pi t} \int\limits_{|\xi - x| = t} u_1(\xi) \, d\xi = t S_t[u_1(x)],$$

where $S_t[u_1(x)]$ is the mean value of $u_1(x)$ over a sphere of radius t with center at x.

Consider now the general case. By formula (26) of Sec. 2.8, we have

$$F[\delta(r - t)] = (2\pi)^{\frac{n}{2}} t^{\frac{n}{2}} \varrho^{1 - \frac{n}{2}} J_{\frac{n}{2} - 1}(t\varrho). \qquad (6)$$

We replace $t\varrho$ by z and $\frac{1}{2}n - 1$ by $m + \frac{1}{2}$ in (6) (so that the subsequent computations are good for odd values of $n \geq 3$). This yields

$$\frac{\varrho^{n - 2}}{t} F[\delta(r - t)] = (2\pi)^{\frac{n}{2}} z^{m + \frac{1}{2}} J_{m + \frac{1}{2}}(z).$$

Now applying formula (23) of Sec. 2.8, we obtain

$$\left(\frac{d}{z \, dz}\right)^m \frac{\varrho^{n - 2}}{t} F[\delta(r - t)] = 2^{\frac{n + 1}{2}} \pi^{\frac{n - 1}{2}} \sin t\varrho.$$

It makes no difference which of the quantities t or ϱ is considered to be constant in the substitution $z = \varrho t$ and which is dependent on z. To facilitate matters when differentiating with respect to z, we shall consider ϱ to be constant. Thus $dz = \varrho \, dt$, $z \, dz = t\varrho^2 \, dt$, and

$$\left(\frac{d}{z \, dz}\right)^m \frac{\varrho^{n - 2}}{t} F[\delta(r - t)] = \frac{1}{\varrho^{2m}} \left(\frac{d}{t \, dt}\right)^m \frac{\varrho^{n - 2}}{t} F[\delta(r - t)]$$

$$= \varrho F\left[\left(\frac{d}{t \, dt}\right)^m \frac{\delta(r - t)}{t}\right] = 2^{\frac{n + 1}{2}} \pi^{\frac{n - 1}{2}} \sin t\varrho.$$

Hence

$$G(t, x) = F^{-1}\left[\frac{\sin t\varrho}{\varrho}\right] = \frac{1}{2^{\frac{n + 1}{2}} \pi^{\frac{n - 1}{2}}} \left(\frac{d}{t \, dt}\right)^m \frac{\delta(r - t)}{t}.$$

19   Shilov

The solution to the Cauchy problem with $\dfrac{\partial u(0, x)}{\partial t} = u_1(x)$ prescribed is given by

$$u(t, x) = G(t, x) * u_1(x) = \frac{1}{2^{\frac{n+1}{2}} \pi^{\frac{n-1}{2}}} \left(\frac{d}{t\, dt}\right)^m \frac{\delta(r - t)}{t} * u_1(x)$$

$$= \frac{1}{1 \cdot 3 \cdots (n - 2)} \left(\frac{d}{t\, dt}\right)^m t^{n-2} S_t[u_1(x)], \tag{7}$$

where $S_t[u_1]$ as before stands for the mean value of $u_1(x)$ over the sphere $|\xi - x| = t$.

Since the solution of the Cauchy problem for the wave equation is obtained through an m-fold differentiation of the initial data, its order of singularity will in general be $m = (n - 3)/2$ more than that of the initial data itself.

Formula (7) shows that at each given instant t, $u(t, x)$ is determined at x by just the values of $u_1(x)$ in the immediate neighborhood of the sphere of radius t centered at x. We wish to emphasize that this result has been obtained for odd n. It actually does not hold for even n. The solution $u(t, x)$ will depend on the values of $u_1(x)$ in the entire ball of radius t when n is even.

By computing the inverse Fourier transform in a somewhat different way, we can derive a formula that is valid for any n, even or odd, though not as transparent. We first note that the classical formula for the inverse Fourier transform of

$$V(t, \sigma) = \frac{\sin t\varrho}{\varrho}$$

cannot be used since the integral is divergent. So we integrate $V(t, \sigma)$ $n - 2$ times with respect to t to obtain

$$W(t, \sigma) = \frac{\cos t\varrho}{\varrho^{n-1}} \left(\text{or} \ \frac{\sin t\varrho}{\varrho^{n-1}}\right).$$

We then form the inverse Fourier transform

$$\frac{1}{(2\pi)^n} \int_{R_n} \frac{\cos t\varrho}{\varrho^{n-1}} e^{-i(\sigma, x)}\, d\sigma.$$

By first integrating over a sphere Ω of radius ϱ and then with respect to ϱ from 0 to ∞, we can show that this integral is now convergent. Indeed, this last integral but with ϱ going from 0 to N has the form

$$I_N = \frac{1}{(2\pi)^n} \int_0^N \frac{\cos t\sigma}{\varrho^{n-1}} \left[\int_\Omega e^{-i(\omega,\,x)\varrho} \, d\omega \right] \varrho^{n-1} \, d\varrho.$$

According to formula (6) of Sec. 2.2,

$$\int_\Omega e^{-(\omega,\,x)\varrho} \, d\omega = \frac{c_n}{r^{n-2}} \int_{-r}^r e^{-i\varrho\xi} \left(r^2 - \xi^2\right)^{\frac{n-3}{2}} \, d\xi.$$

Therefore

$$I_N = \frac{c_n}{(2\pi)^n \, r^{n-2}} \int_0^N \cos t\varrho \left[\int_{-r}^r e^{-i\varrho\xi} \left(r^2 - \xi^2\right)^{\frac{n-3}{2}} \, d\xi \right] d\varrho$$

$$= \frac{c_n}{(2\pi)^n \, r^{n-2}} \int_{-r}^r \left(r^2 - \xi^2\right)^{\frac{n-3}{2}} \left[\int_0^N \cos t\varrho \, (\cos \varrho\xi - i \sin \varrho\xi) \, d\varrho \right] d\xi$$

$$= \frac{c_n'}{r^{n-2}} \int_{-r}^r \left(r^2 - \xi^2\right)^{\frac{n-3}{2}} \left[\frac{\sin N(t + \xi)}{t + \xi} + \frac{\sin N(t - \xi)}{t - \xi} \right] d\xi - \frac{ic_n'}{r^{n-2}}$$

$$\times \int_{-r}^r \left(r^2 - \xi^2\right)^{\frac{n-3}{2}} \left[\frac{1 - \cos N(\xi + t)}{\xi + t} + \frac{1 - \cos N(\xi - t)}{\xi - t} \right] d\xi.$$

The second term tends to zero as $N \to \infty$, while the first has the limit

$$\frac{c}{r^{n-2}} \begin{cases} \left(r^2 - t^2\right)^{\frac{n-3}{2}} & \text{for} \quad t < r, \\ 0 & \text{for} \quad t > r \end{cases}$$

by virtue of the inversion theorem for the Fourier transform. A similar result is obtained if $\cos t\varrho$ is replaced by $\sin t\varrho$. Hence

$$G(t, x) = F^{-1} \left[\frac{\sin t\varrho}{\varrho} \right] = c \frac{\partial^{n-2}}{\partial t^{n-2}} \begin{cases} \dfrac{1}{r} \left(1 - \dfrac{t^2}{r^2}\right)^{\frac{n-3}{2}} & (t < r), \\ 0 & (t > r). \end{cases} \qquad (8)$$

19*

The solution $u(t, x)$ to the Cauchy problem with $u(0, x) = 0$ and $\dfrac{\partial u(0, x)}{\partial t} = u_1(x)$ is given by

$$u(t, x) = G(t, x) * u_1(x) = c_1 \frac{\partial^{n-2}}{\partial t^{n-2}} \int_{r > t} \left(1 - \frac{t^2}{r^2}\right)^{\frac{n-3}{2}} S_r[u_1] \, r^{n-2} \, dr.$$

An equivalent representation can be shown to be

$$u(t, x) = c_2 \frac{\partial^{n-2}}{\partial t^{n-2}} \int_0^t (t^2 - r^2)^{\frac{n-3}{2}} r S_r[u_1] \, dr.$$

When $u_1(x) \equiv 1$, $u(t, x)$ must be identically equal to t. Hence, we find that $c_2 = \dfrac{1}{(n - 2)!}$. Thus our required formula is

$$u(t, x) = \frac{1}{(n - 2)!} \frac{\partial^{n-2}}{\partial t^{n-2}} \int_0^t (t^2 - r^2)^{\frac{n-3}{2}} r S_r[u_1] \, dr.$$

4.7.2. We again consider the general case of a regular equation (1) with index r. The Green's function $G(t, x)$ as we know satisfies the conditions

$$G(0, x) = \cdots = \frac{\partial^{r-2} G(0, x)}{\partial t^{r-2}} = 0, \qquad \frac{\partial^{r-1} G(0, x)}{\partial t^{r-1}} = \delta(x).$$

Now let $G_p(t, x)$ denote the solution of (1) satisfying the conditions

$$G_p(0, x) = \cdots = \frac{\partial^{r-2} G_p(0, x)}{\partial t^{r-2}} = 0, \qquad \frac{\partial^{r-1} G_p(0, x)}{\partial t^{r-1}} = \mathscr{E}_n^{2p}(x), \qquad (9)$$

where

$$\mathscr{E}_n^{2p}(x) = \begin{cases} c_{pn} r^{2p-n}, & \text{for odd } n \text{ or even } n > 2p; \\ c_{pn} r^{2p-n} \log r, & \text{for even } n \leqq 2p, \end{cases}$$

is the fundamental function for the operator Δ^p (Sec. 1.7.2, 9°). Since

$$\Delta^p \mathscr{E}_n^{2p}(x) = \delta(x),$$

we can determine our Green's function $G(t, x)$ through the formula

$$G(t, x) = \Delta^p G_p(t, x). \qquad (10)$$

In some instances when p is sufficiently large, it turns out to be easier to construct the generalized fundamental solution $G_p(t, x)$ and then to obtain $G(t, x)$ using (10).

Our starting point is the plane-wave expansion for r^λ given by (Sec. 2.4, (34))

$$r^\lambda = \frac{1}{2\pi^{\frac{n-1}{2}}} \frac{\Gamma\left(\dfrac{\lambda + n}{2}\right)}{\Gamma\left(\dfrac{\lambda + 1}{2}\right)} \int_\Omega |(\omega, x)|^\lambda \, d\omega$$

and its derivative with respect to λ

$$r^\lambda \log r = \frac{1}{2\pi^{\frac{n-1}{2}}} \frac{\Gamma\left(\dfrac{\lambda + n}{2}\right)}{\Gamma\left(\dfrac{\lambda + 1}{2}\right)} \int_\Omega |(\omega, x)|^\lambda \log |(\omega, x)| \, d\omega$$

$$+ \, c_n(\lambda) \int_\Omega |(\omega, x)|^\lambda \, d\omega.$$

Applying these formulas with $\lambda = 2p - n$, we obtain the following plane-wave expansion for $\mathscr{E}_n^{2p}(x)$:

$$\mathscr{E}_n^{2p}(x) = \begin{cases} c_{np} \int_\Omega |(\omega, x)|^{2p-n} \, d\omega & \text{for odd } n \text{ or } n > 2p; \quad (11) \\[2ex] d_{np} \int_\Omega |(\omega, x)|^{2p-n} \log |(\omega, x)| \, d\omega & \text{for even } n \leq 2p. \quad (12) \end{cases}$$

In the first case, when $n > 2p$, the expression $|(\omega, x)|^{2p-n}$ stands for $\delta^{(n-2p-1)}((\omega, x))$ or $((\omega, x))^{-2p+n}$ depending on whether n is odd or even (Sec. 2.4.6).

The term $c_n(\lambda) \int_\Omega |(\omega, x)|^{2p-n} \, d\omega$ has been omitted in the second case (even $n \leq 2p$) because the operator Δ^p annihilates it.

We can combine (11) and (12) into a single formula:

$$\mathscr{E}_n^{2p}(x) = a_{np} \int_\Omega \mathscr{E}_{2p-n}((\omega, x)) \, d\omega, \qquad (13)$$

where $\mathscr{E}_{2p-n}(\xi)$ is an even generalized function of ξ defined by

$$
\mathscr{E}_k(\xi) = \begin{cases} |\xi|^k, & \text{for odd } k > 0 \text{ or even } k < 0, \\ \delta^{(-k-1)}(\xi), & \text{for odd } k < 0, \\ |\xi|^k \log |\xi|, & \text{for even } k \geq 0. \end{cases}
$$

We also need the q-fold indefinite integrals $\mathscr{E}_{kq}(\xi)$ of $\mathscr{E}_k(\xi)$ which are

$$
\mathscr{E}_{kq}(\xi) = \begin{cases} a_{kq}\xi^{k+q}, & \text{for even } k, \quad k+q < 0, \\ b_{kq}\delta^{(-k-1-q)}(\xi) & \text{for odd } k, \quad k+q < 0, \\ c_{kq}\xi^{k+q} \operatorname{sgn} \xi, & \text{for odd } k, \quad k+q \geq 0, \\ d_{kq}\xi^{k+q} \log |\xi|, & \text{for even } k, \quad k+q \geq 0. \end{cases}
$$

With (13) in mind, we dispense with solving problem (1), (9) and we look for a solution $u_p(\omega, t, x)$ (for fixed ω) of (1) satisfying the conditions

$$
\left. \begin{aligned} u_p(\omega, 0, x) = \cdots &= \frac{\partial^{r-2} u_p(\omega, 0, x)}{\partial t^{r-2}} = 0, \\ \frac{\partial^{r-1} u_p(\omega, 0, x)}{\partial t^{r-1}} &= \mathscr{E}_{2p-n}((\omega, x)). \end{aligned} \right\} \tag{14}
$$

Suppose that a solution $u_p(\omega, t, x)$ has been found which depends continuously on ω (in K') and is m times differentiable with respect to t. Then we assert that

$$
G_p(t, x) = a_{np} \int_\Omega u_p(\omega, t, x) \, d\omega
$$

is the solution to the stated problem. For since the order of differentiation and passage to the limit are interchangeable in K', we have

$$
\frac{\partial^m G_p(t, x)}{\partial t^m} - \sum_{k=0}^{m-1} p_k\left(i\frac{\partial}{\partial x}\right)\frac{\partial^k G_p(t, x)}{\partial t^k}
$$

$$
= a_{np} \int_\Omega \left[\frac{\partial^m u(\omega, t, x)}{\partial t^m} - \sum_{k=0}^{m-1} p_k\left(i\frac{\partial}{\partial x}\right)\frac{\partial^k u(\omega, t, x)}{\partial t^k} \right] d\omega = 0,
$$

$$
G_p(0, x) = \cdots = \frac{\partial^{r-2} G_p(0, x)}{\partial t^{r-2}} = 0,
$$

and

$$\frac{\partial^{r-1} G_p(0, x)}{\partial t^{r-1}} = a_{np} \int_\Omega \frac{\partial^{r-1} u(\omega, 0, x)}{\partial t^{r-1}} d\omega = a_{np} \int_\Omega \mathscr{E}_{2p-n}((\omega, x)) d\omega$$

$$= \mathscr{E}_n^{2p}(x),$$

as required.

Thus, the determination of the Green's function is reduced to solving (1) under the conditions (14). We shall seek a solution to this problem of the form

$$u_p(\omega, t, x) \equiv U_\omega(t, \omega_1 t_1 + \cdots + \omega_n x_n) \equiv U(t, \xi), \quad \xi = (\omega, x). \quad (15)$$

Applying the chain rule for differentiating functions of the type (15) (Sec. 1.7.2, 4°), we find from (1) that

$$\frac{\partial^m U(t, \xi)}{\partial t^m} = \sum_{k=0}^{m-1} p_k \left(i\omega_1 \frac{\partial}{\partial \xi}, \ldots, i\omega_n \frac{\partial}{\partial \xi} \right) \frac{\partial^k U_\omega(t, \xi)}{\partial t^k}. \quad (16)$$

This is an equation involving one space variable ξ (and a parameter ω) which must be solved subject to the conditions

$$U_\omega(0, \xi) = \cdots = \frac{\partial^{r-2} U_\omega(0, \xi)}{\partial t^{r-2}} = 0, \quad \frac{\partial^{r-1} U_\omega(0, \xi)}{\partial t^{r-1}} = \mathscr{E}_{2p-n}(\xi). (17)$$

The equation (16) is a regular equation with index r for almost all ω. To see this, we have to examine the equation

$$\lambda^m = \sum_{k=0}^{m-1} p_k(\omega_1 \alpha, \ldots, \omega_n \alpha) \lambda^k \quad (18)$$

to find out how many of its roots have non-positive real parts. But (18) results from (3) by setting $\sigma_1 = \omega_1 \alpha, \ldots, \sigma_n = \omega_n \alpha$, i.e., for parameter values σ describing a line in R_n. Since (1) is a regular equation with index r, this means that (16) is also regular with the same index for almost all ω. Therefore, the problem (16), (17) has a unique solution for almost all ω. If it depends continuously on ω, then as we have just seen, by integrating it with respect to ω we obtain the solution to our pertinent problem.

4.7.3. Let us examine the case of a homogeneous hyperbolic equation

$$\frac{\partial^m u}{\partial t^m} = \sum_{k_0 + k_1 + \cdots + k_n = m} a_{k_1, \ldots, k_n} i^{k_1 + \cdots + k_n} \frac{\partial^m u(t, x)}{\partial t^{k_0} \partial x_1^{k_1} \cdots \partial x_n^{k_n}}. \quad (19)$$

The corresponding algebraic equation is

$$\lambda^m = \sum_{k_0+k_1+\cdots+k_n=m} a_{k_1\ldots k_n}\sigma_1^{k_1}\ldots\sigma_n^{k_n}\lambda^{k_0}. \tag{20}$$

Its roots $\lambda_0(\sigma)$, ..., $\lambda_{m-1}(\sigma)$ are homogeneous functions of first degree in σ. Take a fixed ray defined by the unit vector $\omega = (\omega_1, \ldots, \omega_n)$. Then on such a ray, $\sigma_j = \varrho\omega_j$, $\lambda_k(\sigma) = \mu_k(\omega)\varrho$, and the quantities μ_k and ω are related by

$$\mu_k^m(\omega) - \sum_{k_0+\cdots+k_n=m} a_{k_1\ldots k_n}\mu_k^{k_0}\omega_1^{k_1}\ldots\omega_n^{k_n} = 0. \tag{21}$$

Equation (19) is called hyperbolic if all $\mu_k(\omega)$ are pure imaginary and strictly hyperbolic if in addition distinct. In this case, the fundamental solution of the Cauchy problem for (16) is (Sec. 4.6, (16))

$$G_\omega(t, \xi) = C \sum_{j=0}^{m-1} B_j(\xi - b_j(\omega)t)^{m-2} \operatorname{sgn}(\xi - b_j(\omega)t), \quad ib_j(\omega) = \mu_j(\omega).$$

The solution of the Cauchy problem (16), (17) can be expressed in the form

$$U_\omega(t, \xi) = C \sum_{j=0}^{m-1} B_j(\xi - b_j(\omega)t)^{m-2} \operatorname{sgn}(\xi - b_j(\omega)t) * \mathscr{E}_{2p-n}(\xi). \tag{22}$$

In order to simplify the convolution, we note that

$$\frac{d^{m-1}}{d\xi^{m-1}}(\xi - b_j(\omega)t)^{m-2} \operatorname{sgn}(\xi - b_j(\omega)t) = C_1\delta(\xi - b_j(\omega)t).$$

On the other hand, if we use the indefinite integrals of $\mathscr{E}_{2p-n}(\xi)$, we can write

$$\mathscr{E}_{2p-n}(\xi) = \frac{d^{m-1}}{d\xi^{m-1}}\mathscr{E}_{2p-n,\,m-1}(\xi).$$

Transferring the operator $\dfrac{d^{m-1}}{d\xi^{m-1}}$ from the second factor of the convolution in (22) to the first, we arrive at

$$U_\omega(t, \xi) = C_2 \sum_{j=0}^{m-1} B_j\delta(\xi - b_j(\omega)t) * \mathscr{E}_{2p-n,\,m-1}(\xi)$$

$$= C_2 \sum_{j=0}^{m-1} B_j\mathscr{E}_{2p-n,\,m-1}(\xi - b_j(\omega)t).$$

Hence

$$G(t, x) = c_{pn}\Delta^p \int_\Omega U_\omega(t, (\omega, x))\, d\omega$$

$$= c\Delta^p \int_\Omega \sum_{j=0}^{m-1} B_j \mathscr{E}_{2p-n,\, m-1}((\omega, x) - b_j(\omega)\, t)\, d\omega. \qquad (23)$$

The operator Δ can be expressed in terms of $\dfrac{\partial^2}{\partial t^2}$ by virtue of the relations

$$\frac{\partial^2}{\partial t^2} f((\omega, x) - b_j t) = b_j^2 f''((\omega, x) - b_j t)$$

and

$$\Delta f((\omega, x) - b_j t) = \sum_{k=1}^{n} \frac{\partial^2}{\partial x_k^2} f((\omega, x) - b_j t)$$

$$= \sum_{k=1}^{n} f''((\omega, x) - b_j t)\, \omega_k^2 = f''((\omega, x) - b_j t),$$

which are valid for any $f(\xi)$ (generalized as well).

Thus we find that

$$G(t, x) = c\Delta^p \int_\Omega \sum_{j=0}^{m-1} B_j \mathscr{E}_{2p-n,\, m-1}((\omega, x) - b_j(\omega)\, t)\, d\omega$$

$$= c\, \frac{\partial^{2p}}{\partial t^{2p}} \int_\Omega \sum_{j=0}^{m-1} B_j b_j^{-2p} \mathscr{E}_{2p-n,\, m-1}((\omega, x) - b_j(\omega)\, t)\, d\omega. \qquad (24)$$

If equation (19) is spherically symmetric, i.e., differentiation with respect to x occurs only as a Laplacian, then the roots $b_j(\omega)$ are actually independent of ω. We can then apply the formula (6) of Sec. 2.2.7 for an integral over a sphere of a function of the argument (ω, x). Accordingly,

$$G(t, x) = c_1 \frac{\partial^{2p}}{\partial t^{2p}} \int_{-1}^{1} \sum_{j=0}^{m-1} B_j b_j^{-2p} \mathscr{E}_{2p-n,\, m-1}(rh - b_j t)\, (1 - h^2)^{\frac{n-3}{2}}\, dh$$

$$= c_1 \frac{\partial^{2p}}{\partial t^{2p}} \frac{1}{r^{n-2}} \int_{-r}^{r} (r^2 - \xi^2)^{\frac{n-3}{2}} \sum_{j=0}^{m-1} B_j b_j^{-2p} \mathscr{E}_{2p-n,\, m-1}(\xi - b_j t)\, d\xi. \qquad (25)$$

The number p is temporaily still arbitrary. We could choose it so as to simplify the form of $\mathscr{E}_{2p-n,\,m-1}(\xi - b_j t)$.

For example, suppose we are considering the wave equation

$$\frac{\partial^2 u}{\partial t^2} = \varDelta u.$$

Here

$$m = 2,\ \lambda^2 = -\sigma^2,\ \lambda_{0,\,1} = \pm i|\sigma|,\ \mu_{0,\,1} = \pm i,\ b_{0,\,1} = \pm 1,$$

$$B_{0,\,1} = \pm \frac{1}{2}.$$

If n is odd and greater than one, we set $2p = n - 1$ and obtain

$$G(t, x) = c\,\frac{\partial^{n-1}}{\partial t^{n-1}}\,\frac{1}{r^{n-2}}\int_{-r}^{r}(r^2 - \xi^2)^{\frac{n-3}{2}}\,\chi_{[-t,\,t]}(\xi)\,d\xi,$$

where $\chi_{[-t,\,t]}(\xi)$ equals 1 for $|\xi| \leqq t$ and 0 for $|\xi| > t$. Thus

$$G(t, x) = c\,\frac{\partial^{n-1}}{\partial t^{n-1}}\,\frac{1}{r^{n-2}} \begin{cases} \displaystyle\int_{-t}^{t}(r^2 - \xi^2)^{\frac{n-3}{2}}\,d\xi & \text{for } r > t. \\[2ex] \displaystyle\int_{-r}^{r}(r^2 - \xi^2)^{\frac{n-3}{2}}\,d\xi & \text{for } r < t, \end{cases}$$

Carrying out one differentiation, we arrive at

$$G(t, x) = c\,\frac{\partial^{n-2}}{\partial t^{n-2}}\,\frac{1}{r^{n-2}} \begin{cases} (r^2 - t^2)^{\frac{n-3}{2}} & \text{for } r > t, \\[2ex] 0 & \text{for } r < t, \end{cases}$$

which coincides with (8).

In the general case the integral (24) over the sphere Ω can be converted into an integral over the surface

$$H(\sigma) \equiv 1 + \sum_{j=0}^{m-1} p_j(\sigma) = 0.$$

The corresponding representation (a variant of the Herglotz-Petrovsky formula) is given by

$$G(t, x) = c \frac{\partial^{2p}}{\partial t^{2p}} \int_{H=0} \frac{\mathscr{E}_{2p-n,\, m-1}((\xi, x) - t)\, dH(\xi)}{\text{grad } H(\xi) \text{ sgn } (\xi) \text{ grad } H(\xi)}, \tag{26}$$

where $dH(\xi)$ is the surface element of $H = 0$ at ξ. The derivation of (26) from (24) may be found in a number of books.†

4.7.4. The next general case we consider is where (19) is a regular equation. This means that the roots of

$$\mu^m = \sum_{k_0 + \cdots + k_n = m} a_{k_1 \ldots k_n} \mu^{k_0} \omega_1^{k_1} \ldots \omega_n^{k_n}$$

satisfy the inequalities

$$\text{Re } \mu_0 \leqq \cdots \leqq \text{Re } \mu_{q-1} < \text{Re } \mu_q = \cdots = \text{Re } \mu_{r-1}$$
$$= 0 < \text{Re } \mu_r \leqq \cdots \leqq \text{Re } \mu_{m-1},$$

where q and r are independent of ω and $q + r = m$. As initial data, we may assign $u(0, x), \ldots, \dfrac{\partial^{r-1} u(0, x)}{\partial t^{r-1}}$. We shall assume in addition that $\mu_0, \ldots, \mu_{m-1}$ are distinct. In this case, the solution of (16) under the initial conditions (17) assumes the form (cf. (23) of Sec. 4.6)

$$U_\omega(t, \xi) = C \left\{ \sum_{j=0}^{r-1} B_j(-\mu_j t + i\xi)^{r-2} \log (-\mu_j t + i\xi) \right.$$
$$\left. + (-1)^{r-1} \sum_{s=q}^{m-1} A_s (\mu_s t - i\xi)^{r-2} \log (\mu_s t - i\xi) \right\} * \mathscr{E}_{2p-n}(\xi). \tag{27}$$

When $2p < n$, $\mathscr{E}_{2p-n}(\xi)$ is a convolutor in the space $\mathscr{H}$ (as is $G_\omega(t, \xi)$) and the convolution in (27) is therefore well-defined.

We wish to use the same method applied to a hyperbolic equation and this necessitates differentiating $G_\omega(t, \xi)$. Leibniz's rule easily leads to the following formula for repeated differentiation:

$$(z^\lambda \log z)^{(l)} = \begin{cases} z^{\lambda-l} [a_{\lambda l} \log z + b_{\lambda l}] & (l \leqq \lambda), \\ b_{\lambda l} z^{\lambda-l} & (l > \lambda). \end{cases}$$

† F. John, *Plane Waves and Spherical Means Applied to Partial Differential Equations*, Tract. 2, Interscience, New York, 1955; I. M. Gelfand and G. E. Shilov, *Generalized Functions*, Vol. 1, *Properties and Operations*, Academic Press, New York, 1964.

We can therefore write

$$
U_\omega(t, \xi) = c_r \left\{ \sum_{j=0}^{r-1} B_j(- \mu_j t + i\xi)^{r-2-l} \right.
$$

$$
\times [a_{r-2, l} \log (- \mu_j t + i\xi) + b_{r-2, l}]
$$

$$
+ (-1)^{r+l-1} \sum_{s=q}^{m-1} A_s(\mu_s t - i\xi)^{r-2-l}
$$

$$
\left. \times [a_{r-2, l} \log (\mu_s t - i\xi) + b_{r-2, l}] \right\} * \mathscr{E}_{2p-n, l}(\xi), \qquad (28)
$$

with the constants $a_{r-2, l}$ vanishing for $l > r - 2$.

At this point, we impose on p and l the condition $2p + l = n - 1$.

For n odd, we have $\mathscr{E}_{2p-n, l}(\xi) = b\delta(\xi)$ and for n even, $\mathscr{E}_{2p-n, l}(\xi) = a\xi^{-1}$. Accordingly, for n odd,

$$
U_\omega(t, \xi) = bc_r \left\{ \sum_{j=0}^{r-1} B_j(- \mu_j t + i\xi)^{2p+r-n-1} \right.
$$

$$
\times [a_{r-2, l} \log (- \mu_j t + i\xi) + b_{r-2, l}]
$$

$$
+ (- 1)^{r+n} \sum_{s=q}^{m-1} A_s(\mu_s t - i\xi)^{2p+r-n-1}
$$

$$
\left. \times [a_{r-2, l} \log (\mu_j t - i\xi) + b_{r-2, l}] \right\}. \qquad (29)
$$

Hence

$$
G(t, x) = c\Delta^p \int_\Omega \left\{ \sum_{j=0}^{r-1} B_j(- \mu_j t + i(\omega, x))^{2p+r-n-1} \right.
$$

$$
\times [a_{r-2, l} \log (- \mu_j t + i(\omega, x)) + b_{r-2, l}]
$$

$$
+ (- 1)^{r+n} \sum_{s=q}^{m-1} A_s(\mu_s t - i(\omega, x))^{2p+r-n-1}
$$

$$
\left. \times [a_{r-2, l} \log (\mu_j t - i(\omega, x)) + b_{r-2, l}] \right\} d\omega, \qquad (30)
$$

where $a_{r-2, l} = 0$ for $l > r - 2$.

For n even, we have to work out the convolution with $a\xi^{-1}$. The convolution of any generalized function $f(\xi)$ with $a\xi^{-1}$ may be determined for example as follows: find the Fourier transform $g(\sigma)$ of $f(\xi)$, multiply it by

$$
F[a\xi^{-1}] = a\pi i \operatorname{sgn} \sigma
$$

and then take the inverse Fourier transform to obtain

$$f(\xi) * \xi^{-1} = F^{-1}[g(\sigma) \cdot \pi i \operatorname{sgn} \sigma].$$

In our case,

$$g(\sigma) = V(\omega, t, \sigma) = (i\sigma)^l \begin{cases} \dfrac{1}{\sigma^{r-1}} \displaystyle\sum_{j=0}^{r-1} B_j e^{\mu_j \sigma t} & \text{for } \sigma > 0, \\[3mm] \dfrac{1}{\sigma^{r-1}} \displaystyle\sum_{s=q}^{m-1} A_s e^{\mu_s \sigma t} & \text{for } \sigma < 0. \end{cases}$$

Therefore

$$a\pi g(\sigma) \operatorname{sgn} \sigma = a\pi(i\sigma)^l \begin{cases} \dfrac{1}{\sigma^{r-1}} \displaystyle\sum_{j=0}^{r-1} B_j e^{\mu_j \sigma t} & \text{for } \sigma > 0, \\[3mm] -\dfrac{1}{\sigma^{r-1}} \displaystyle\sum_{s=q}^{m-1} A_s e^{\mu_s \sigma t} & \text{for } \sigma < 0. \end{cases}$$

Hence

$$U_\omega(t, \xi) = C\pi \left(\frac{d}{d\xi}\right)^l \Bigg\{ \int_0^\infty \frac{1}{\sigma^{r-1}} \sum_{j=0}^{r-1} B_j e^{\mu_j \sigma t - i\sigma\xi} \, d\sigma$$

$$- \int_{-\infty}^0 \frac{1}{\sigma^{r-1}} \sum_{s=q}^{m-1} A_s e^{\mu_j \sigma t - i\sigma\xi} \, d\sigma \Bigg\}$$

$$= aC_r\pi \left(\frac{d}{d\xi}\right)^l \Bigg\{ \sum_{j=0}^{r-1} B_j(i\xi - \mu_j t)^{r-2} \log (i\xi - \mu_j t)$$

$$+ (-1)^r \sum_{s=q}^{m-1} A_j(\mu_j t - i\xi)^{r-2} \log (\mu_j t - i\xi) \Bigg\}$$

$$= a\pi \hat{C}_r \Bigg\{ \sum_{j=0}^{r-1} B_j(-\mu_j t + i\xi)^{2p+r-n-1}[a_{r-2,\,l} \log (-\mu_j t + i\xi)$$

$$+ b_{r-2,\,l}] - (-1)^{r+n} \sum_{s=q}^{m-1} A_s(\mu_j t - i\xi)^{2p+r-n-1}$$

$$\times [a_{r-2,\,l} \log (\mu_j t - i\xi) + b_{r-2,\,l}] \Bigg\}. \tag{31}$$

Finally our desired Green's function $G(t, x)$ is given by

$$
G(t, x) = c\Delta^p \int_\Omega \left\{ \left[\sum_{j=0}^{r-1} B_j(-\mu_j t + i(\omega, x))^{2p+r-n-1} \right. \right.
$$

$$
\times [a_{r-2, l} \log(-\mu_j t + i(\omega, x)) + b_{r-2, l}]
$$

$$
+ (-1)^{r+n-1} \sum_{s=q}^{m-1} A_s(\mu_s t - i(\omega, x))^{2p+r-n-1}
$$

$$
\left. \times [a_{r-2, l} \log(\mu_j t - i(\omega, x)) + b_{r-2, l}] \right\} d\omega. \qquad (32)
$$

The constants $a_{r-2, l}$ likewise vanish here for $l > r - 2$. The exponent p is still arbitrary (except that $2p < n$) and can be used to further simplify (30) and (32) in the various possible particular cases.

4.8. An Equation with Right-Hand Side

4.8.1. In this section, we shall examine the question of the existence of well-posed boundary value problems for an equation with a right-hand side and the form they take. Consider the equation

$$
\frac{\partial^m u(t, x)}{\partial t^m} - \sum_{k=0}^{m-1} P_k\left(i \frac{\partial}{\partial x} \right) \frac{\partial^k u(t, x)}{\partial t^k} = f(t, x). \qquad (1)
$$

As before, our objective is to clarify what initial conditions determine a well-posed problem for (1), i.e., one whose solution $u(t, x)$ belongs to $\mathcal{H}$ for each $t \geq 0$, increases there no faster than a power of t as $t \to \infty$, is unique in this class, and depends continuously on the initial conditions and $f(t, x)$. Of course, we shall assume that $f(t, x)$ itself satisfies the conditions set forth for the solution. It is to belong to $\mathcal{H}$ for $t \geq 0$ and to increase no faster than a power of t as $t \to \infty$. A more precise condition will be formulated below.

Taking Fourier transforms with respect to x in (1), we obtain the equation

$$
\frac{d^m v(t, \sigma)}{dt^m} - \sum_{k=0}^{m-1} P_k(\sigma) \frac{d_k v(t, \sigma)}{dt^k} = g(t, \sigma), \qquad (2)
$$

where $g(t, \sigma)$ belongs to H for each $t \geq 0$ and has growth there no faster than a power of t has $t \to \infty$.

To illustrate the various possibilities that exist here, we shall drop the parameter σ tentatively and consider the case of an ordinary differential equation and for simplicity one of first order:

$$\frac{dv(t)}{dt} - av(t) = g(t), \quad |g(t)| \leq C(1 + t)^h. \tag{3}$$

Its solution under the initial condition $v(0) = v_0$ is given by

$$v(t) = e^{at}v_0 + \int_0^t e^{a(t-\theta)}g(\theta) \, d\theta = e^{at}\left[v_0 + \int_0^t e^{-a\theta}g(\theta) \, d\theta\right]. \tag{4}$$

Suppose first that $a \leq 0$. Let us show that the solution $v(t)$ increases no faster than a power of t as $t \to \infty$ for arbitrary v_0. In this case,

$$|v(t)| \leq |v_0| + \int_0^t |g(\theta)| \, d\theta \leq |v_0| + Ct(1 + t)^h,$$

as asserted.

Suppose now that $a > 0$. Then the integral

$$I(g) = \int_0^\infty e^{-a\theta}g(\theta) \, d\theta$$

is convergent and we can write the solution (4) in the form

$$v(t) = e^{at}[v_0 + I(g)] - \int_t^\infty e^{a(t-\theta)} g(\theta) \, d\theta. \tag{5}$$

The second term has power growth at the most since

$$\int_t^\infty e^{a(t-\theta)}g(\theta)d\theta = \int_0^\infty e^{-a\theta}g(\theta + t) \, d\theta \tag{6}$$

and

$$\left|\int_t^\infty e^{a(t-\theta)}g(\theta) \, d\theta\right| \leq C \int_0^\infty e^{-a\theta}(1 + \theta + t)^h \, d\theta = O(t^h).$$

If $v_0 + I(g) \neq 0$, then $v(t)$ is clearly exponentially increasing as $t \to \infty$ Hence

The condition $v_0 + I(g) = 0$ is necessary and sufficient for the existence of a solution having power growth in t.

Thus when $a > 0$, we cannot impose any additional condition on v_0 because v_0 is uniquely determined by $I(g)$.

An equation of the type (3) involving a parameter σ comes about by taking Fourier transforms with respect to x in the following first order equation in t:

$$\frac{\partial u(t,x)}{\partial t} - p\left(i\,\frac{\partial}{\partial x}\right) u(t,x) = f(t,x).$$

The polynomial $p(\sigma)$ plays the role of the coefficient a. We can now state the following result. For values of σ where $p(\sigma) \leq 0$, $v_0(\sigma)$ may be assigned in an arbitrary way. For values of σ where $p(\sigma) > 0$, it cannot be so assigned and is on the contrary entirely determined by the condition

$$v_0(\sigma) + \int_0^\infty e^{-p(\sigma)\theta} g(\theta, \sigma)\,d\theta = 0. \tag{7}$$

For each fixed σ, the solution $v(t, \sigma)$ is of the form

$$v(t, \sigma) = \begin{cases} e^{tp(\sigma)} v_0(\sigma) + \displaystyle\int_0^t e^{(t-\theta)p(\sigma)} g(\theta, \sigma)\,d\theta & \text{for} \quad p(\sigma) \leq 0; \tag{8} \\[2ex] -\displaystyle\int_0^\infty e^{-\theta p(\sigma)} g(t + \theta, \sigma)\,d\theta & \text{for} \quad p(\sigma) > 0. \tag{9} \end{cases}$$

The expression for $v(t, \sigma)$ when $p(\sigma) \leq 0$ is an element of H having there at most power growth in t together with $g(t, \sigma)$. Indeed, if

$$|g(t, \sigma)| \leq C(1 + t)^h (1 + \sigma^2)^q \psi(\sigma), \quad \psi(\sigma) \in L_2,$$

then for $p(\sigma) \leq 0$ we have

$$|v(t, \sigma)| \leq |v_0(\sigma)| + C \int_0^t |g(\theta, \sigma)|\,d\theta$$

$$\leq |v_0(\sigma)| + C(1 + \sigma^2)^q (1 + t)^h |\psi(\sigma)|.$$

For $p(\sigma) > 0$, the analogous assertion requires the imposition of additional assumptions on $g(t, \sigma)$. For instance, we might assume that $g(t, \sigma)$ decreases fast enough in H as $t \to \infty$ so as to assure the fulfillment of the estimate

$$|g(t, \sigma)| \leq C(1 + t)^h (1 + \sigma^2)^q \psi(\sigma)$$

with $h < -1$. For then

$$|v(t, \sigma)| \leq C(1 + \sigma^2)^q |\psi(\sigma)| \int_0^\infty (1 + t + \theta)^h\,d\theta$$

$$= C_1(1 + \sigma^2)^q |\psi(\sigma)| (1 + t)^{h+1}.$$

A weaker condition is simply the assumption that the integral (9) belongs to H for $t \geq 0$ and has at most power growth in t.

We shall prove the following theorem for the general equation (1):

THEOREM. *Let $\lambda_0(\sigma), ..., \lambda_{m-1}(\sigma)$ be the roots of*

$$\lambda^m(\sigma) - \sum_{k=0}^{m-1} p_k(\sigma)\lambda^k(\sigma) = 0$$

and G_r the set of all $\sigma \in R_n$ for which exactly r of the $\lambda_j(\sigma)$ have a non-positive real part $(r = 0, 1, ..., m - 1)$.

Further let $f(t, x)$ be a function in $\mathcal{H}$ for each $t \geq 0$ having there at most power growth in t and such that on G_r each of the integrals

$$\int_0^\infty e^{-\theta\lambda_r(\sigma)}\theta^k |g(\theta + t, \sigma)| \, d\theta \quad (k = 0, 1, ..., m - 1) \qquad (10)$$

is in H and has there at most power growth in t.

Then the following initial value problem is well-posed for equation (1): Assign $v_{k-1}(\sigma)$ almost everywhere on G_r, where $v_{k-1}(\sigma)$ belongs to $H(G_{k-1})$ and is the Fourier transform of $\dfrac{\partial^{k-1}u(t, x)}{\partial t^{k-1}}$ at $t = 0$ $(k = 1, 2, ..., r; r = 0, ..., m - 1).$

The proof will be carried out under the assumption that $\lambda_0(\sigma), ..., \lambda_{m-1}(\sigma)$ are all distinct almost everywhere.

4.8.2. Consider a non-homogeneous linear system of m first order ordinary differential equations involving m unknowns $v_0(t), ..., v_{m-1}(t)$ with constant coefficients given in vector form by

$$\frac{dv(t)}{dt} - Pv(t) = g(t), \qquad (11)$$

where

$$|g(t)| \leq C(1 + t)^h. \qquad (12)$$

The solution to (11) with initial data $v(0) = v_0$ can be written in the similar form to (4),

$$v(t) = e^{Pt}v(0) + \int_0^t e^{(t-\theta)P}g(\theta) \, d\theta = e^{Pt}\left[v(0) + \int_0^t e^{-\theta P}g(\theta) \, d\theta\right]. \quad (13)$$

Suppose that the linear operator defined by the matrix P has r eigenvalues with a non-positive real part and $m - r$ eigenvalues with a posi-

20   Shilov

tive real part. The space Q in which P is an operator splits up into a direct sum of invariant subspaces Q^- and Q^+, Q^- being generated by the eigenvectors (and root vectors) corresponding to the eigenvalues of the first group and Q^+ by the eigenvectors (and root vectors) corresponding to the eigenvalues of the second group. The vectors $v(t)$, v_0, and $g(t)$ have corresponding decompositions

$$v(t) = v^-(t) + v^+(t), \quad v_0 = v_0^- + v_0^+, \quad g(t) = g^-(t) + g^+(t).$$

Since P and e^{tP} are invariant in Q^- and Q^+, we have

$$v^-(t) = e^{tP}v_0^- + \int_0^t e^{(t-\theta)P}g^-(\theta)\, d\theta \tag{14}$$

and

$$v^+(t) = e^{tP}v_0^+ + \int_0^t e^{(t-\theta)P}g^+(\theta)\, d\theta. \tag{15}$$

It is asserted that $v^-(t)$ has no more than power growth in Q^-. The proof of this merely requires the application of inequality (9) of Sec. 4.2 which states that

$$\| e^{tP} \| \leqq e^{t\Lambda}\left(1 + 2t\| P \| + \cdots + \frac{(2t)^{m-1}}{(m-1)!}\| P \|^{m-1}\right), \tag{16}$$

where Λ is the maximum of the real parts of the eigenvalues of P. Denote the eigenvalues of P in Q^- by $\lambda_0, \ldots, \lambda_{r-1}$. Then

$$\operatorname{Re} \lambda_0 \leqq \cdots \leqq \operatorname{Re} \lambda_{r-1} \leqq 0.$$

Therefore by (16) we have

$$\| e^{tP} \| \leqq C(1 + t)^{m-1} \tag{17}$$

in Q^-. Hence

$$\| v^-(t) \| \leqq \| e^{tP} \|\, \| v_0^- \| + \int_0^t \| e^{(t-\theta)P} \|\, \| g^-(\theta) \|\, d\theta$$

$$\leqq C_1 \| e^{tP} \|\, \| v_0 \| + C_1 \int_0^t \| e^{(t-\theta)P} \|\, \| g(\theta) \|\, d\theta$$

$$\leqq C_2(1 + t)^{m-1} + C_3 \int_0^t (1 + t - \theta)^{m-1}(1 + \theta)^h\, d\theta = O(t^{m+1+h}),$$

as asserted.

We now write the expression (15) for $v^+(t)$ in the form

$$v^+(t) = e^{tP}\left\{v_0^+ + \int_0^t e^{-\theta P}g^+(\theta)\,d\theta\right\} \tag{18}$$

and we show that the integral

$$I = \int_0^\infty e^{-\theta P}g^+(\theta)\,d\theta \tag{19}$$

is convergent. Denote the eigenvalues of P in Q^+ by $\lambda_r, ..., \lambda_{m-1}$ and suppose that

$$\mathrm{Re}\,\lambda_r \leqq \cdots \leqq \mathrm{Re}\,\lambda_{m-1}.$$

By hypothesis $\mathrm{Re}\,\lambda_r > 0$. The eigenvalues of the operator $-P$ in Q^+ are $-\lambda_r, ..., -\lambda_{m-1}$ and

$$\mathrm{Re}\,(-\lambda_{m-1}) \leqq \cdots \leqq \mathrm{Re}\,(-\lambda_r) < 0.$$

Therefore applying the inequality (16) and the estimate (12), we have

$$\left\|\int_0^\infty e^{-\theta P}g^+(\theta)\,d\theta\right\| \leqq \int_0^\infty \|e^{-\theta P}\|\,\|g^+(\theta)\|\,d\theta$$

$$\leqq C\int_0^\infty (1+\theta)^{m-1}\,e^{-\theta\,\mathrm{Re}\,\lambda_r}(1+\theta)^h\,d\theta < \infty.$$

The vector I obviously belongs to Q^+. We incorporate it in (18) obtaining

$$v^+(t) = e^{tP}[v_0^+ + I] - \int_t^\infty e^{(t-\theta)P}g^+(\theta)\,d\theta.$$

By virtue of (12) and (16), an estimate for the norm of this last integral in Q is

$$\left\|\int_t^\infty e^{(t-\theta)P}g^+(\theta)\,d\theta\right\| = \left\|\int_0^\infty e^{-\theta P}g^+(\theta+t)\,d\theta\right\|$$

$$\leqq C\int_0^\infty e^{-\theta\,\mathrm{Re}\,\lambda_r}(1+\theta)^{m-1}(1+\theta+t)^h\,d\theta = O(t^h).$$

It therefore increases in Q no faster than a power of t.

At the same time the expression $e^{tP}[v_0^+ + I]$ is afortiori exponentially increasing if $v_0^+ + I \neq 0$. One merely has to observe that e^{tP} operates on each basis vector e_s of P in the subspace Q^+ like the factor $e^{t\lambda_s}$.

20*

Hence it follows that a necessary condition for the existence of a solution to the system (11) with growth in Q no faster than a power of t is that

$$v_0^+ + I = v_0^+ + \int_0^\infty e^{-\theta P} g^+(\theta)\, d\theta = 0. \tag{20}$$

Whenever condition (20) holds, (11) has a solution

$$v(t) = v^-(t) + v^+(t)$$
$$= e^{tP}\left\{ v_0^- + \int_0^t e^{-\theta P} g^-(\theta)\, d\theta - \int_t^\infty e^{-\theta P} g^+(\theta)\, d\theta \right\} \tag{21}$$

increasing in Q, as we have seen, no faster than a power of t as $t \to \infty$. Thus,

Condition (20) is necessary and sufficient for the existence of a solution $v(t)$ with growth in Q no faster than a power of t as $t \to \infty$.

In particular,

The component v_0^- of the initial vector v_0 uniquely determines the solution $v(t)$ and can be chosen arbitrarily (in Q^-). The component v_0^+ is determined by $g(t)$ through (20).

Suppose now that the system (11) comes from a single m-th order equation

$$\frac{d^m v(t)}{dt^m} - \sum_{k=0}^{m-1} p_k \frac{d^k v(t)}{dt^k} = g(t) \tag{22}$$

by the customary substitution $v(t) = v_0(t), \dfrac{dv}{dt} = v_1(t), ..., \dfrac{d^{m-1}v(t)}{dt^{m-1}}$
$= v_{m-1}(t)$. In this case, we can state that the arbitrary assignment of $v(0), ..., \dfrac{d^{r-1}v(0)}{dt^{r-1}}$ will uniquely determine a solution having at most power growth at infinity.

Indeed, let $v_0 = v_0^- + v_0^+$ and $v_0^+ = (\xi_0, ..., \xi_{r-1}, ..., \xi_{m-1})$. If $v(0) = \eta_0, ..., \dfrac{d^{r-1}v(0)}{dt^{r-1}} = \eta_{r-1}$ are prescribed, then the first r components $\eta_0 - \xi_0, ..., \eta_{r-1} - \xi_{r-1}$ of the vector $v_0^- = v_0 - v_0^+$ are known. But then by Sec. 4.3.2, v_0^- is uniquely determined.

If (11) comes from taking Fourier transforms in the system of partial differential equations

$$\frac{\partial u(t, x)}{\partial t} - P\left(i\frac{\partial}{\partial x}\right) u(t, x) = f(t, x),$$ (23)

then the vector σ will appear as a parameter (in the matrix P and the right-hand side $g = g(t, \sigma)$). Condition (20) now assumes the form

$$v_0^+(\sigma) + I(\sigma) \equiv v_0^+(\sigma) + \int_0^\infty e^{-tP(\sigma)} g^+(\theta, \sigma) \, d\theta = 0.$$ (24)

As before, it is a necessary and sufficient condition for the system (11) to have a proper solution for each value of σ.

Considered for all σ, the condition (24) is also necessary and sufficient for the existence of a solution $v(t, \sigma)$ to (23) having for each σ power growth in t at the most. But as a function of σ, the resulting solution (21) does not belong to H apriori. In what follows, we shall show that for the system (23) corresponding to the single equation

$$\frac{\partial^m u(t, x)}{\partial t^m} - \sum_{k=0}^{m-1} P_k\left(i\frac{\partial}{\partial x}\right) \frac{\partial^k u(t, x)}{\partial t^k} = f(t, x),$$ (25)

the fulfillment of (24) almost everywhere in σ is a necessary and sufficient condition for a problem with initial data $v_0(\sigma)$ to be well-posed.

4.8.3. To prove this fact, we need to analyze more thoroughly the solution $v(t)$ of the system (11) of ordinary differential equations corresponding to the case of a single equation and particularly to derive an expression for $v(t)$ in terms of the data $v_0, ..., v_{r-1}$.

Thus consider again a single m-th order equation (22). It is equivalent to the system

$$\left.\begin{array}{l} \dfrac{dv_0}{dt} - v_1 = 0, \\[2ex] \dfrac{dv_1}{dt} - v_2 = 0, \\[2ex] \cdots \cdots \cdots \cdots \cdots \cdots \cdots \cdots \cdots \\[2ex] \dfrac{dv_{m-1}}{dt} - p_0 v_0 - p_1 v_1 - p_2 v_2 - \cdots - p_{m-1} v_{m-1} = g(t) \end{array}\right\}$$ (26)

The coefficient matrix

$$P = \begin{Vmatrix} 0 & 1 & \cdots & 0 \\ 0 & 0 & \cdots & 0 \\ \cdot & \cdot & \cdot & \cdot \\ 0 & 0 & \cdots & 1 \\ p_0 & p_1 & \cdots & p_{m-1} \end{Vmatrix}$$

and has the eigenvectors

$$e_j = (1, \lambda_j, \lambda_j^2, \ldots, \lambda_j^{m-1}) \quad (j = 0, 1, \ldots, m-1)$$

(see Sec. 4.3.2), where $\lambda_0, \ldots, \lambda_{m-1}$ are the roots of

$$\lambda^m - \sum_{k=0}^{m-1} p_k \lambda^k = 0.$$

As we already know, the solution $v(t)$ can be uniquely expressed just in terms of $v_0 = v(0), \ldots, v_{r-1} = \dfrac{d^{r-1}v(0)}{dt^{r-1}}$. We wish to find an explicit expression for $v(t)$. For simplicity, we shall assume that $\lambda_0, \ldots, \lambda_{m-1}$ are distinct.

A vector $\xi = (\xi_0, \ldots, \xi_{m-1}) \in Q$ can be represented in terms of the basis vectors $e_0, e_1, \ldots, e_{m-1}$ by

$$\xi = \xi^0 e_0 + \xi^1 e_1 + \cdots + \xi^{m-1} e_{m-1}. \tag{27}$$

Written out in coordinate form, (27) is equivalent to the system

$$\left. \begin{aligned} \xi_0 &= \xi_0 & + \xi^1 & + \cdots + \xi^{m-1} \\ \xi_1 &= \xi^0 \lambda_0 & + \xi^1 \lambda_1 & + \cdots + \xi^{m-1} \lambda_{m-1} \\ & \cdots \cdots \cdots \cdots \cdots \cdots \cdots \cdots \cdots \\ \xi_{m-1} &= \xi^0 \lambda_0^{m-1} + \xi^1 \lambda_1^{m-1} + \cdots + \xi^{m-1} \lambda_{m-1}^{m-1} \end{aligned} \right\}. \tag{28}$$

Solving this system for the unknowns ξ^j, we find that

$$\xi^j = \frac{1}{W} \begin{vmatrix} 1 & \cdots & \xi_0 & \cdots & 1 \\ \lambda_0 & \cdots & \xi_1 & \cdots & \lambda_{m-1} \\ \cdot & \cdot & \cdot & \cdot & \cdot \\ \lambda_0^{m-1} & \cdots & \xi_{m-1} & \cdots & \lambda_{m-1}^{m-1} \end{vmatrix} \tag{29}$$

$$= \frac{(-1)^j}{W} [W_{j0}\xi_0 - W_{j1}\xi_1 + \cdots + (-1)^{m-1} W_{j,\,m-1}\xi_{m-1}].$$

Here W is the Vandermonde determinant formed from $\lambda_0, \ldots, \lambda_{m-1}$ and W_{jk} is the minor of the element in the $(j+1)$-st column and $(k+1)$-st row $(j, k = 0, 1, \ldots, m-1)$. In particular, if the vector is

$$V_0 = (v_0, v_1, \ldots, v_{m-1}) = \sum_{j=0}^{m-1} v^j e_j$$

we have

$$v^j = \frac{(-1)^j}{W} [W_{j0}v_0 - W_{j1}v_1 + \cdots + (-1)^{m-1} W_{j,m-1}v_{m-1}]. \qquad (30)$$

Similarly, for

$$G(t) = (0, 0, \ldots, g(t)) = \sum_{j=0}^{m-1} g^j(t)e_j$$

we obtain

$$g^j(t) = \frac{(-1)^{j+m-1}}{W} W_{j,m-1}g(t). \qquad (31)$$

If we resolve (18) into components with respect to the vectors $e_r, \ldots, e_{m-1}$ and use the fact that e^{tP} maps e_s into $e^{t\lambda_s}e_s$, we can express the condition (20) in the equivalent form

$$v^s + \int_0^\infty e^{-\theta\lambda_s}g^s(\theta)\, d\theta = 0 \quad (s = r, \ldots, m-1). \qquad (32)$$

Substituting (30) and (31) in (32) and simplifying, we obtain

$$\frac{W_{s0}}{W_{s,m-1}} v_0 - \frac{W_{s1}}{W_{s,m-1}} v_1 + \cdots + (-1)^{m-1} \frac{W_{s,m-1}}{W_{s,m-1}} v_{m-1}$$

$$+ (-1)^{m-1} \int_0^\infty e^{-\theta\lambda_s} g(\theta)\, d\theta = 0, \quad (s = r, \ldots, m-1). \qquad (33)$$

We may regard (33) as an equivalent expression (in the original coordinate form) for the basic condition (20) assuring that the problem in question is well-posed.

Now set

$$A_j = \int_0^t e^{-\theta\lambda_j}g(\theta)\, d\theta, \qquad B_j = \int_t^\infty e^{-\theta\lambda_j}g(\theta)\, d\theta,$$

and

$$C_j = \int_0^\infty e^{-\theta\lambda_j}g(\theta)\, d\theta = A_j + B_j.$$

Formula (32), which is equivalent to (33), determines $v^r, ..., v^{m-1}$ in terms of $C_r, ..., C_{m-1}$. Specifically,

$$v^s = - \int_0^\infty e^{-\theta \lambda_s} g^s(\theta) \, d\theta = \frac{(-1)^{s+m}}{W} W_{s, m-1} C_s \quad (s \geqq r). \qquad (34)$$

Let us find $v^0, ..., v^{r-1}$. To this end, we use the first r equations of the system (28) with $\xi = V_0$:

$$v_0 = v^0 \qquad + v^1 \qquad + \cdots + v^{m-1},$$
$$v_1 = v^0 \lambda^0 \qquad + v^1 \lambda_1 \qquad + \cdots + v^{m-1} \lambda_{m-1},$$
$$\cdots \cdots \cdots \cdots \cdots \cdots \cdots \cdots$$
$$v_{r-1} = v^0 \lambda_0^{r-1} + v^1 \lambda_1^{r-1} + \cdots + v^{m-1} \lambda_{m-1}^{m-1}.$$

For simplicity, we take $v_0 = v_1 = \cdots = v_{r-1} = 0$. Transposing the unknowns to the left-hand side and the knowns to the right-hand side, we have

$$v^0 \qquad + v^1 \qquad + \cdots + v^{r-1} \qquad = - \sum_{s \geqq r} v^s,$$
$$v^0 \lambda_0 \qquad + v^1 \lambda_1 \qquad + \cdots + v^{r-1} \lambda_{r-1} = - \sum_{s \geqq r} v^s \lambda_s,$$
$$\cdots \cdots \cdots \cdots \cdots \cdots \cdots \cdots \cdots$$
$$v^0 \lambda_0^{r-1} + v^1 \lambda_1^{r-1} + \cdots + v^{r-1} \lambda_{r-1}^{r-1} = - \sum_{s \geqq r} v^s \lambda_s^{r-1}.$$

Using Cramer's rule to solve this system, we obtain

$$v^j = - \frac{1}{W_r} \sum_{s \geqq r} W_{r(s, j)} v^s,$$

where W_r is the Vandermonde determinant formed from $\lambda_0, ..., \lambda_{r-1}$ and $W_{r(s, j)}$ is obtained from W_r by replacing the column $1, \lambda_j, ..., \lambda_j^{r-1}$ by $1, \lambda_s, ..., \lambda_s^{r-1}$. Thus $W_{r(s, j)}$ is the Vandermonde determinant formed from $\lambda_0, ..., \lambda_{j-1}, \lambda_s, \lambda_{j+1}, ..., \lambda_{r-1}$. Substituting the known values (34) of $v^r, ..., v^{m-1}$ in this last equation, we find that

$$v^j = \sum_{s \geqq r} (-1)^{s+m-1} \frac{W_{s, m-1}}{W} \frac{W_{r(s, j)}}{W_r} C_s. \qquad (35)$$

The coefficient of C_s can be converted into a simpler form. Using the value of the Vandermonde determinant, we can write

$$\frac{W_{s,m-1}}{W}\cdot\frac{W_{r(s,j)}}{W_r}=\frac{\prod_{j<k}^{m-1}{}^{(s)}(\lambda_k-\lambda_j)\prod_{i<l}^{r-1}{}^{(s,j)}(\lambda_l-\lambda_i)}{\prod_{j<k}^{m-1}(\lambda_k-\lambda_j)\prod_{i<l}^{r-1}(\lambda_l-\lambda_i)}.$$

The first product in the numerator does not involve λ_s while the second product has λ_j replaced by λ_s. Canceling, we obtain

$$\frac{W_{s,m-1}}{W}\cdot\frac{W_{r(s,j)}}{W_r}=\frac{1}{(\lambda_s-\lambda_0)\dots(\lambda_s-\lambda_{s-1})(\lambda_{s+1}-\lambda_s)\dots(\lambda_{m-1}-\lambda_s)}$$

$$\times\frac{(\lambda_s-\lambda_0)\dots(\lambda_s-\lambda_{j-1})(\lambda_{j+1}-\lambda_s)\dots(\lambda_{r-1}-\lambda_s)}{(\lambda_j-\lambda_0)\dots(\lambda_j-\lambda_{j-1})(\lambda_{j+1}-\lambda_j)\dots(\lambda_{r-1}-\lambda_j)}$$

$$=\frac{(-1)^{m-s-1}}{\prod_{k=r}^{m-1}{}'(\lambda_s-\lambda_k)\prod_{i=0}^{r-1}{}'(\lambda_j-\lambda_i)(\lambda_s-\lambda_j)}.$$

Thus

$$v^j=\sum_{s=r}^{m-1}\frac{C_s}{\prod_{k=r}^{m-1}{}'(\lambda_s-\lambda_k)\prod_{i=0}^{r-1}(\lambda_j-\lambda_i)(\lambda_s-\lambda_j)}. \tag{36}$$

Consider again the solution (21) of our problem under condition (20). We may write it in the form

$$V(t)=(v_0(t),\dots,v_{m-1}(t))=\sum_{j=0}^{m-1}v^j(t)e_j$$

$$=\sum_{j=0}^{r-1}\left[v^j+\int_0^t e^{-\theta\lambda_j}g^j(\theta)\,d\theta\right]e^{t\lambda_j}e_j-\sum_{s=r}^{m-1}\left[\int_t^\infty e^{-\theta\lambda_s}g^s(\theta)\,d\theta\right]e^{t\lambda_s}e_s. \tag{37}$$

What interests us is the first component of $V(t)$,

$$v(t)=v_0(t)=\sum_{j=0}^{m-1}v^j(t)$$

$$=\sum_{j=0}^{r-1}\left[v^j+\int_0^t e^{-\lambda\theta_j}g^j(\theta)\,d\theta\right]e^{t\lambda_j}-\sum_{s=r}^{m-1}\left[\int_t^\infty e^{-\theta\lambda_s}g^s(\theta)\,d\theta\right]e^{t\lambda_s}. \tag{38}$$

Substituting the computed values of v^j from (36) and the known values of g^j and g^s from (31), we find that

$$
v(t) = \sum_{j=0}^{r-1} e^{t\lambda_j} \left[\sum_{s=r}^{m-1} \frac{C_s}{\prod\limits_{k=r}^{m-1}{}' (\lambda_s - \lambda_k) \prod\limits_{i=0}^{r-1}{}' (\lambda_j - \lambda_i)(\lambda_s - \lambda_j)} + (-1)^{j+m-1} \frac{W_{j,m-1} A_j}{W} \right]
$$

$$
- \sum_{s=r}^{m-1} (-1)^{s+m-1} \frac{e^{t\lambda_s} W_{s,m-1} B_s}{W}
$$

$$
= \sum_{j=0}^{r-1} e^{t\lambda_j} \left[\sum_{s=r}^{m-1} \frac{A_s}{\prod\limits_{k=r}^{m-1}{}' (\lambda_s - \lambda_k) \prod\limits_{i=0}^{r-1}{}' (\lambda_j - \lambda_i)(\lambda_s - \lambda_j)} + (-1)^{j+m-1} \frac{W_{j,m-1} A_j}{W} \right]
$$

$$
+ \sum_{s=r}^{m-1} B_s \left[\sum_{j=0}^{r-1} \frac{e^{t\lambda_j}}{\prod\limits_{k=r}^{m-1}{}' (\lambda_s - \lambda_k) \prod\limits_{i=0}^{r-1}{}' (\lambda_j - \lambda_i)(\lambda_s - \lambda_j)} + (-1)^{s+m} \frac{W_{s,m-1} e^{t\lambda_s}}{W} \right]
$$

$$
= \int_0^t \left\{ \sum_{j=0}^{r-1} e^{t\lambda_j} \left[\sum_{s=r}^{m-1} \frac{e^{-\theta\lambda_s}}{\prod\limits_{k=r}^{m-1}{}' (\lambda_s - \lambda_k) \prod\limits_{i=0}^{r-1}{}' (\lambda_j - \lambda_i)(\lambda_s - \lambda_j)} + (-1)^{j+m-1} \frac{W_{j,m-1} e^{-\theta\lambda_j}}{W} \right] \right\}
$$

$$
\times g(\theta)\, d\theta
$$

$$
+ \int_t^\infty \left\{ \sum_{s=r}^{m-1} e^{-\theta\lambda_s} \left[\sum_{j=0}^{r-1} \frac{e^{t\lambda_j}}{\prod\limits_{k=r}^{m-1}{}' (\lambda_s - \lambda_k) \prod\limits_{i=0}^{r-1}{}' (\lambda_j - \lambda_i)(\lambda_s - \lambda_j)} \right. \right.
$$

$$
\left. \left. + (-1)^{s+m} \frac{W_{s,m-1} e^{t\lambda_s}}{W} \right] \right\} g(\theta)\, d\theta. \tag{39}
$$

The braces in the first integral can be reduced to the form

$$
\sum_{j=0}^{r-1} \frac{1}{\prod\limits_{i=0}^{r-1}{}' (\lambda_j - \lambda_i)} \left[\sum_{s=r}^{m-1} \frac{e^{t\lambda_j - \theta\lambda_s}}{\prod\limits_{k=r}^{m-1}{}' (\lambda_s - \lambda_k)(\lambda_s - \lambda_j)} + \frac{e^{t\lambda_j - \theta\lambda_j}}{\prod\limits_{k=r}^{m-1} (\lambda_j - \lambda_k)} \right]
$$

$$
= \sum_{j=0}^{r-1} \frac{1}{\prod\limits_{i=0}^{r-1}{}' (\lambda_j - \lambda_i)} \sum_{s=j,r,\ldots,m-1} \frac{e^{t\lambda_j - \theta\lambda_s}}{\prod\limits_{k=j,r,\cdots,m-1}{}' (\lambda_s - \lambda_k)}. \tag{40}
$$

Similarly, the braces in the second integral can be converted to

$$\sum_{s=r}^{m-1} \frac{1}{\prod\limits_{k=r}^{m-1}{}'(\lambda_s - \lambda_k)} \left[\sum_{j=0}^{r-1} \frac{e^{t\lambda_j - \theta\lambda_s}}{\prod\limits_{i=0}^{r-1}{}'(\lambda_j - \lambda_i)(\lambda_s - \lambda_j)} - \frac{e^{t\lambda_s - \theta\lambda_s}}{\prod\limits_{i=0}^{r-1}(\lambda_s - \lambda_i)} \right]$$

$$= - \sum_{s=r}^{m-1} \frac{1}{\prod\limits_{k=r}^{m-1}{}'(\lambda_s - \lambda_k)} \sum_{j=0,\ldots,r-1,s} \frac{e^{t\lambda_j - \theta\lambda_s}}{\prod\limits_{i=0,\ldots,r-1,s}{}'(\lambda_j - \lambda_i)}. \qquad (41)$$

The expression (40) is the compound divided difference (Sec. 4.2) for the function $F(z, \zeta) = e^{tz - \theta\zeta}$ of the form $F(\lambda_0, \ldots, \lambda_{r-1}; \lambda_r, \ldots, \lambda_{m-1})$. Therefore by (26) of Sec. 4.2, we have

$$|F(\lambda_0, \ldots, \lambda_{r-1}; \lambda_r, \ldots, \lambda_{m-1})| \leq Ct^{m-1} |e^{tz_0 - \theta\zeta_0}|, \qquad (42)$$

with z_0 a point of the convex polygon spanned by $\lambda_0, \ldots, \lambda_{r-1}$ and ζ_0 a point of the convex polygon spanned by $z_0, \lambda_r, \ldots, \lambda_{m-1}$. Since all $\operatorname{Re} \lambda_j \leq 0$ for $j \leq r - 1$, $\operatorname{Re} z_0 \leq 0$ and since $\operatorname{Re} \lambda_s > 0$ for $s \geq r$, $\operatorname{Re} \zeta_0 \geq \operatorname{Re} z_0$. Therefore for $0 \leq \theta \leq t$ we can write

$$\operatorname{Re}(t\zeta_0 - \theta z_0) = t \operatorname{Re} z_0 - \theta \operatorname{Re} \zeta_0$$

$$= (t - \theta) \operatorname{Re} z_0 + \theta (\operatorname{Re} z_0 - \operatorname{Re} \zeta_0) \leq 0.$$

Finally, we obtain the estimate

$$\left| \sum_{j=0}^{r-1} \frac{1}{\prod\limits_{i=0}^{r-1}{}'(\lambda_j - \lambda_i)} \sum_{s=j,r,\ldots,m-1} \frac{e^{t\lambda_j - \theta\lambda_s}}{\prod\limits_{k=j,r,\ldots,m-1}{}'(\lambda_s - \lambda_k)} \right| \leq Ct^{m-1} \qquad (43)$$

for our compound divided difference (40).

The expression (41) is the compound divided difference of $F(z, \zeta) = e^{t\zeta - \theta z}$ of the form $F(\lambda_r, \ldots, \lambda_{m-1}; \lambda_0, \ldots, \lambda_{r-1})$. Therefore again by the estimate (26) of Sec. 4.2,

$$\left| F(\lambda_r, \ldots, \lambda_{m-1}; \lambda_0, \ldots, \lambda_{r-1}) \right| \leq C\theta^{m-1} |e^{t\zeta_0 - \theta z_0}|, \qquad (44)$$

where z_0 is a point of the convex polygon spanned by $\lambda_r, \ldots, \lambda_{m-1}$ and ζ_0 is a point of the convex polygon spanned by $\lambda_0, \ldots, \lambda_{r-1}, z_0$.

Since $\operatorname{Re} \lambda_s > 0$ for $s \geq r$, we have $\operatorname{Re} z_0 \geq \operatorname{Re} \lambda_r$ and since $\operatorname{Re} \lambda_j \leq 0$ for $j < r$, $\operatorname{Re} \zeta_0 \leq \operatorname{Re} z_0$. Therefore when $0 \leq t \leq \theta$,

$$\operatorname{Re}(tz_0 - \theta\zeta_0) = t \operatorname{Re} \zeta_0 - \theta \operatorname{Re} z_0$$

$$= (t - \theta) \operatorname{Re} z_0 + t (\operatorname{Re} \zeta_0 - \operatorname{Re} z_0) \leq (t - \theta) \operatorname{Re} \lambda_r.$$

Thus the second braces in (39) does not exceed

$$C\theta^{m-1}e^{(t-\theta)\operatorname{Re}\lambda_r}$$

in magnitude. As a result we obtain the inequality

$$|v(t)| \leq Ct^{m-1}\int_0^t |g(\theta)|\,d\theta + C\int_t^\infty \theta^{m-1}e^{(t-\theta)\operatorname{Re}\lambda_r}|g(\theta)|\,d\theta$$

$$= Ct^{m-1}\int_0^t |g(\theta)|\,d\theta + C\int_0^\infty (\theta+t)^{m-1}e^{-\theta\operatorname{Re}\lambda_r}|g(\theta+t)|\,d\theta \qquad (45)$$

with constants depending just on r and m.

4.8.4. We are now in a position to consider the equation

$$\frac{\partial^m u(t,x)}{\partial t^m} - \sum_{k=0}^{m-1} p_k\left(i\frac{\partial}{\partial x}\right)\frac{\partial^k u(t,x)}{\partial t^k} = f(t,x). \qquad (46)$$

The solution $v(t,\sigma)$ of the dual equation

$$\frac{d^m v(t,\sigma)}{dt^m} - \sum_{k=0}^{m-1} p_k(\sigma)\frac{d^k v(t,\sigma)}{dt^k} = g(t,\sigma) \qquad (47)$$

under the conditions

$$v(0,\sigma) = \cdots = \frac{d^{r-1}v(0,\sigma)}{dt^{r-1}} = 0$$

can be estimated by means of (45). We merely have to introduce the parameter σ appropriately:

$$|v(t,\sigma)| \leq Ct^{m-1}\int_0^t |g(\theta,\sigma)|\,d\theta$$

$$+ C\int_0^\infty (\theta+t)^{m-1}e^{-\theta\operatorname{Re}\lambda_r(\sigma)}|g(\theta+t,\sigma)|\,d\theta. \qquad (48)$$

The fact that $g(t,\sigma)$ *belongs to H and has at most power growth in t* is expressible by the inequality

$$|g(t,\sigma)| \leq c_0(1+t)^h(1+\sigma^2)^q\,\psi(\sigma), \quad \psi(\sigma)\in L_2. \qquad (49)$$

Inserting (49) in (48), we obtain

$$|v(t,\sigma)| \leq c_1 t^{m-1}(1+\sigma^2)^q\,\psi(\sigma)(1+t)^{h+1}$$

$$+ \sum_{k=0}^{m-1} C_k t^{m-k-1}\int_0^\infty \theta^k e^{-\theta\operatorname{Re}\lambda_r(\sigma)}|g(\theta+t,\sigma)|\,d\theta.$$

By virtue of (10), $v(t, \sigma)$ is an element of H for each $t \geq 0$ and has growth there no faster than a power of t when $t \to \infty$, as required.

Now consider the general case where the data $v_0(\sigma), \ldots, v_{r-1}(\sigma)$ are arbitrary elements of H.

By Sec. 4.4.2, the equation

$$\frac{d^m w(t, \sigma)}{dt^m} - \sum_{k=0}^{m-1} p_k(\sigma) \frac{d^k w(t, \sigma)}{dt^k} = 0$$

has a solution $w(t, \sigma)$, in H for each $t \geq 0$, having power growth in t at the most, and satisfying the initial conditions

$$w(0, \sigma) = v_0(\sigma), \ldots, \frac{d^{r-1} w(0, \sigma)}{dt^{r-1}} = v_{r-1}(\sigma).$$

Denote by $V(t, \sigma)$ the solution of (47) we constructed above satisfying the initial conditions

$$V_0(0, \sigma) = \cdots = V_{r-1}(0, \sigma) = 0.$$

The function

$$v(t, \sigma) = V(t, \sigma) + w(t, \sigma)$$

is again a solution of (47) belonging to H for each $t \geq 0$ and having there power growth in t at the most. It satisfies the initial conditions

$$v(0, \sigma) = v_0(\sigma), \ldots, \frac{d^{r-1} v(0, \sigma)}{dt^{r-1}} = v_{r-1}(\sigma).$$

Thus the required solution exists and our theorem has been completely proved.

REMARK. As in Sec. 4.4.5 the conditions of the problem may be modified. It is possible to look for a solution in the class of functions $u(t, x) \in \mathscr{H}$ which behave like exponential-power functions of the form $t^h e^{\alpha t}$ at infinity for some fixed α. The above theorem holds in its entirety for this case with one difference. The number $r = r(\sigma)$ is determined by the condition

$$\operatorname{Re} \lambda_0(\sigma) \leq \cdots \leq \operatorname{Re} \lambda_{r-1}(\sigma) \leq \alpha < \operatorname{Re} \lambda_r(\sigma) \leq \cdots \leq \operatorname{Re} \lambda_{m-1}(\sigma).$$

Suppose for instance that (46) is proper in the sense of Petrovsky. This means that the real parts of all eigenvalues are uniformly bounded, say by the constant α. Owing to our theorem, the Cauchy problem for this equation is well-posed for any initial data $u_0(x), \ldots, u_{m-1}(x)$ in the following sense. A solution exists in $\mathcal{H}$ with growth no faster than $t^h e^{\alpha t}$ for $t \to \infty$. It is unique in this class and depends continuously on the initial data.

4.9.   Mixed problems

4.9.1. In this section we shall examine certain well-posed mixed problems for the equation

$$\frac{\partial^m u(t, x)}{\partial t^m} = \sum_{k=0}^{m-1} p_k \left(i \frac{\partial}{\partial x} \right) \frac{\partial^k u(t, x)}{\partial t^k} \tag{1}$$

in the region $t \geq 0$, $x \geq 0$.

A problem that we wish to show is well-posed is an extension to generalized functions of the following classical problem: Find a solution to (1) ratisfying the given boundary conditions

$$w_0(t) = u(t, 0), \; w_1(t) = \frac{\partial u(t, 0)}{\partial x}, \ldots, \; w_{p-1}(t) = \frac{\partial^{p-1} u(t, 0)}{\partial x^{p-1}} \tag{2}$$

and given initial conditions

$$u_0(x) = u(0, x), \; u_1(x) = \frac{\partial u(0, x)}{\partial t}, \ldots, \; u_{r-1}(x) = \frac{\partial^{r-1} u(0, x)}{\partial t^{r-1}}. \tag{3}$$

The number p is the order of (1) relative to x. The significance of r will be indicated below.

In Sec. 1.5.6 of Chapt. 1, we considered initial value problems for ordinary differential equations involving generalized functions. Let us recall the result deduced there. With each ordinary differential equation

$$\sum_{j=0}^{p} a_j y^{(j)}(x) = f(x) \quad (x \geq 0) \tag{4}$$

(in ordinary functions) and set of initial conditions

$$y(0) = y_0, \ldots, \; y^{(p-1)}(0) = y_{p-1} \tag{5}$$

there is associated an equation in generalized functions

$$\sum_{j=0}^{p} a_j [y^{(j)}(x) - \sum_{i=0}^{j-1} y_i \delta^{(j-1-i)}(x)] = F(x), \tag{6}$$

where $F(x)$ equals $f(x)$ for $x > 0$ and 0 for $x < 0$.

Every solution of (4) satisfying the conditions (5) when regarded as a generalized function equaling 0 for $x < 0$ is a solution of (6). Conversely, every solution of (6) equaling 0 for $x < 0$ is an ordinary function solving (4) for $x > 0$ and satisfying the conditions (5).

We can handle equation (1) with its additional t-derivatives in an analogous fashion. We first rewrite (1) in the form

$$\frac{\partial^m u(t,x)}{\partial t^m} = \sum_{k=0}^{m-1} p^k \left(i \frac{\partial}{dx} \right) \frac{\partial^k u}{\partial t^k} = \sum_{j=0}^{p} Q_j \left(\frac{\partial}{\partial t} \right) \frac{\partial^j u(t,x)}{\partial x^j} \tag{7}$$

and then replace each ordinary x-derivative by an x-derivative in K' plus a correction term consisting of δ-functions and their derivatives as in (6).

As a result we obtain the equation

$$\frac{\partial^m u(t,x)}{\partial t^m} = \sum_{j=0}^{p} Q_j \left(\frac{\partial}{\partial t} \right) \left[\frac{\partial^j u(t,x)}{\partial x^j} - \sum_{i=0}^{j-1} w_i(t)\delta^{(j-i-1)}(x) \right]. \tag{8}$$

The solution of this equation equaling 0 for $x < 0$ is what we shall define to be the solution of problem (1), (2).

For an ordinary differential equation involving an ordinary function, we established that every solution of (6) is actually an ordinary function (having some particular jump at $x = 0$). The partial differential equation (8) generally speaking has a solution of a more complicated nature. Nevertheless the passage from (1), (2) to (8) is a natural one in the sense that if a solution $u(t, x)$ of (8) vanishing for $x < 0$ is an ordinary function having a sufficient number of continuous derivatives for $x \geq 0$, then it will also be a solution of problem (1), (2) in the usual sense and conversely.

The proof will be carried out under slightly more general circumstances. Instead of equation (1), we consider a system of first order equations in x

$$A \left(\frac{\partial}{\partial t} \right) u(t, x) = B \left(\frac{\partial}{\partial t} \right) \frac{\partial u(t, x)}{\partial x}, \tag{9}$$

and instead of (8) a system of the form

$$A \left(\frac{\partial}{\partial t} \right) u(t, x) = B \left(\frac{\partial}{\partial t} \right) \left[\frac{\partial u(t, x)}{\partial x} - u(t,0)\delta(x) \right]. \tag{10}$$

Suppose that $u(t, x)$ is a generalized vector function equaling the solution $u(t, x)$ of (9) for $x > 0$ and 0 for $x < 0$. Denote by $u_1(t, x)$ the generalized vector function equaling $\dfrac{\partial u(t, x)}{\partial x}$ for $x > 0$ and 0 for $x < 0$. Equation (9) can then be written as

$$A\left(\frac{\partial}{\partial t}\right)u(t, x) = B\left(\frac{\partial}{\partial t}\right)u_1(t, x). \tag{11}$$

The function $u_1(t, x)$ is not the derivative of $u(t, x)$ (in K'); just as in Sec. 1.5 of Chapt. 1, we have

$$\frac{\partial u(t, x)}{\partial x} = u_1(t, x) + u(t, 0)\,\delta(x). \tag{12}$$

Substituting $u_1(t, x)$ as given by (12) in (11), we obtain (10) as required. By reversing the order of our argument, we can prove the converse statement.

We now wish to make our assumptions about the nature of the solution $u(t, x)$ more precise.

Let $\mathscr{H}^\beta_+$ be the space of functions $u(x)$ which when multiplied by $e^{-\beta x}$ are square integrable for $x \geq 0$ plus all their derivatives (in K'). A solution $u(t, x)$ is assumed to be in $\mathscr{H}^\beta_+$ for each $t \geq 0$. β depends on the equation (1) and its value will be specified below. Denote by H^β_+ the collection of Fourier transforms of all functions in $\mathscr{H}^\beta_+$. In Sec. 4.9.3, the class H^β_+ will be shown to consist of (ordinary) functions $v(s)$ which are analytic in the half-plane Im $s > \beta$ and which after division by some power of s are square integrable along each line Im $s = \tau$. Moreover, for each $v(s)$ the integral

$$\|v(s)\|^2_q = \int\limits_{-\infty}^{\infty} \frac{|v(s)|^2}{|s|^{2q}}\,dv$$

is uniformly bounded in τ. We point out that this class possesses the following important monotonicity property. If $v_0(s) \in H^\beta_+$ and $|v_1(s)| \leq C|v_0(s)|$, where $v_1(s)$ is analytic, then $v_1(s) \in H^\beta_+$ also. Therefore the class H^β_+ admits multiplication by analytic functions having power growth for $|s| \to \infty$.

As already stated, the required solution $u(t, x)$ is to be an element of $\mathscr{H}^\beta_+$ depending on t as a parameter.

At $t = 0$, the generalized function $u(t, x)$ and its t-derivatives up to order $r - 1$ are to take on the prescribed values $u_0(x), \ldots, u_{r-1}(x)$ likewise generalized functions.

We now specify the admissible behavior of $u(t, x)$ for $t \to \infty$. We shall assume that $u(t, x)$ has at most power growth in $\mathscr{H}_+^{\beta}$, i.e.,

$$\|u(t, x)\|_{\beta,q}^2 \equiv \int_{-\infty}^{\infty} \frac{|u(t, s)|^2}{|s|^{2q}} \, d\sigma \leq Ct^h \tag{13}$$

for some q and h. The number β and the number r of assignable initial conditions are determined by the following considerations. Let

$$\lambda^m = \sum_{k=0}^{m-1} P_k(s) \lambda^k \tag{14}$$

be the equation resulting from (1) by the replacement of $\frac{\partial}{\partial t}$ by λ and $i\frac{\partial}{\partial x}$ by s. For each complex s, equation (14) has m complex roots $\lambda_0(s), \ldots, \lambda_{m-1}(s)$. These roots are analytic functions of s everywhere except at a finite number points (the branch points of the algebraic function defined by (14)). We shall consider these roots to be in the half-plane $\operatorname{Im} s > \beta$ where β is so chosen that all singularities of the roots lie in the half-plane $\operatorname{Im} s < \beta$. We shall also assume that none of the roots $\lambda_0(s), \ldots, \lambda_{m-1}(s)$ coincide identically. Thus β may be selected so that all of the roots are everywhere distinct from each other in the half-plane $\operatorname{Im} s > \beta$. By the same token, $\lambda_0(s), \ldots, \lambda_{m-1}(s)$ are single-valued analytic functions in this domain. We now separate all the roots into two groups by the following rule. The first contains the roots $\lambda_0(s), \ldots, \lambda_{r-1}(s)$ which have a non-positive real part everywhere in $\operatorname{Im} s > \beta$ and the second the roots $\lambda_r(s), \ldots, \lambda_{m-1}(s)$ which have a positive real part for at least one point in $\operatorname{Im} s > \beta$. This then determines r.

We assume in addition that the functions

$$w_0(t) = u(t, 0), \quad w_1(t) = \frac{\partial u(t, 0)}{\partial x}, \ldots, w_{p-1} = \frac{\partial^{p-1} u(t, 0)}{\partial x^{p-1}}$$

are continuous, are $m - 1$ times differentiable in t (i.e., as many times as formula (8) requires), and increase together with their corresponding

20a

derivatives no faster than Ct^h as $t \to \infty$ for some h. We form the function

$$g(t, s) = \sum_{j=1}^{p} Q_j(\partial/\partial t) \, (w_{j-1}(t)$$

$$- isw_{j-2}(t) + \cdots + (-is)^{j-1}w_0(t)). \tag{15}$$

Owing to the conditions imposed on the functions $w_j(t)$, the integrals

$$\Phi_{k,q}(t, s) = \int_0^\infty e^{-\theta\lambda_q(s)}\theta^k \, |g(t + \theta, s)| \, d\theta \quad (k \leqq m - 2) \tag{16}$$

exist for each s such that Im $s > \beta$, where Re $\lambda_q(s) > 0$. We shall assume that every analytic function $v(t, s)$ satisfying the inequality

$$|v(t, s)| \leqq \Phi_{k,q}(t, s) \tag{17}$$

for Im $s > \beta$ (for at least one q and one k) belongs to the class H_+^β and increases there no faster than a power of t. This condition is in the final analysis a condition on $w_0(t), \ldots, w_{p-1}(t)$. It is satisfied for instance whenever the integrals

$$\int_0^\infty \theta^k|w_j^{(i)}(\theta)| \, d\theta \quad (k \leqq m - 2, \ i \leqq m - 1) \tag{18}$$

all exist.

The following theorem will be proved:

THEOREM 1. *Let* (1) *be given together with the conditions* (2) *and* (3) *with* β *and* r *determined as in the above. Let the prescribed functions* $w_0(t), \ldots, w_{p-1}(t)$ *satisfy the fore-mentioned conditions. Then a necessary and sufficient condition for the existence (and uniqueness) of a solution to the problem* (1)–(3) *in* $\mathscr{H}_+^\beta$ *is that the functions*

$$G_q(s) = \int_0^\infty e^{-\theta\lambda_q(s)}g(\theta, s) \, d\theta \quad (q = r, \ldots, m - 1), \tag{19}$$

initially defined only in the domain Re $\lambda_q(s) > 0$, Im $s > \beta$, *admit analytic continuation to the entire half-plane* Im $s > \beta$ *as elements of* H_+^β.

4.9.2. Let us first consider a few examples.

a) Suppose that each of the eigenvalues $\lambda_0(s), \ldots, \lambda_{m-1}(s)$ has a non-positive real part in the half-plane Im $s > \beta$. Then $r = m$ and the well-posed problem entails assigning all of the functions $u_0(x), \ldots, u_{m-1}(x)$, $w_0(t), \ldots, w_{p-1}(t)$. The simplest equation evincing this possibility is

$$\frac{\partial u}{\partial t} + \frac{\partial u}{\partial x} = 0. \tag{20}$$

It has a general solution $u = f(t - x)$ whose level curves are lines parallel to the bisector of the first quadrant. Clearly, the arbitrary assignment of $u(0, x)$ and $u(t, 0)$ results in a well-posed problem. In view of our theory, $\lambda = is$ and the regions $\operatorname{Re} \lambda < 0$ and $\operatorname{Im} s > 0$ coincide. Thus $\lambda = is$ belongs to the first group of roots. Hence both $u_0(x) = u(0, x)$ and $w_0(t) = u(t, 0)$ figure in the well-posed problem for (20).

Slightly more general is a homogeneous equation

$$\sum_{k=0}^{m} a_k \frac{\partial^m u(t, x)}{\partial t^k \partial x^{m-k}} = 0 \tag{21}$$

for which the roots of

$$\sum_{k=0}^{m} a_k \lambda^k = 0 \tag{22}$$

are real and non-negative. Finally, we may add a group of lower order terms to the left-hand side of (21).

b) Suppose that the eigenvalues $\lambda_0(s), \ldots, \lambda_{r-1}(s)$ each have a non-positive real part in the half-plane $\operatorname{Im} s > \beta$ while the remaining ones $\lambda_r, \ldots, \lambda_{m-1}$ have a positive real part everywhere in this domain. Then the condition that the integrals (19) be analytically continuable drops out and the well-posed problem amounts to assigning $u_0(x), \ldots, u_{r-1}(x)$, $w_0(t), \ldots, w_{m-1}(t)$. As an example consider the equation

$$\frac{\partial u}{\partial t} - \frac{\partial u}{\partial x} = 0$$

with a general solution $u = f(t + x)$ whose level curves are parallel to the line $x + t = 0$. Clearly, giving just $w_0(t)$ is sufficient to determine the solution. In view of our theory, $\lambda = -is$ and the regions $\operatorname{Re} \lambda < 0$ and $\operatorname{Im} s < 0$ coincide so that $\lambda = -is$ belongs to the second group of roots. Hence only $w_0(t)$ should figure in the well-posed problem.

A second example is the wave equation

$$\frac{\partial^2 u}{\partial t^2} - \frac{\partial^2 u}{\partial x^2} = 0.$$

The eigenvalues satisfy the equation $\lambda^2 = -s^2$ and so $\lambda_{0,1} = \pm is$. One of these has a negative real part in $\operatorname{Im} s > 0$ and the other a positive real part. In view of our general theory, the well-posed problem consists in the assignment of $w_0(t)$, $w_1(t)$, and $u_0(x)$. This is the classical mixed initial-boundary value problem for the wave equation.

In general, if each of the roots $\lambda_j(s)$ has a fixed sign in $\operatorname{Im} s \geq \beta$, then equation (1) is called proper in the sense of Petrovsky relative to $\dfrac{\partial}{\partial x}$.

Some of the branches of the algebraic function $\lambda = \lambda(s)$ map the half-plane $\operatorname{Im} s \geq \beta$ into the half-plane $\operatorname{Re} \lambda \geq 0$ and others map it into the half-plane $\operatorname{Re} \lambda \leq 0$. But this means that all the branches of the inverse function $s = s(\lambda)$ map the line $\operatorname{Re} \lambda = 0$ into the half-plane $\operatorname{Im} s \leq \beta$. If we replace $\dfrac{\partial}{\partial x}$ by $\dfrac{\partial}{\partial t'}$, and $\dfrac{\partial}{\partial t}$ by $\dfrac{\partial}{\partial x'}$, in (1) and s and λ by $-i\lambda'$ and is', respectively, we obtain the more customary condition for properness (cf. Sec. 4.4): The function $\lambda' = \lambda'(s)$ takes on values for $\operatorname{Im} s' = 0$ such that the real parts of $\lambda'(\sigma)$ are uniformly bounded from above (by β). Equations of this type were considered by Sobolev. Included among them are the parabolic and hyperbolic equations, one such being the homogeneous equation (21) for which all roots of (22) are real.

c) Somewhat more complicated in nature are the equations for which the real parts of $\lambda_r(s)$, ..., $\lambda_{m-1}(s)$ do not have a fixed sign in the half-plane $\operatorname{Im} s \geq \beta$. Pertinent here is the generalized Cauchy-Riemann equation

$$\frac{\partial u}{\partial t} = ai \frac{\partial u}{\partial x}, \quad \operatorname{Im} a = 0.$$

Any solution of this equation for $a = 1$ is obviously an analytic function of the argument $t + ix$. To assign $w_0(t)$ arbitrarily so that it is analytically continuable into the quarter-plane $t \geq 0$, $x \geq 0$ is now far from being possible. In view of our theory, we have $\lambda = s$ and the region $\operatorname{Re} \lambda > 0$ is the half-plane $\operatorname{Re} s > 0$. The well-posed problem is to assign a function $w_0(t)$ such that the integral

$$\int_{0_i}^{\infty} e^{-ts} w_0(t) \, dt$$

is analytically continuable from the half-plane Re $s > 0$ (where it is defined a priori) to the domain Im $s > 0$ as an element of the class H_+^β.

When $a = -1$, we have

$$\frac{\partial u}{\partial t} = -i \frac{\partial u}{\partial x},$$

which is again an equation with analytic solutions. The well-posed problem consists in assigning a function $w_0(t)$ for which

$$\int_{0}^{\infty} e^{ts} w_0(t) \, dt$$

is analytically continuable from the half-plane Re $s < 0$ (where it is defined a priori) to Im $s > 0$ as an element of H_+^β.

Consider next Laplace's equation

$$\frac{\partial^2 u}{\partial t^2} + \frac{\partial^2 u}{\partial x^2} = 0.$$

The eigenvalues satisfy the equation $\lambda^2 = s^2$ and thus $\lambda = \pm s$. The well-posed problem is to assign $w_0(t)$ and $w_1(t)$ satisfying the following conditions:

$$\int_{0}^{\infty} e^{-ts}[w_1(t) - isw_0(t)] \, dt \quad \text{and} \quad \int_{0}^{\infty} e^{ts}[w_1(t) - isw_0(t)] \, dt \qquad (23)$$

should be analytically continuable from their domains of definition (Re $s > 0$ in the first case and Re $s < 0$ in the second) to the half-plane Im $s > 0$ as elements of H_+^0. Since

$$\left. \begin{aligned} s \int_{0}^{\infty} e^{-ts} w_0(t) \, dt &= \int_{0}^{\infty} e^{-ts} w_0'(t) \, dt + w_0(0), \\[2mm] s \int_{0}^{\infty} e^{ts} w_0(t) \, dt &= - \int_{0}^{\infty} e^{ts} w_0'(t) \, dt - w_0(0), \end{aligned} \right\} \qquad (24)$$

we can rewrite the integrals (23) in the form

$$
\left.
\begin{aligned}
\int_0^\infty e^{-ts}[w_1(t) - iw_0'(t)]\, dt - iw_0(0), \\[2ex]
\int_0^\infty e^{ts}[w_1(t) + iw_0'(t)]\, dt + iw_0(0).
\end{aligned}
\right\}
\tag{25}
$$

In view of the preceding example, we can express the condition for a well-posed problem in the following curious form: $w_1(t) - w_0'(t)$ should be continuable to the domain $t > 0$, $x > 0$ as an analytic function and $w_1(t) + iw_0'(t)$ should be continuable to this domain as an anti-analytic function.

Similarly, for any homogeneous equation (21) with eigenvalues $\lambda_j = \mu_j s$ such that μ_j is not pure imaginary, we can deduce the following corresponding condition. Each of the integrals

$$
\int_0^\infty e^{\mu_k s \theta} g(\theta, s)\, d\theta =
$$

$$
= \int_0^\infty e^{\mu_k s \theta} \sum_{j=1}^{m-1} a_j \frac{\partial^j}{\partial \theta^j} (w_{j-1}(\theta) - is w_{j-2}(\theta) + \cdots + (-is)^{j-1} w_0(\theta))\, d\theta
\tag{26}
$$

should be analytically continuable from its domain of definition (Re $\mu_k s < 0$) to the half-plane Im $s > 0$ as an element of H_+^0. The integral (26) may be rewritten in the form

$$
\int_0^\infty e^{\mu_k s \theta} \sum_{j=1}^{m-1} a_j \frac{\partial^j}{\partial \theta^j} \left(w_{j-1}(\theta) + \cdots + \frac{1}{(-i\mu_k)^{j-1}} w_0^{(j-1)}(\theta) \right) d\theta + C
$$

and our condition expressed in the following way: The function

$$
\sum_{j=1}^{m-1} a_j \frac{\partial^j}{\partial t^j} \left(w_{j-1}(t) + \cdots + \frac{1}{(-i\mu_k)^{j-1}} w_0^{(j-1)}(t) \right)
$$

should be continuable from the half-line $t > 0$ to the quarter-plane $t > 0$, $x > 0$ as a μ_k-analytic function, i.e., as a solution of the equation

$$
\frac{\partial u}{\partial t} + i\mu_k \frac{\partial u}{\partial x} = 0.
$$

4.9.3. Denote by $\mathscr{H}_{+}^{\beta,0}$ the space of all (complex-valued) functions $u(x)$ satisfying the condition

$$\int_0^\infty |u(x)|^2\, e^{-2\beta x}\, dx < \infty. \tag{27}$$

Each $u(x) \in \mathscr{H}_{+}^{\beta,0}$ has a one-sided Fourier transform

$$v(s) = \int_0^\infty u(x)\, e^{ixs}\, dx, \tag{28}$$

which exists in the mean for all $s = \sigma + i\tau$ such that $\tau \geq \beta$. In addition, by Plancherel's theorem,

$$\int_{-\infty}^\infty |v(\sigma + i\tau)|^2\, d\sigma = \frac{1}{2\pi} \int_{-\infty}^\infty |u(x)|^2\, e^{-2\tau x}\, dx. \tag{29}$$

The transforms $v(s)$ admit the following intrinsic characterization: *Each $v(s)$ is an analytic function for $\tau \geq \beta$ such that*

$$\int_{-\infty}^\infty |v(\sigma + i\tau)|^2\, d\sigma$$

exists for each τ and is bounded throughout the region $\beta \leq \tau < \infty$ (Paley-Wiener lemma). The necessity follows from (29) since

$$\int_0^\infty |u(x)|^2\, e^{-2\tau x}\, dx \leq \int_0^\infty |u(x)|^2\, e^{-2\beta x}\, dx.$$

To prove the sufficiency, we start from the inverse Fourier transform of $v(\sigma + i\tau)\, e^{\tau x}$. We have

$$u(x) = \frac{1}{2\pi} \int_{-\infty}^\infty v(\sigma + i\tau)\, e^{-i(\sigma + i\tau)x}\, d\sigma,$$

or

$$u(x)\, e^{-\tau x} = \frac{1}{2\pi} \int_{-\infty}^\infty v(\sigma + i\tau)\, e^{-i\sigma x}\, d\sigma,$$

and so by Plancherel's theorem,

$$\int_{-\infty}^\infty |u(x)|^2\, e^{-2\tau x}\, dx = \frac{1}{2\pi} \int_{-\infty}^\infty |v(\sigma + i\tau)|^2\, d\sigma. \tag{30}$$

This implies that $u(x)$ must vanish almost everywhere for $x < 0$, For, if $u(x) \neq 0$ on a set of positive measure for $x < 0$, then

$$\int_{-\infty}^{0} |u(x)|^2 \, e^{-2\tau x} \, dx \to \infty$$

as $\tau \to \infty$. But this is impossible since the right-hand side of (30) is by assumption bounded. Thus $u(x) = 0$ for $x < 0$ and we have

$$\int_{0}^{\infty} |u(x)|^2 \, e^{-2\tau x} \, dx = \frac{1}{2\pi} \int_{-\infty}^{\infty} |v(\sigma + i\tau)|^2 \, d\sigma \leq C$$

for all $\tau \geq \beta$. Letting $\tau \to \beta$, we conclude that

$$\int_{0}^{\infty} |u(x)|^2 \, e^{-2\beta x} \, dx \leq C.$$

Thus $u(x)$ belongs to $\mathscr{H}_{+}^{\beta,0}$. The initial function $v(s)$ is the Fourier transform of $u(x)$ by the uniqueness theorem. This completes the proof of the Paley-Wiener lemma.

Let $\mathscr{H}_{+}^{\beta,q}$ denote the collection of (generalized) functions $u(x)$ which result from the elements of $\mathscr{H}_{+}^{\beta,0}$ by differentiations of order not exceeding q:

$$u(x) = \sum_{k=0}^{q} \frac{d^k u_k(x)}{dx^k}, \quad u_k(x) \in \mathscr{H}_{+}^{\beta,0}.$$

The class $H_{+}^{\beta,q}$ of Fourier transforms of functions $u(x) \in \mathscr{H}_{+}^{\beta,q}$ obviously consists of functions $v(s) \in H_{+}^{\beta,0}$ multiplied by polynomials in s of degree not exceeding q.

The class $H_{+}^{\beta} = \bigcup_{q=0}^{\infty} H_{+}^{\beta,q}$ includes any function $v(s)$ analytic for Im $s > \beta$ which has at most power growth at infinity. This is because the division of each such $v(s)$ by a sufficiently high power of s will yield a function belonging a fortiori to $H_{+}^{\beta,0}$. Furthermore, H_{+}^{β} has the monotonicity property mentioned in Sec. 4.9.1: if $v_0(s) \in H_{+}^{\beta}$ and $v_1(s)$ is an analytic function not exceeding $v_0(s)$ in absolute value, then $v_1(s) \in H_{+}^{\beta}$ also. Hence it follows that multiplication by analytic functions with at most power growth at infinity is admissible in H_{+}^{β}.

4.9.4. The necessity part of Theorem 1 will be proved in the next two subsections. However it is more convenient to consider the first order system in t

$$\frac{\partial \mathbf{u}(t, x)}{\partial t} = P\left(i \frac{\partial}{\partial x}\right) \mathbf{u}(t, x), \quad ((t, x) \geq 0), \tag{31}$$

rather than the single equation (1). Here $\mathbf{u}(t, x)$ is an m-component vector function and $P\left(i \dfrac{\partial}{\partial x}\right)$ is an $m \times m$ matrix whose elements are polynomials of the operator $i \dfrac{\partial}{\partial x}$ of maximum degree p.

In addition to (31), we have the (vector) initial condition

$$\mathbf{u}(0, x) = \mathbf{u}_0(x) \tag{32}$$

and (vector) boundary conditions

$$\mathbf{u}(t, 0) = \mathbf{w}_0(t), \ldots, \frac{\partial^{p-1} \mathbf{u}(t, 0)}{\partial x^{p-1}} = \mathbf{w}_{p-1}(t). \tag{33}$$

In order to pass to generalized functions, we shall utilize (7) and (8) as in Sec. 4.9.1 but interpreting them here to be m-component vector equations. We first write (31) in the form

$$\frac{\partial \mathbf{u}(t, x)}{\partial t} = \sum_{k=0}^{p} a_k \left(i \frac{\partial}{\partial x}\right)^k \mathbf{u}(t, x), \tag{34}$$

where the a_k are matrices with numerical entries. Then according to (8), (34) turns into the system in generalized functions,

$$\frac{\partial \mathbf{u}(t, x)}{\partial t} = \sum_{k=0}^{p} a_k \left[\left(i \frac{\partial}{\partial x}\right)^k \mathbf{u}(t, x) - \mathbf{w}_{k-1}(t) \, \delta(x) - \cdots \right.$$
$$\left. - \mathbf{w}_0(t) \delta^{(k-1)}(x)\right].$$

We shall assume that all generalized functions being considered belong as before to $\mathscr{H}^{\beta}_{+}$. Taking Fourier transforms, we obtain

$$\frac{d\mathbf{v}(t, s)}{dt} = P(s) \mathbf{v}(t, s) + \mathbf{g}(t, s), \tag{35}$$

where $\mathbf{g}(t, s)$ is a vector function given by

$$\mathbf{g}(t, s) = \sum_{k=1}^{p} a_k(\mathbf{w}_{k-1}(t) - is\mathbf{w}_{k-2}(t) + \cdots + (-is)^{k-1}\mathbf{w}_0(t)),$$

and $\mathbf{v}(t, s)$ is a vector function with components in H_+^β . The initial condition (32) goes over into the condition

$$\mathbf{v}(0, s) = \mathbf{v}_0(s). \tag{36}$$

For given s, the solution to problem (35), (36) is

$$\mathbf{v}(t, s) = e^{tP(s)}\mathbf{v}_0(s) + \int_0^t e^{(t-\theta)P(s)} \mathbf{g}(\theta, s) \, d\theta. \tag{37}$$

For s fixed, $\mathbf{v}(t, s)$, $\mathbf{v}_0(s)$, and $\mathbf{g}(t, s)$ are defined in m-dimensional complex space $R = R_s$ endowed with the usual Euclidean metric. Thus, for any vector $\xi = (\xi_1, ..., \xi_m)$,

$$|\xi|_s^2 = \sum_{j=1}^{m} |\xi_j|^2.$$

For fixed s, the linear operator defined in R_s by the matrix $P(s)$ has a certain number $\varrho = \varrho(s)$ of eigenvalues with non-positive real parts $(0 \leq \varrho \leq m)$ and $m - \varrho$ eigenvalues with positive real parts.

Accordingly, R_s may be decomposed into a direct sum of invariant subspaces under $P(s)$ which we denote by R_s^+ and R_s^-. The vectors $\mathbf{v}(t, s)$, $\mathbf{v}_0(s)$, and $\mathbf{g}(t, s)$ have the corresponding decompositions,

$$\mathbf{v}(t, s) = \mathbf{v}^-(t, s) + \mathbf{v}^+(t, s) \quad \mathbf{v}_0(s) = \mathbf{v}_0^-(s) + \mathbf{v}_0^+(s),$$

$$\mathbf{g}(t, s) = \mathbf{g}^-(t, s) + \mathbf{g}^+(t, s).$$

Since the operators $P(s)$ and $e^{tP(s)}$ are invariant in the subspaces R_s^- and R_s^+, we have

$$\mathbf{v}^-(t, s) = e^{tP(s)} \mathbf{v}_0^-(s) + \int_0^t e^{(t-\theta)P(s)}\mathbf{g}^-(\theta, s) \, d\theta,$$

and

$$\mathbf{v}^+(t, s) = e^{tP(s)}\mathbf{v}_0^+(s) + \int_0^t e^{(t-\theta)P(s)}\mathbf{g}^+(\theta, s) \, d\theta.$$

As in Sec. 4.8, it is easy to verify that $\mathbf{v}^-(t, s)$ increases in R_s^- no faster than a power of t.

The expression for $\mathbf{v}^+(t, s)$ can be written as

$$\mathbf{v}^+(t, s) = e^{tP(s)} \left\{ \mathbf{v}_0^+(s) + \int_0^t e^{-\theta P(s)} \mathbf{g}^+(\theta, s) \, d\theta \right\}. \tag{38}$$

As in Sec. 4.8, it can be shown that the integral

$$\mathbf{I}(s) = \int_0^\infty e^{-\theta P(s)} \mathbf{g}^+(\theta, s) \, d\theta$$

is convergent. It obviously belongs to R_s^+. Incorporating it in (38), we obtain

$$\mathbf{v}^+(t, s) = e^{tP(s)} [\mathbf{v}_0^+(s) + \mathbf{I}(s)] - \int_t^\infty e^{(t-\theta)P(s)} \mathbf{g}^+(\theta, s) \, d\theta.$$

As in Sec. 4.8, it is easy to show that this last integral increases in R_s no faster than a power of t. At the same time, the expression $e^{tP(s)}[\mathbf{v}_0^+(s) + \mathbf{I}(s)]$ is a fortiori exponentially increasing if $\mathbf{v}_0^+(s) + \mathbf{I}(s) \neq 0$. Hence it follows that

A necessary condition for the existence of a solution to the problem (35), (36) *which for given s increases no faster than a power of t is the fulfillment of the relation*

$$\mathbf{v}_0^+(s) + \mathbf{I}(s) \equiv \mathbf{v}_0^+(s) + \int_0^\infty e^{-\theta P(s)} \mathbf{g}^+(\theta, s) \, d\theta = 0. \tag{39}$$

4.9.5. Let us examine the form of (39) in the case of our single equation

$$\frac{\partial^m u}{\partial t^m} = \sum_{k=0}^{m-1} P_k \left(i \frac{\partial}{\partial x} \right) \frac{\partial^k u}{\partial t^k}. \tag{40}$$

This equation is equivalent to the system

$$\left. \begin{aligned} \frac{\partial u_0}{\partial t} &= u_1, \\[2mm] \frac{\partial u_1}{\partial t} &= u_2, \\[2mm] &\cdots\cdots\cdots\cdots\cdots\cdots \\[2mm] \frac{\partial u_{m-1}}{\partial t} &= P_0 \left(i \frac{\partial}{\partial x} \right) u_0 + P_1 \left(i \frac{\partial}{\partial x} \right) u_1 + \cdots + P_{m-1} \left(i \frac{\partial}{\partial x} \right) u_{m-1}, \end{aligned} \right\} \tag{41}$$

$m - 1$. The roots themselves are analytic functions of s in the half-plane Im $s > \beta$ which grow no faster than $|s|^p$ as $|s| \to \infty$. $v_0(s), ..., v_{m-1}(s)$ belong to H_+^β by assumption. Therefore the left-hand side of (46) is an analytic function of s for Im $s > \beta$ which also belongs to H_+^β.

Each of the quantities

$$(-1)^m \int_0^\infty e^{-\theta \lambda_v(s)} g(\theta, s)\, d\theta = G_v(s) \quad (v = r, ..., m - 1) \tag{47}$$

is apriori defined (as a function of s) just in the region Re $\lambda_v(s) > 0$, Im $s > \beta$ and is analytic there. We thus see that

A necessary condition for problem (1)–(3) to have a solution is that each $G_v(s)$ be analytically continuable from the region where it is defined by (47) (i.e. from Re $\lambda_v(s) > 0$, Im $s > \beta$) to the entire half-plane Im $s > \beta$ as an element of H_+^β.

The necessity part of Theorem 1 is thus proved.

4.9.6. We now proceed to prove the sufficiency.

Suppose that for equation (1) and the boundary conditions (2), $g(s,t)$ has been constructed according to equation (15) and suppose that all

$$G_v(s) = (-1)^m \int_0^\infty e^{-\theta \lambda v(s)} g(\theta, s)\, d\theta \quad (v = r, ..., m - 1)$$

are analytically continuable into the half-plane Im $s > \beta$ as elements of H_+^β. We wish to show that the problem (1)–(3) has a solution that satisfies all of the stated conditions.

Our first objective is to establish that the relations of (46), which we now write in the form

$$Q_{v0}v_0 - Q_{v1}v_1 + \cdots + (-1)^{m-1} Q_{v,m-1} = G_v(s) \quad (v = r, ..., m-1) \tag{48}$$

$(Q_{vk} = W_{v,k} / W_{v,m-1})$, make it possible to uniquely determine $v_r(s), ..., v_{m-1}(s)$ in terms of $v_0(s), ..., v_{r-1}(s)$, and $G_v(s)$ for each s. First of all, (45) determines $v^r(s), ..., v^{m-1}(s)$ in terms of $G(\lambda_r), ..., G(\lambda_{m-1})$ through

$$v^v(s) = -\int_0^\infty e^{-\theta \lambda v(s)} g^v(\theta, s)\, d\theta = \frac{(-1)^v}{W(\lambda)} W_{v,m-1} G_v(s).$$

Since the algebraic function $W_{v,m-1}(s)/W(\lambda(s))$ has at most power growth at infinity, $v^r(s) \in H^\beta_+$ together with $G_v(s)$.

Let us find $v^0, ..., v^{r-1}$. To this end, we use the first r equations of the system (42) with $\boldsymbol{\xi} = \mathbf{v}_0$ obtaining

$$v_0 = v_0 \quad\;\; + v^1 \quad\; + \cdots + v^{m-1}$$
$$v_1 = v^0\lambda_0 \;\; + v^1\lambda_1 \;\; + \cdots + v^{m-1}\lambda_{m-1}$$
$$\cdots\cdots\cdots\cdots\cdots$$
$$v_{r-1} = v^0\lambda_0^{r-1} + v^1\lambda_1^{r-1} + \cdots + v^{m-1}\lambda_{m-1}^{r-1}$$

The determinant of this system of r equations in the r unknowns $v^0, ..., v^{r-1}$ is non-vanishing for Im $s > \beta$ (the eigenvalues are distinct in this region). Hence, $v^0, ..., v^{r-1}$ are uniquely determined in terms of $v_0, ..., v_{r-1}$ and $v^r, ..., v^{m-1}$, or in the final analysis, in terms of $v_0,...,v_{r-1}$ and $G(\lambda_r), ..., G(\lambda_{r-1})$. The determinant is an algebraic function of s. Its inverse increases no faster than a power of s as $|s| \to \infty$. Hence, $v^0,..., v^{r-1}$ belong to H^β_+ together with $v_0, ..., v_{r-1}$. The last $m - r$ equations of the system (42) determine the required $v_r, ..., v_{m-1}$ in terms of $v_0, ..., v_{r-1}$, $G_r(s), ..., G_{m-1}(s)$; the results once more obviously belong to H^β_+.

As we know, for each s the solution of the problem is uniquely determined in terms of $v_0(s), ..., v_{m-1}(s)$, and $g(t, s)$ through formula (37) or

$$\mathbf{v}(t, s) = e^{tP(s)}[\mathbf{v}_0(s) + \int_0^t e^{-\theta P(s)}\, \mathbf{g}(\theta, s)\, d\theta]. \tag{49}$$

If (39) is taken into account, then

$$\mathbf{v}(t, s) = e^{tP(s)}[\mathbf{v}_0^-[s] + \int_0^t e^{-\theta P(s)}\, \mathbf{g}^-(\theta, s)\, d\theta] - \int_t^\infty e^{(t-\theta)P(s)}\, \mathbf{g}^+(\theta, s)\, d\theta. \tag{50}$$

Formula (49) shows that $\mathbf{v}(t, s)$ is an analytic function for each t and each s in the half-plane Im $s > \beta$. We must show that it belongs to H^β_+ and increases there no faster than Ct^h. To do this, we apply (50) and the main result of Sec. 4.8. The following was established there. Suppose that the initial data $v_0(\sigma) = v(0, \sigma)$, $v_1(\sigma) = \dfrac{dv(0, \sigma)}{dt}, ..., v_{m-1}(\sigma) = \dfrac{d^{m-1}v(0, \sigma)}{dt^{m-1}}$

for the equation

$$\frac{d^m v}{dt^m} - \sum_{k=0}^{m-1} P_k(\sigma)\frac{d^k v}{dt^k} = g(t, \sigma) \tag{51}$$

22*

are assigned on a domain of values of σ where the real parts of $\lambda_0(\sigma), ..., \lambda_{m-1}(\sigma)$ have fixed signs (so that for instance $\operatorname{Re} \lambda_j(\sigma) \leqq 0$ for $j \leqq r - 1$ and $\operatorname{Re} \lambda_\nu(\sigma) > 0$ for $\nu \geqq r$). Suppose that the data are square integrable after division by a certain power of $|\sigma| + 1$ and satisfy the relations

$$Q_{\nu_0}v_0 - Q_{\nu_1}v_1 + \cdots + (-1)^{m-1} Q_{\nu, m-1}v_{m-1} = (-1)^m \int_0^\infty e^{-\theta\lambda_\nu(\sigma)}g(\theta, \sigma)\, d\theta.$$

Then there exists a solution $\mathbf{v}(t, \sigma)$ satisfying the initial conditions and the inequality (cf. (45) of Sec. 4.8)

$$|\mathbf{v}(t, s)| \leqq Ct^{m-1} \int_0^t |g(\theta, s)|\, d\theta + C \int_0^\infty (\theta + t)^{m-1} e^{-\theta \operatorname{Re}\lambda_r(s)}|g(\theta + t, s)|\, d\theta$$

$$+ C(1 + t)^m (1 + |s|)^{m(p-1)} \sum_{k=0}^{r-1} |v_k(s)|, \tag{52}$$

where $\lambda_r(s)$ is the root with the smallest positive real part.

Since the half-plane $\operatorname{Im} s > \beta$ can be represented as a finite union of regions in which the real parts of the eigenvalues have fixed signs, the estimate (52) is valid in the entire half-plane $\operatorname{Im} s > \beta$. But in that event, the estimate implies by hypothesis that $\mathbf{v}(t, s)$ belongs to H_+^β and increases there no faster than a power of t, as required.

REMARK 1. As in Sec. 4.4, the conditions of the problem may be modified. It is possible to look for a solution in the class of functions $u(t, x) \in \mathscr{H}_+^\beta$ increasing no faster than $t^h e^{\alpha t}$ for some fixed α. One then has to change the definition of r in the hypotheses of the theorem as follows. The eigenvalues $\lambda_0(s), ..., \lambda_{r-1}(s)$ should each have a real part not exceeding α throughout the half-plane $\operatorname{Im} s > \beta$. This condition holds particularly for equations that are proper in the sense of Petrovsky relative to $\dfrac{\partial}{\partial x}$, i.e., for which $\operatorname{Im} s_j(\lambda)$ remains uniformly bounded for $\operatorname{Re} \lambda = 0$. If we denote the bound by μ, we can conclude as we did in Example b) of Sec. 4.9.2 that $\operatorname{Re} \lambda_0(s), ..., \operatorname{Re} \lambda_{r-1}(s)$ do not exceed α in the half-plane $\operatorname{Im} s > 0$. Thus the results of Example b) can be extended with the indicated change to this general case.

REMARK 2. The results obtained for a quarter-plane may be used to specify well-posed problems for the quarter-space $t \geqq 0$, $x_1 \geqq 0$, $-\infty < x_2 < \infty, ..., -\infty < x_n < \infty$ for the equation

$$\frac{\partial^m u}{\partial t^m} = \sum_{k=0}^{m-1} p_k\left(i\frac{\partial}{\partial x_1}, ..., i\frac{\partial}{\partial x_n}\right)\frac{\partial^k u}{\partial t^k}. \tag{53}$$

The equation reduces to an equation of the form (1) with parameters $\sigma_2, ..., \sigma_n$ after taking Fourier transforms with respect to $x_2, ..., x_n$. The initial data $u(0, x_1, ..., x_n)$, $\dfrac{\partial u(0, x_1, ..., x_n)}{\partial t}, ..., \dfrac{\partial^{m-1} u(0, x_1, ..., x_n)}{\partial t^{m-1}}$ go over into

$$\tilde{u}(0, x_1, \sigma_2, ..., \sigma_n), ..., \dfrac{\partial^{m-1} \tilde{u}(0, x_1, \sigma_2, ..., \sigma_n)}{\partial t^{m-1}}.$$

Suppose for example that (53) is strictly proper in the sense of Petrovsky relative to $\dfrac{\partial}{\partial x_1}$. This means that for fixed arbitrary $\sigma_2, ..., \sigma_n$ the roots of

$$\lambda^m = \sum_{k=0}^{m-1} p_k(s_1, \sigma_2, ..., \sigma_n) \lambda^k$$

have either a positive or a non-positive real part throughout a certain half-plane Im $s_1 > \beta$ and that the number r of roots with a non-positive real part is independent of $\sigma_2, ..., \sigma_n$. Then the well-posed problem consists in assigning the data

$$u(t, 0, x_2, ..., x_n), ..., \dfrac{\partial^{p-1} u(t, 0, x_2, ..., x_n)}{\partial x_1^{p-1}},$$

$$u(0, x_1, x_2, ..., x_n), ..., \dfrac{\partial^{r-1} u(0, x_1, , x_2, ... x_n)}{\partial t^{r-1}}$$

so as to satisfy conditions analogous to those given in the examples of Sec. 4.9.2.

The condition for an equation to be strictly proper in the sense of Petrovsky relative to $\dfrac{\partial}{\partial x_1}$ has a more customary equivalent formulation:
If Re $\lambda = 0$, then all the roots $s_1 = s_1(\lambda_1, \sigma_2, ..., \sigma_n)$ should have a non-positive real part. As in Remark 1, one may consider equations that are proper in the sense of Petrovsky. This means that for Re $\lambda = 0$, the real parts of the roots $s_1(\lambda_1, \sigma_2, ..., \sigma_n)$ are bounded by a fixed constant α and thus the growth in t of the solution to a well-posed problem is restricted to being of the exponential-power form $t^h e^{\alpha t}$.

Bibliographical Comments

Although the premises underlying the formation of the theory of generalized functions are rooted deep in classical mathematics, the physicists were actually the first to introduce and use the concepts [1]. The first mathematically rigorous treatment of a generalized function as a functional is due to Sobolev [2]. A systematic development of the theory involving numerous conclusive examples was carried out by Schwartz [3]. Almost instantly after the appearance of Schwartz's book, the concepts spread to a large part of analysis where they helped clear up many old facts and enabled new general relationships to be found. The theory was further developed in the series of books by Gelfand et al. entitled *Generalized Functions* [4–8]. Besides contributing to the classical branches of analysis, these books touch on such new fields as general measure theory in linear topological spaces, representations of classical groups, and the theory of stochastic processes.

With the growth of generalized function theory, it has become possible to construct a general theory of partial differential equations encompassing any type of equation of any order. Although the role of fundamental solutions in the classical problems has been known certainly as far back as the past century, only the advent of generalized functions has made it possible to give a precise definition of a fundamental function (solution) of a differential operator and to study the question of its existence. The first proof of the existence of a fundamental function for a general linear partial differential operator with constant coefficients is due to Malgrange [9]. The class of hypoelliptic equations was first characterized by Hörmander [10]. The first general investigation of fundamental functions for non-hypoelliptic equations was carried out by Borovikov [11]. Radon's method for constructing fundamental solutions of elliptic and hyperbolic equations was systematically applied by John [12]. The general problem of characterizing well-posed problems for a half-space was formulated and solved by Dikopolov, Palamodov, and Shilov [13–16].

This book has described but a very small portion of the entire gamut of ideas, problems, and results in the general theory of partial differential equations. Readers who wish to obtain more complete information about the field are referred to the review paper of Vishik and Shilov presented at the Fourth All-Union Mathematical Conference [17] in which an extensive bibliography is cited. We mention also the two new interesting books of Hörmander [18] and Trèves [19].

[1] P. A. M. DIRAC, *The Principles of Quantum Mechanics*, Oxford University Press, 3rd ed., 1944.
[2] S. L. SOBOLEV, *Méthode nouvelle à résoudre le probléme de Cauchy pour les équations linéaires hyperboliques normales*, Mat. Sb., No. 1 (43), 1936, pp. 39–72.
[3] L. SCHWARTZ, *Théorie des distributions*, Vols. 1 and 2, Hermann et Cie., Paris, 1950–1951.

[4] I. M. GELFAND and G. E. SHILOV, *Generalized Functions, Vol.* 1: *Properties and Operations*, Academic Press, New York, 1964.

[5] I. M. GELFAND and G. E. SHILOV, *G. F., Vol.* 2: *Spaces of Test and Generalized Functions*, Fizmatgiz, 1958 (English translation in preparation, Academic Press).

[6] I. M. GELFAND and G. E. SHILOV, *G. F., Vol.* 3: *Special Topics in the Theory of Differential Equations*, Fizmatgiz, 1958 (English translation in preparation, Academic Press).

[7] I. M. GELFAND and N. Ya. VILENKIN, *G. F. Vol.* 4: *Applications of Harmonic Analysis. Rigged Hilbert Spaces*, Academic Press, New York, 1965.

[8] I. M. GELFAND, N. Ya. VILENKIN, and M. I. GRAEV, *G. F., Vol.* 5: *Integral Geometry and Representation Theory*, Academic Press, New York, 1965.

[9] B. MALGRANGE, *Équations aux derivées partielles à coefficients constants*, I. *Solution élémentaire*, C. R. Acad. Sci., 237, No. 25, 1953, pp. 1620–1622.

[10] L. HÖRMANDER, *On the theory of general partial differential operators*, Acta Math., 94, 1955, pp. 161–248.

[11] V. A. BOROVIKOV, *Fundamental solutions of linear partial differential equations with constant coefficients*, Trudy Moskov. Mat. Obshch., Vol. 8, 1959, pp. 199–258 (In Russian).

[12] F. JOHN, *Plane Waves and Spherical Means Applied to Partial Differential Equations*, Interscience, New York, 1955.

[13] G. V. DIKOPOLOV and G. E. SHILOV, *On well-posed boundary value problems for partial differential equations in a half-space*, Izv. Akad. Nauk SSSR Ser. Mat., Vol. 24, 1960, pp. 369–380 (In Russian).

[14] V. P. PALAMODOV, *On well-posed boundary value problems for partial differential equations in a half-space*, Izv. Akad. Nauk SSSR Ser. Mat., Vol. 24, 1960, pp. 381–386 (In Russian).

[15] G. V. DIKOPOLOV and G. E. SHILOV, *On well-posed boundary value problems in a half-space for partial differential equations with a right-hand side*, Sibirsk. Mat. Zh., Vol. 2, No. 1, 1960, pp. 45–61 (In Russian).

[16] G. E. SHILOV, *On boundary value problems in a quarter-plane for partial differential equations with constant coefficients*, Sibirsk. Mat. Zh., Vol. 2, No. 1, 1961, pp. 144–160 (In Russian).

[17] M. I. VISHIK and G. E. SHILOV, *The general theory of partial differential equations and certain questions in the theory of boundary value problems*, Trans. Fourth All-Union Mathematical Conference, Leningrad, 1961, Vol. 1, Izd-vo Akad. Nauk SSSR, Leningrad, 1963, pp. 55–85 (In Russian).

[18] L. HÖRMANDER, *Linear Partial Differential Operators*, Academic Press, New York, 1963.

[19] F. TRÈVES, *Lectures on Linear Partial Differential Equations with Constant Coefficients*, Notas de Matimatika, No. 7, Rio de Janeiro, 1961.

Index